PSYCHOLOGY OF
ACADEMIC CHEATING

UNM-GALLUP   DUP

3 7996 1006 7574 9

PSYCHOLOGY OF ACADEMIC CHEATING

EDITORS
ERIC M. ANDERMAN
TAMERA B. MURDOCK

AMSTERDAM · BOSTON · HEIDELBERG · LONDON
NEW YORK · OXFORD · PARIS · SAN DIEGO
SAN FRANCISCO · SINGAPORE · SYDNEY · TOKYO

Academic Press is an imprint of Elsevier

ELSEVIER

Elsevier Academic Press
30 Corporate Drive, Suite 400, Burlington, MA 01803, USA
525 B Street, Suite 1900, San Diego, California 92101-4495, USA
84 Theobald's Road, London WC1X 8RR, UK

This book is printed on acid-free paper. ∞

Copyright © 2007, Elsevier Inc.

No part of this publication may be reproduced or transmitted in any form or by any
means, electronic or mechanical, including photocopy, recording, or any information
storage and retrieval system, without permission in writing from the publisher.

The following articles are copyright by their respective authors: Foreword, by Alfie Kohn,
Cheating on Tests: Prevalance, Detection, and Implications for On-Line Testing, by
Walter M. Haney and Michael J. Clarke.

Permissions may be sought directly from Elsevier's Science & Technology Rights
Department in Oxford, UK: phone: (+44) 1865 843830, fax: (+44) 1865 853333, E-mail:
permissions@elsevier.co.uk. You may also complete your request on-line via the Elsevier
homepage (http://elsevier.com), by selecting "Customer Support" and then "Obtaining
Permissions."

Library of Congress Cataloging-in-Publication Data

Psychology of academic cheating/editors, Eric M. Anderman, Tamera B. Murdock.
    p. cm.
  Includes index.
  ISBN-13: 978-0-12-372541-7 (alk.paper)
  ISBN-10: 0-12-372541-0 (alk.paper)
  1. Cheating (Education)   I. Anderman, Eric M.   II. Murdock, Tamera Burton.
  LB3609.P79 2006
  371.5'8—dc22

                                                              2006019372

British Library Cataloguing in Publication Data
A catalogue record for this book is available from the British Library

ISBN 13: 978-0-12-372541-7
ISBN 10: 0-12-372541-0

For all information on all Elsevier Academic Press publications
visit our Web site at www.books.elsevier.com

Printed in the United States of America
07 08 09 10 9 8 7 6 5 4 3 2 1

Working together to grow
libraries in developing countries

www.elsevier.com | www.bookaid.org | www.sabre.org

ELSEVIER   BOOK AID
            International   Sabre Foundation

CONTENTS

3

PART II

ACHIEVEMENT MOTIVATION AND CHEATING

4

5

6

7

12

THE PRESSURE TO CHEAT IN A HIGH-STAKES TESTING ENVIRONMENT 289

SHARON L. NICHOLS AND DAVID C. BERLINER

Contributors

Editors:

Eric M. Anderman, University of Kentucky Lexington, KY 40506-0001

Tamera B. Murdock, University of Missouri-Kansas City, Department of Psychology, Kansas City, MO 64110-0238

Chapter Contributors:

Eric M. Anderman, University of Kentucky Lexington, KY 40506-0001

Lynley Anderman, University of Kentucky Lexington, KY 40504-0017

David C. Berliner, Arizona State University Tempe, AZ 85281

David Callahan, Demos, New York, NY 10001

Leslie Christensen, University of Missouri-Kansas City, Kansas City, MO 64110

Michael J. Clarke, Merkert Chemistry Center Boston College Chestnut Hill, MA 02167-3860

Tierra M. Freeman, University of Missouri-Kansas City Kansas City, MO 64110-2499

Linda Garavalia, UMKC Department of Psychology Kansas City, MO 64110

Hunter Gehlbach, University of Connecticut Teachers for a New Era Storrs, CT 06269-2064

Courtney J. Golant, University of Illinois at Chicago, Department of Educational Psychology, Chicago, IL 60607-7133

Walter Haney, Boston College Chestnut Hill, MA 02467

Alfie Kohn

Fred Kuch, UNLV Las Vegas, NV 89154

Stephen Lehman, Utah State University, Department of Psychology Logan, UT 84322-2810

Trish Lehman, University of Colorado, Boulder School of Education Boulder, CO 80309-0249

Matthew T. McCrudden, University of North Florida College of Education and Human Services Jacksonville, FL 3222402645

Angela D. Miller

Christian E. Mueller, University of Memphis Counseling, Educational Psychology and Research, Memphis, TN 38152-3570

Tamera B. Murdock, University of Missouri-Kansas City, Department of Psychology, Kansas City, MO 64110-0238

Sharon L. Nichols, University of Texas at San Antonio San Antonio, TX 78209

Lori Olafson, UNLV Las Vegas, NV 89154

Elizabeth Olson, Division of Counseling and Educational Psychology University of Missouri-Kansas City, Kansas City, MO 64110

Amy L. Poindexter

David A. Rettinger, University of Mary Washington, Psychology Department, Fredericksburg, VA, 22407

L. Dale Richesin, University of Alaska, Fairbanks-Bristol Bay campus, Togiak, AK 99678

Emily Russell, Division of Counseling and Educational Psychology University of Missouri-Kansas City Kansas City, MO 64110

Gregory Schraw, UNLV Las Vegas, NV 89154

Jason M. Stephens, University of Connecticut Storrs, CT 06269-2064

Theresa A. Thorkildsen, Department of Educational Psychology, University of Illinois at Chicago Chicago, IL 60607-7133

FOREWORD

It has been three decades since Lee Ross of Stanford University coined the term "fundamental attribution error" to describe the tendency to invoke an individual's personality or character to explain what is actually due to the social environment. Researchers as well as ordinary folks routinely succumb to this temptation, and while there are surely examples to be found anywhere, it seems to be particularly prevalent in a society where individualism is both a descriptive reality and a cherished ideal. We Americans are apt to assume that people who commit crimes are morally deficient, that the have-nots in our midst are lazy (or at least insufficiently resourceful), that children who fail to learn simply aren't studying hard enough (or have unqualified teachers). We treat each instance of illegality, poverty, or academic difficulty as if it had never happened before, as if the individual in question was acting out of sheer perversity. In so doing we overlook the social structures—the historical, political, and economic realities—that affect human behavior. After all, to acknowledge the importance of these forces is inconvenient because we might feel an obligation to do something about them.

It's not surprising, then, that discussions about academic cheating are typically characterized by condemnations of the students in question, attempts to ascertain what features of their personalities or backgrounds led them to break the rules, and ideas for deterring (or catching) them. The condemnation may not be explicit, particularly in essays written by researchers, but if the point is to figure out what's wrong with these young people and how they can be stopped, then the structural issues that could help us make sense of why cheating takes place are still being neglected.

Copyright © 2007 by Alfie Kohn
All rights of reproduction in any form reserved.

Fortunately, many of the chapters in this anthology do describe evidence dealing with structural issues, or at least with the classroom practices that tend to be associated with cheating: academic tasks that are too difficult or are experienced by students as unengaging; teachers who are indifferent to, or unskilled at, helping students to learn, or who fail to develop any sort of connection with them; instructional practices that heighten the salience of grades (including honor rolls and incentives for good marks— that is, rewards for rewards) or otherwise lead students to become more focused on how well they are doing than on *what* they are doing.

That last distinction is particularly useful in thinking not only about the priorities of instructors but about the culture and goals of schools. "Only extraordinary education is concerned with learning," the writer Marilyn French (1985, p. 387) once observed, whereas "most is concerned with achieving: and for young minds, these two are very nearly opposites." Rather than responding to an idea by wondering, "What does that mean?", many students are more likely to ask "Do we have to know this?" (or "Will it be on the test?"). Such a reaction reflects the influence of an institutional emphasis on levels of achievement rather than on layers of understanding. Less interest in learning for its own sake is one consequence of such schooling. Cheating is another. The prevalence of both phenomena will vary, of course, since all students do not behave identically in the same environment, but to lose sight of the primary contribution of that environment— and others to which they've been exposed—is to do exactly what Lee Ross warned us about.

The distinction between raising achievement and promoting learning is particularly sharp, at least for elementary and secondary students, when achievement is defined as higher scores on standardized tests. If there is pressure to raise those scores, lower quality learning may again be accompanied by a higher frequency of cheating, but this time on the part of adults. First, and most familiar, we find *explicit cheating*, in which students are supplied with answers or hints during the exam or their answer sheets are changed after the fact. The second version is *hidden cheating*, in which educators contrive to raise a school's overall performance profile by (a) retaining low-scoring students in grade—not because this is in their best interest (which it almost never is) but because they will be a year older when they take the test, (b) classifying more students as "special needs" in order to exclude their scores from school averages, or (c) lavishing attention on students who are close to passing, ignoring those who are sure to do well and those likely to fail. Finally, what might be called *legal cheating* is more commonly known as teaching to the test. When real instruction gives way to extensive exam preparation, scores can be raised without improving learning at all. At best, the results are then meaningless because the more pressure there is to ratchet up the scores, the less valid the tests are as measures of anything other than the extent of test preparation.

However, higher test scores may actually be worrisome if more important kinds of instruction have been sacrificed in order to raise test scores. Indeed, we might say that students are cheated out of a decent education—in the name of "higher standards" and "accountability."[1]

Regardless of whether it is teachers or students who break the rules, we need to attend to the larger context in which their actions take place. That context, as I've suggested, may involve classroom structures that emphasize performance or achievement more than learning, and it may involve external pressure to improve that performance. Worst of all, though, is an environment in which the pressure is experienced in terms of one's standing relative to others. Competition is perhaps the single most toxic ingredient in a classroom, and it is also a reliable predictor of cheating. Eric Anderman and others (Anderman et al., 1998; Anderman and Midgley, 2004) discovered a significant relationship between competitive educational settings and the extent to which students cheat, including students who acknowledge it is wrong to do so. The precise respects in which competition is harmful may describe independent paths to that outcome. Competition typically has an adverse impact on relationships; it often contributes to a loss of intrinsic motivation; it undermines academic self-confidence (even for winners); and it interferes with the development of higher-order thinking (Kohn, 1992). How ironic, then, that many of the same people who deplore cheating also support academic contests, awards, or the practice of grading on a curve, all of which make it more likely that cheating will take place.

Cheating is properly viewed less as a discrete set of behaviors than as a point of entry for understanding what promotes, and what impedes, meaningful learning. List the classroom practices that nourish a disposition to

---

[1] They may also be cheated out of a diploma. (For evidence that high-stakes testing raises the dropout rate, particularly among the most vulnerable students, see Jacob [2001], Reardon and Galindo [2002], and Marchant and Paulson [2005].) The problems inherent to the design of standardized tests, moreover, raise questions about their use not only in classrooms but in research. Perhaps we ought to hesitate not only before conducting studies, but even before *citing* studies, that evaluate educational practices on the basis of standardized test results. First, these tests are so fundamentally flawed (Kohn, 2000) that advocating a given intervention solely because it helps students to perform better on those tests may not constitute a persuasive defense of its use. This is particularly true of norm-referenced tests, like the Stanfords, Iowas, or Terra Novas, which were designed to maximize response variance—that is, to create a broad range of scores for the purpose of sorting students efficiently—rather than to gauge whether a given teaching strategy is effective. Such tests are not merely inappropriate as a strategy to *change* teaching (that is, as an accountability tool) but problematic when employed to *measure* teaching. Second, every time a study is published—particularly in a reputable journal or by a reputable researcher—that uses standardized test scores as the primary dependent variable, those tests gain further legitimacy. If we are not keen on bolstering their reputation and perpetuating their use in schools, we would want to avoid relying on them even in the course of pursuing other objectives and investigating other topics.

find out about the world, the teaching strategies that are geared not to covering a prefabricated curriculum but to *dis*covering the significance of ideas, the pedagogical approaches that invite students to play an active role in designing the direction that a course will take (rather than expecting them to listen passively to lectures and memorize material for the purpose of passing a test), and you will have enumerated the conditions under which cheating is much less likely to occur.

To some observers it may seem objectionable, or at least perplexing, that instructors do not work more vigorously to detect and prevent cheating. But a broader perspective reminds us that there may be other objectives that matter just as much, if not more. Suppose that cheating could be at least partly curtailed by tightly monitoring and regulating students or by repeatedly announcing the dire penalties that await anyone who breaks the rules. Would this result be worth the cost of creating a climate of mistrust, undermining a sense of community, and perhaps leading students to become less enthusiastic about learning? Rebecca Moore Howard (2001), who teaches writing at Syracuse University, put it this way: "In our stampede to fight what some call a 'plague' of plagiarism, we risk becoming the enemies rather than the mentors of our students; we are replacing the student-teacher relationship with the criminal-police relationship. . . . Worst of all, we risk not recognizing that our own pedagogy needs reform . . . [if it] encourages plagiarism because it discourages learning." This is not to say that all instructors will choose to strike the same balance between cheating and other considerations. But it's important to make sure we're asking the right questions and seeing the whole picture.

That picture, moreover, includes more than individual classrooms. If schools focus on relative achievement and lead students to do the same, it may be because they exist in a society where education is sometimes conceived as little more than a credentialing ritual. Schools then become, as educational historian David Labaree (1997, pp. 258, 32) put it, "a vast public subsidy for private ambition" in which "self-interested actors [seek] opportunities for gaining educational distinctions at the expense of each other." And if the point is just to get ahead, individuals may seek "to gain the highest grade with the minimum amount of learning," Labaree (1997, p. 259) continues. Cheating could be seen as a rational choice in a culture of warped values.

A deep analysis of cheating may lead us to investigate not only the structures that give rise to it but the process by which we come to decide what will be classified as cheating in the first place. Even a critical analysis of the social context usually assumes that cheating is, by definition, unethical. But perhaps the situation is more complicated. If cheating is defined as a violation of the rules, then it would seem to make sense to ask whether those rules are reasonable, who devised them, and who stands to benefit by them. In practice, however, these questions are rarely asked; the premises are almost always taken for granted.

To be sure, some kinds of cheating involve actions that are indisputably objectionable. Plagiarism is one example. While it's not always clear in practice where to draw the line between an idea that has been influenced by the work of other writers and one that clearly originated with someone else (and ought to be identified as such),[2] we should be able to agree that it is wrong to use a specific concept or a verbatim passage from another source without giving credit, the objective being to deceive the reader about its origin. But more interesting, and perhaps just as common, are those cases where what is regarded as cheating actually consists of a failure to abide by restrictions that may be arbitrary and difficult to defend. It's not just that questionable educational practices may *cause* students to cheat, in other words; it's that such practices are responsible for *defining* certain behaviors as cheating. In the absence of those practices and the ideology supporting them, such behaviors would not be regarded as illegitimate.

This admittedly unsettling possibility enjoys a prima facie plausibility because of other phenomena whose very existence turns out to be dependent on social context. Sportsmanship, for example, is an artificial concept that would not exist except for competition: Only in activities where people are attempting to defeat one another is it meaningful to talk about doing so in a graceful or virtuous fashion. (Participants in cooperative games do not require reminders to be "good sports" because they are working *with* one another toward a common goal.) Likewise, theft does not exist in cultures where there is no private property—not because people refrain from stealing but because the idea literally has no meaning if people's possessions are not off-limits to one another. There is no such thing as leisure unless work is experienced as alienating or unfulfilling. You cannot commit blasphemy unless you believe there is a God to be profaned. Finally, jaywalking is a meaningless concept in Boston, where I live, because there is simply no expectation that pedestrians should cross only at intersections.

On what, then, does the concept of cheating depend for *its* existence? One answer was supplied by a scandal at the Massachusetts Institute of Technology in the early 1990s. More than seventy students were punished for "cheating" because they worked in small groups to write computer programs for fear that they would otherwise be unable to keep up with their

---

[2] "Encouraged by digital dualisms, we forget that plagiarism means many different things: downloading a term paper, failing to give proper credit to the source of an idea, copying extensive passages without attribution, inserting someone else's phrases or sentences—perhaps with small changes—into your own prose, and forgetting to supply a set of quotation marks. If we ignore these distinctions, we fail to see that most of us have violated the plagiarism injunctions in one way or another, large or small, intentionally or inadvertently, at one time or another. The distinctions are just not that crisp" (Howard, 2001).

class assignments. "Many feel that the required work is clearly impossible to do by straightforward [that is, solitary] means," observed the faculty member who chaired MIT's Committee on Discipline (quoted in Butterfield, 1991). The broader issue here is that cooperative learning, beyond its efficiency at helping students deal with an overwhelming workload, provides a number of benefits when compared with individual or competitive instructional models. By working together, students not only are able to exchange information and divide up tasks but typically end up engaging in more sophisticated problem-solving strategies, which, in turn, results in more impressive learning on a range of measures. Structured cooperation in the classroom also proves beneficial in terms of self-esteem, relationships, and motivation to learn (Johnson and Johnson, 1989; Kohn, 1992).

The problem, however, is that, aside from the occasional sanctioned group project, the default condition in most American classrooms—particularly where homework and testing are concerned—is reflected in that familiar injunction heard from elementary school teachers: "I want to see what *you* can do, not what your neighbor can do." (Or, if the implications were spelled out more precisely, "I want to see what you can do all by yourself, deprived of the resources and social support that characterize most well-functioning real-world environments, rather than seeing how much more you and your neighbors could accomplish together.") Whether, and under what circumstances, it might make more sense to have students learn, and to assess their performance, in groups is an issue ripe for discussion and disagreement. But as a rule schools are characterized by an *absence* of discussion about this, to the point that collaboration is referred to as "cheating."

By the same token, students may be disciplined if they consult reference sources during any sort of assessment in which the teacher has forbidden this. Rather than merely investigating how often this happens, drawing an elaborate demographic profile of the students most likely to do this, or searching for ways to stop them, it would seem more appropriate to begin by questioning the premise. What does it say about the instructor, and the larger education system, that assessment is geared largely to students' ability to memorize? What pedagogical purpose is served by declaring that students will be judged on this capacity and must therefore spend a disproportionate amount of time attempting to cram dates, definitions, and other facts into their short-term memory? To what other educational purposes might that time have been put? And what is the purpose of this sort of assessment? Is information being collected about students' proficiency for the purpose of helping them to learn more effectively—or is the exercise more about sorting them (comparing students to one another) or controlling them (by using assessment as an extrinsic motivator)?

It may well be that students who use "unauthorized" materials or assistance thereby compromise the teacher's assessment plan. But perhaps this should lead us to question the legitimacy of that plan and ask why those materials have been excluded. Similarly, if "cheating hinders standardization" (Garavalia et al., 2007), should our primary response be to condemn the cheaters or to question the value of a standardized education? Again, we can expect lively debate on these questions; but again, what is troubling is the absence of such debate—the result of uncritically accepting conventional definitions and assumptions. Consulting a reference source during an exam (or working with one's peers on an assignment) will be classified as cheating in one classroom, with all the repercussions and grave moral implications attendant on that label, while it will be seen as appropriate, even admirable, in another. Students unlucky enough to find themselves in the first classroom stand condemned of cheating, with little attention paid to the nature of the rules they broke. To that extent, *their actions have violated a purely* conventional *set of prohibitions but they are treated as though guilty of a* moral *infraction*.

Moreover, any student who offered just such a defense, arguing that her action is actually less problematic than the instructor's requirements, that what she did was more analogous to entering a lecture hall through a door marked "exit" than to lying or stealing, would likely be accused of engaging in denial, attempting to displace responsibility for what she has done, or trying to rationalize her behavior. Once we have decided that someone's action is morally wrong, her efforts to challenge that premise, no matter how well-reasoned, merely confirm our view of her immorality.

In 2006, a front-page story in the *New York Times* (Glater, 2006) described how instructors and administrators are struggling to catch college students who use ingenious high-tech methods of cheating. In every example cited in the article, the students were figuring out ways to consult their notes during exams; in one case, a student was "caught" using a computer spell-check program. The implication is that students at the university level were being tested primarily on their capacity to memorize, a fact noted neither by the reporter nor by any of his sources. Only a single sentence dealt with the nature of the assessments: "Several professors said they tried to write exams on which it was hard to cheat, posing questions that outside resources would not help answer." Even here, the intent appeared to be foiling cheaters rather than improving the quality of assessment and instruction. To put it differently, the goal was to find ways to prevent students from being *able* to cheat rather than addressing the reasons they *wanted* to cheat. Detecting or deterring cheating does nothing to address the "educational damage" caused by whatever systemic forces have taught students that "the final product takes precedence over learning" (Renard 1999/2000, p. 41).

It is sometimes said that students who take forbidden shortcuts with their homework will just end up "cheating themselves" because they will not derive any intellectual benefits from doing the assignment. This assertion, too, is often accepted on faith rather than prompting us to ask just how likely it is that the assignment really would prove valuable if it had been completed in accordance with instructions. A review of the available evidence on the effects of homework fails to support widely held beliefs about its benefits (Kohn, 2006). To that extent, students' violations of the instructor's rules not only may fail to constitute a moral infraction but also may not lead to any diminution of learning. The outraged condemnation of cheating, at least in such instances, may turn out to have more to do with power than with ethics or pedagogy. Perhaps what actually elicits that outrage is not a lack of integrity on the part of students so much as a lack of conformity.

A penetrating analysis of cheating will at least raise these possibilities even if it may not always lead to these conclusions. It will invite us to reexamine what comes to be called cheating and to understand the concept as a function of the context in which the label is used. Even if the reality of cheating is unquestioned, however, its causes will lead us to look at the actions of teachers as well as the (re)actions of students, and at classroom and cultural structures as well as individual behaviors. Such a perspective reminds us that how we educate students is the dog; cheating is just the tail.

REFERENCES

Anderman, E. M., Griesinger, T., & Westerfield, G. (1998). Motivation and cheating during early adolescence. *Journal of Educational Psychology, 90*, 84–93.

Anderman, E. M., & Midgley, C. (2004). Changes in self-reported academic cheating across the transition from middle school to high school. *Contemporary Educational Psychology, 29*, 499–517.

Butterfield, F. (1991). Scandal over cheating at M.I.T. stirs debate on limits of teamwork. *New York Times*, May 22, A23.

French, M. (1985). *Beyond power: On women, men, and morals*. New York: Summit.

Garavalia, L., Olson, E., Russell, E., & Christensen, L. (2007). How do students cheat? This volume.

Glater, J. D. (2006). Colleges chase as cheats shift to higher tech. *New York Times*, May 18, A1, A24.

Howard, R. M. (2001). Forget about policing plagiarism. Just *teach*. *Chronicle of Higher Education*, November 16, B24.

Jacob, B. A. (2001). Getting tough? The impact of high school graduation exams. *Educational Evaluation and Policy Analysis, 23*, 99–121.

Johnson, D. W., & Johnson, R. T. (1989). *Cooperation and competition: Theory and research*. Edina, MN: Interaction Books.

Kohn, A. (1992). *No contest: The case against competition*. Rev. ed. Boston: Houghton Mifflin.

Kohn, A. (2000). *The case against standardized testing: Raising the scores, ruining the schools*. Portsmouth, NH: Heinemann.

Kohn, A. (2006). *The homework myth*. Cambridge, MA: Da Capo Press.

Labaree, D. F. (1997). *How to succeed in school without really learning: The credentials race in American education.* New Haven: Yale University Press.

Marchant, G. J., & Paulson, S. E. (2005). The relationship of high school graduation exams to graduation rates and SAT scores. *Education Policy Analysis Archives, 13.* Available at: http://epaa.asu.edu/epaa/v13n6.

Reardon, S. F., & Galindo, C. (2002). Do high-stakes tests affect students' decisions to drop out of school? Evidence from NELS. Working Paper 03-01, Population Research Institute, The Pennsylvania State University.

Renard, L. (1999/2000). Cut and paste 101: Plagiarism and the net. *Educational Leadership,* December/January, 38–42.

1

THE PSYCHOLOGY OF ACADEMIC CHEATING

ERIC M. ANDERMAN AND TAMERA B. MURDOCK

Academic cheating is extremely common in educational institutions. Cheating undermines the use of assessment data as both indicators of student learning and as sources of feedback to teachers for instructional planning. Although cheating appears to increase as students move through the K–12 school system, no age group is exempt from acts of academic dishonesty. For example, research indicates that cheating occurs among elementary school children (Kanfer & Duerfeldt, 1968), in middle and high school by adolescents (Anderman, Griessinger, & Westerfield, 1998; Murdock, Hale, & Weber, 2001), and as late as college (Newstead, Franklyn-Stokes, & Armstead, 1996) and even graduate school (Baldwin, Daugherty, Rowley, & Schwarz, 1996). Although most individuals associate cheating with students, research indicates that educators may also become involved in forms of cheating, in order to bolster their students' scores in high-stakes assessment systems (Kane & Staiger, 2002).

Academic cheating can be viewed from a number of disciplinary and theoretical perspectives. Indeed, it has been studied from the domains of education (Cizek, 1999), sociology (Black, 1962), philosophy (Green, 2004), and economics (Kerkvliet, 1994). Whereas each of these vantages offers something in terms of understanding how and why cheating occurs, in the final analysis, when individuals engage in any type of cheating behavior, they are making the decision to engage in that behavior. This decision,

Copyright © 2007 by Academic Press, Inc.
All rights of reproduction in any form reserved.

which occurs within the mind of the individual, is inherently *psychological* in nature. This book therefore focuses on the psychological aspects of academic cheating.

PSYCHOLOGY AND CHEATING

Cheating on academic work involves a diverse array of psychological phenomena, including learning, development, and motivation. These phenomena form the core of the field of educational psychology.

From the perspective of learning, cheating is a strategy that serves as a cognitive shortcut. Whereas effective learning often involves the use of complex self-regulatory and cognitive strategies, cheating precludes the need to use such strategies. Thus students may choose to cheat either because they do not know how to use effective learning strategies or simply because they do not want to invest the time in using such strategies.

From a developmental perspective, cheating may occur in different quantities and qualities depending on students' levels of cognitive, social, and moral development. Whereas cheating tends to occur less in younger children than in adolescents (Miller, Murdock, Anderman, & Poindexter, this volume), these developmental differences are due to changes both in students' cognitive abilities and in the social structures of the educational contexts in which children and adolescents interact. For example, cheating may be more likely to occur in middle and high school classrooms than in elementary school classrooms because the instructional practices used in middle schools and high schools are more focused on grades and ability than is the case in elementary schools (Anderman & Midgley, 2004; Anderman & Turner, 2004).

From a motivational perspective, learners report many different reasons for engaging in academic cheating (Murdock & Anderman, in press). Some students cheat because they are highly focused on extrinsic outcomes such as grades; others cheat because they are concerned with maintaining a certain image to themselves or to their peers; still others cheat because they lack the requisite self-efficacy to engage in complex tasks or because of the types of attributions they have developed.

In the present volume, we have invited leading researchers from the field of educational psychology to report on various psychological factors that are related to academic cheating. All of the authors have been involved either with research directly related to academic cheating or with research in areas that influence the psychological causes of cheating. More specifically, we have disaggregated cheating into four subcomponents. We have framed each of these subcomponents as a general section.

1. The Anatomy of Cheating
2. Achievement Motivation and Cheating
3. Moral and Social Motivations for Dishonesty
4. Prevention and Detection of Cheating

In the first section of this book (The Anatomy of Cheating), the authors examine some of the descriptive characteristics of academic cheaters and academic cheating. In Chapter 2, Angela Miller and her colleagues specifically examine the descriptive characteristics of academic cheaters. As readers will see, defining the characteristics of the typical cheater is a difficult task: the majority of students cheat at some point in their academic careers, and the characteristics that are typically associated with increased cheating, such as low self-efficacy, are highly influenced by the demands of the learning situation and the students' larger social context. Furthermore, as detailed by Linda Garavalia and her colleagues in Chapter 3, students have developed innumerable ways of beating the system (i.e., cheating) that reflect both individuals' goals and the changing context of learning. Increasingly, for example, students are relying on technological ways of cheating that were not part of the educational landscape a decade ago. Understanding students' decisions to cheat requires recognition of the heterogeneity of students' definitions of what behaviors constitute dishonesty and the assessed severity of these various behaviors.

The chapters in the second section of the book (Achievement Motivation and Cheating) explore students' reasons for engaging in cheating behaviors, from a variety of psychological perspectives. The first three of these chapters utilize theories of achievement motivation as frameworks for student cheating. Consistent across these chapters is a focus on the relation between students' academic dishonesty and the psychological meaning that students assign to the learning task. When tasks are seen as valuable in their own right, cheating occurs less frequently than when they are completed for other reasons. This perspective is elucidated in Chapter 4 by Gregory Schraw and his colleagues in an examination of the relations between "interest" and engagement in cheating behaviors as well as in Chapter 5 by Eric Anderman, who examines the relations between one of the most widely cited theories of academic motivation (goal orientation theory) and academic cheating. The next chapter, by Jason Stephens and Hunter Gehlbach, more specifically examines motivational profiles of students who report cheating. Working within the social-psychological framework of decision theory, in Chapter 7 David Rettinger develops an elaborate model of the steps leading to students' decisions to cheat. These chapters continually remind us that while the decisions to cheat may be psychological in nature, modification to both the curriculum and the classroom task structure may help reduce dishonesty.

In the next section, Theresa Thorkildsen and her colleagues and Lynley Anderman and her colleagues examine the relations of various moral and social motivational factors to academic cheating. In line with both self-determination theory from psychology and social bond theory from sociology, in Chapter 9 Anderman et al. address the question, "Does having a strong sense of belonging in one's school environment lessen the likelihood that students will cheat?" In contrast, Thorkildsen et al. take a more cognitive approach, focusing on the ways schools foster students' moral engagement. Both of these chapters remind us that engaging in cheating is more than a means of accomplishing an achievement goal: there are social and ethical components to dishonesty that influence students' decisions to engage in these behaviors.

In Chapter 10, Tamera Murdock and Jason Stephens explore how students can frame cheating so that it is viewed as more or less wrong, and for which they are more or less responsible. Given the strong relations between these belief systems and students' actual dishonest behavior, understanding students' reasoning processes when they decide to cheat might help develop social-cognitive interventions.

The final section of the book examines the prevalence and detection of cheating. These include responses by instructors, parents, and the larger society. The chapter written by Walter Haney and Michael Clarke (Chapter 11) examines the psychological aspects of online testing, which is a current and important trend that has seldom been considered in the cheating literature. Sharon Nichols and David Berliner's chapter, Chapter 12, examines cheating within the context of high-stakes testing.

These authors delineate the ways in which the numerous contextual factors that appear to be linked to increased cheating by students in our society are also creating more temptations and rewards for teachers themselves to be dishonest.

REFERENCES

Anderman, E. M., Griessinger, T., & Westerfield, G. (1998). Motivation and cheating during early adolescence. *Journal of Educational Psychology, 90*, 84–93.

Anderman, E. M., & Midgley, C. (2004). Changes in self-reported academic cheating across the transition from middle school to high school. *Contemporary Educational Psychology, 29*, 499–517.

Anderman, E. M., & Turner, J. C. (2004). Changes in academic cheating across the transition to middle school. Paper presented at the American Educational Research Assocation, San Diego, CA.

Baldwin, D. C., Daugherty, S. R., Rowley, B. D., & Schwarz, M. D. (1996). Cheating in medical school: A survey of second-year students at 31 schools. *Academic Medicine, 71*(3), 267–273.

Black, D. B. (1962). The falsification of reported examination marks in a senior university education course. *Journal of Educational Sociology, 35*(8), 346–354.

Cizek, G. J. (1999). *Cheating on tests: How to do it, detect it, and prevent it.* Mahwah, NJ: Lawrence Erlbaum.

Green, S. P. (2004). Cheating. *Law and Philosophy, 23*(2), 137–185.

Kane, T. J., & Staiger, D. O. (2002). The promise and pitfalls of using imprecise school accountablity measures. *Journal of Economic Perspectives, 16*(4), 91–114.

Kanfer, F. H., & Duerfeldt, P. H. (1968). Age, class standing, and commitment as determinants of cheating in children. *Child Development, 39*(2), 545–557.

Kerkvliet, J. (1994). Cheating by economics students: A comparison of survey results. *Journal of Economic Education, 25*(2), 121–133.

Miller, A., Murdock, T., Anderman, E. M., & Poindexter, A. L. (this volume). Who are all these cheaters? Characteristics of academically dishonest students. In E. M. Anderman & T. B. Murdock (Eds.), *Psychology of Academic Cheating.* Boston: Elsevier.

Murdock, T. B., & Anderman, E. M. (in press). Motivational perspectives on student cheating: Towards an integrated model of academic dishonesty. *Educational Psychologist.*

Murdock, T. B., Hale, N. M., & Weber, M. J. (2001). Predictors of cheating among early adolescents: Academic and social motivations. *Contemporary Educational Psychology, 26*(1), 96–115.

Newstead, S. E., Franklyn-Stokes, A., & Armstead, P. (1996). Individual differences in student cheating. *Journal of Educational Psychology, 88,* 229–241.

PART

I

THE ANATOMY
OF CHEATERS

2

WHO ARE ALL THESE CHEATERS? CHARACTERISTICS OF ACADEMICALLY DISHONEST STUDENTS

ANGELA D. MILLER, TAMERA B. MURDOCK, ERIC M. ANDERMAN, AND AMY L. POINDEXTER

INTRODUCTION

In the 2000 John Stockwell film, *Cheaters*, students from an underprivileged, underfunded, rundown public school steal a test in order to win an academic decathlon title that has consistently been defended by an uppity private school. Although talent got them to the state competition, they resort to cheating in this final competition to guarantee a win and then are plagued with investigations, media, and a lawsuit, all in an attempt to obtain glory for the underdog. If the ability to win was apparent, why would students resort to cheating? Certainly this act is aggrandized and sensationalized for the big screen; however, cheating is more than a film topic and far too serious and extensive for it to be considered as only a subject of entertainment. In life, as in the film, many students who are capable of achieving high grades through their own efforts are cheating as a way to get ahead.

Cheating is also in the news, garnering special segments on major network news programs explaining the elaborate schemes and technological aids used by cheaters. With the arrival of David Callahan's (2004) book *The Cheating Culture*, discussion of a moral compass and a cheating society has emerged across the country from the classroom to the boardroom.

Copyright © 2007 by Academic Press, Inc.
All rights of reproduction in any form reserved.

Cheating can be considered an epidemic according to many statistics available on the prevalence of cheating behaviors; one-third of elementary-age students admit to cheating (Cizek, 1999), and approximately 60 percent of middle school students cite cheating as a major problem in schools (Evans & Craig, 1990). Seventy-four percent of high school students admit to cheating on tests (McCabe, 2001), and among college students cheating rates are as high as 95 percent (McCabe & Trevino, 1997). Cheating rates appear to be increasing because these numbers are substantially higher than those in earlier studies; however, it could be that students are just more willing to admit to cheating these days, perhaps owing to changes in social norms.

This chapter examines the demographic, academic, behavioral, and personality characteristics of students who are either self-identified or observed cheaters in recent research. Studies that examine the personal characteristics of students who engage in cheating behaviors constitute a large portion of the research on academic cheating. Following a large-scale study of academic integrity (Bowers, 1964), most of the research in this area has focused on students' individual factors including gender, GPA, competition, self-esteem, and work ethic. Beginning in 1990, the literature shifted to studies that focused more on contextual factors such as classroom characteristics (Murdock, Miller, & Kohlhardt, 2005; McCabe & Trevino, 1993; McCabe, Trevino, & Butterfield, 2002) and peer group norms (Robinson, Amburgey, Swank, & Faulker, 2004). Across all of the extant research, the majority of the studies have concentrated on students at the college level, including a comprehensive review of the literature in which both prevalence and correlates of cheating were examined (Whitley, 1998).

Reports of cheating behavior vary greatly; Whitley (1998) reported prevalence statistics from 5 to 95 percent across 46 studies of cheating. The variation in these numbers is most likely due to varied definitions of cheating as well as to measurement methods; however, across the studies reviewed, Whitley found that although cheating behavior could be predicted by certain contextual variables, individual characteristics also increased the odds of cheating. Another, more recent, review of the literature reached similar conclusions: even though much of the research has turned to contextual factors, there are still many individual difference variables that are related to cheating behaviors (McCabe, Trevino, & Butterfield, 2001). These variables, which have been prominent in the last decade of cheating research, are the focus of this chapter.

Each of the first four sections of this chapter discusses one of the categories of characteristics that have been examined in relation to cheating: demographic, academic, motivational, and personality characteristics. Within these sections, we selected those factors that either have been the most studied (i.e., age) or are just starting to emerge in the literature as having promise for predicting cheating (i.e., impulsivity). Our fifth and final section provides suggestions for educators and future research in the field.

DEMOGRAPHICS

Individual differences in student cheating have been studied in relation to multiple demographic factors including age, gender, ethnicity, socioeconomic status, and school type. In this review we focus on two of the most commonly studied factors, gender and age, as well as on cultural differences in cheating, which is an emerging area of research. A short section on other less frequently studied demographic characteristics is also included as evidence of the breadth of demographic variables that have been studied in relationship to cheating behaviors.

GENDER

Numerous studies have specifically investigated gender differences in cheating behaviors. Most of these studies have operationalized cheating behaviors based on student self-report and have found that males report more cheating than females (Calabrese & Cochran, 1990; Davis, Grover, Becker, & McGregor, 1992; Michaels & Miethe, 1989; Newstead, Franklyn-Stokes, & Armstead, 1996). Some exceptions to this finding include Haines, Diekhoff, LaBeff, and Clark (1986), who found no differences in reported cheating between genders, and Jacobson, Berger, and Millhan (1970), who found that females cheated more than did males. Moreover, gender has been infrequently found to be a significant predictor of self-reported cheating behavior when it was included in studies as a control variable rather than the focus of the investigation (Anderman, Griesinger, & Westerfield, 1998; Anderman & Midgley, 2004; Genereux & McLeod, 1995; McCabe & Trevino, 1997).

One possible explanation for these inconsistent relations between gender and cheating is that gender differences may be moderated by both the context in which the cheating took place and the way in which cheating was assessed in a given study. For example, Calabrese and Cochran (1990) found that females admit to cheating as frequently as males when the cheating is done to help another (altruism). In a meta-analytic review of college and university cheating studies, Whitley, Nelson, and Jones (1999) found that although the effect size for gender attitude differences was medium sized, the actual effect size for behavior was very small. That is, although men and women expressed different attitudes and opinions about cheating, they were almost equally likely to engage in the behavior. Thus, whether males are actually cheating more than females is speculative; the reported differences might be more a result of the self-reporting procedures used rather than actual cheating occurrences.

AGE AND GRADE-LEVEL DIFFERENCES

Whereas age refers to a characteristic of students, grade level is often treated as a contextual factor that affects dishonesty. For the purposes of this chapter, we are reviewing the two variables simultaneously because the strong correlation between chronological age and progression through the educational system makes it difficult to disentangle these effects.

Age appears to have a curvilinear rather than a linear relation to academic cheating; as students progress through the K–12 system, cheating rates increase, but these rates then decline throughout college and again in graduate and professional programs (Franklyn-Stokes & Newstead, 1995; Haines, Diekhoff, LaBeff, & Clark, 1986; Jensen, Arnett, Feldman, & Cauffman, 2002; Sheard, Markham, & Dick, 2003). For example, in the only longitudinal study of cheating behavior, Anderman and Midgley (2004) found that self-reported rates of cheating in math classes increased across the transition from middle school (grade 8) to high school (grade 9). In this instance, the authors explain this increase as being related to contextual changes in students' learning environments. Specifically, cheating only increased for those students whose grade-related transition resulted in their moving to a setting that was more competitive and grade-focused than the one they had experienced during middle school (see Anderman, this volume).

At the same time, several studies report that younger students are more likely to cheat than older students when comparisons are made between high school students and college students (Jensen et al., 2002). Moreover, studies that focus only on college-age students, comparing underclassmen to upperclassmen (Franklyn-Stokes & Newstead, 1995; Haines et al., 1986) and undergraduates to graduates (Sheard et al., 2003), find that cheating declines with age. However, as the authors of these studies note, there are many confounds in trying to tease out age effects at the college level. For example, among college-age populations, age is usually highly correlated with year in school. Due to "weeding out" processes, fewer weaker students are left in the junior and senior years versus the freshman year.

Rates of cheating are not the only differences we see as a function of age (grade level); the way in which students cheat also changes. For example, Jensen et al. (2002) found that high school students cheat more than college students in four specific ways, including copying another student's work, copying from someone else's exam, and letting another student copy in the same two situations. Comparisons among undergraduate and graduate students suggest that, although rates of cheating change across schooling level, the reasons for students' dishonesty are largely constant. Graduate versus undergraduate students in an information technology program admitted to fewer incidences of cheating. However, across both groups, pressure due to limited time and worry about potential failure were the most common reasons stated for engaging in cheating behaviors (Sheard et al., 2003). In contrast, although nontraditional (and typically older) students cheat less often than traditional students, Franklyn-Stokes and Newstead (1995) argue that these differences are less a reflection of age than of the fact that that older students have different reasons for being in school (i.e., educational goals) than their younger peers. In short, the context of students' lives and of their schooling changes as they go through school. Schooling itself may become harder and more competitive at higher grade levels, thus leading to a prediction of more cheating. At the same time, with maturity, students may be developing their own internalized reasons for getting an education; as such, their behavior may be less influenced by the context of which they are a part. As seen here, the study of demographic variables such as age and grade level is undeniably intertwined with contextual factors of cheating as well.

In addition to contextual changes that make it difficult to determine the meaning of age effects, there are also individual characteristics that covary with age and that may be responsible for age-linked trends in dishonesty. Some argue, for example, that age effects may actually be related to changes in levels of moral development (Newstead et al., 1996), though the links between moral maturity and cheating are also quite inconclusive (see Murdock & Stephens, this volume). Age has also been related to the development of self-control, which is negatively correlated with cheating behavior (discussed later on in this chapter). In short, although age and cheating are related, many potential developmental and contextual reasons might explain these relations, and our understanding of the factors is quite limited.

Finally, whereas cheating rates tend to decline in general across the transition from high school to college, and continue to decline with students' age, it also appears that students who begin cheating earlier in life are likely to continue to engage in this behavior in college. In one of the first major studies of cheating (Bowers, 1964), 64 percent of the students who cheated before entering the university also cheated at the university level, while 67 percent of the students who had no prior cheating experience did not cheat at the university level. Three decades later, Davis and Ludvigson (1995) found that 98 percent of students who cheated before entering the university also continued to cheat while at the university. Thus, while age may be a correlate of cheating, it also appears that there is stability across age when the unit of analysis is the individual over time.

CULTURAL DIFFERENCES

Many studies have attempted to compare and contrast demographic variables across cultures, with a goal of examining the similarities and differences of "the portrait of a cheater" across cultural boundaries. For

example, one of the most frequently cited studies across the cheating literature conducted by Newstead and colleagues (1996) found that the typical
college student who cheats is a younger male of lower ability in science
courses. This data was collected in England and parallels most of the
studies carried out in the United States. Other cross-cultural studies
attempt to determine the extent to which cheating reflects the type of
educational system in which a student studies and the societal values in
which the student is raised. In general, these studies suggest that cheating
is universal and occurs in all educational systems, although the levels of
cheating are not constant and the perception of its severity and consequences varies according to cultural and societal differences.

Eleventh grade students (or equivalent) in Germany reported less cheating and, consequently, viewed cheating as a less serious problem in their
school than either Costa Rican or American students (Evans, Craig, &
Mietzal, 1993). German students also expressed more liberal views of
passive cheating behavior, including sharing of information, than did students from the other two countries. This may reflect the fact that the educational environment in Germany is less competitive at the high school
Gymnasium level than in the earlier years when students are competing to
get into a particular program of study. Therefore, helping one another
illegally (i.e., cheating) may just be a more integrated part of the cooperative environment of the Gymnasium.

A second cross-cultural study compared cheating at the college level
between American and Australian students (Davis, Noble, Zak, & Dreyer,
1992). Higher rates of cheating were identified among American students.
The authors hypothesized that these differences reflected cultural differences in the extent to which the groups of students value learning versus
grades. Although they demonstrate that Americans do have higher grade
orientations than Australian students, these results were based on different
samples, and thus links between culture, goals, and cheating could not be
firmly established.

Even though cross-cultural differences appear to exist in the type and
amount of cheating students engage in, it is also evident that no culture
that has been studied is completely free from academic dishonesty.

OTHER DEMOGRAPHIC VARIABLES

In this last section, we acknowledge the many other demographic characteristic variables. An excellent review paper by Whitley (1998) details
the effect sizes for a myriad of variables that have been investigated in
relation to cheating. Here we briefly note some of the more interesting
findings related to ethnicity, socioeconomic status (SES), marital or work
status, and religion.

Calabrese and Cochran (1990) studied cheating at the high school level
and found that Caucasian students are more likely to cheat than their

Hispanic or Asian counterparts. This study also found that high SES private school students reported more cheating than students from the comparison public high school. However, overall very few differences in cheating behaviors have been found between ethnic groups as reported in studies using ethnicity as a control variable.

Diekhoff and colleagues (1996) found that the number of hours worked was negatively related to cheating. This finding seems to be counterintuitive to the time pressure excuse preferred by students in studies of college cheating. In this study, the researchers also found that unmarried students were more likely to cheat than married students, a finding that had also been reported in an earlier study (Haines et al., 1986).

Finally, an additional demographic variable that has received some attention in relationship to cheating is religion. The findings on cheating and religion have been diverse. A recent natural setting experiment conducted with college students enrolled in a dual religious and liberal arts curriculum found less cheating in religious study courses than in liberal arts courses (Rettinger & Jordan, 2005). This study also maintained a negative correlation between religiosity and cheating while controlling for motivation and attitudes toward cheating. Two earlier studies examining moral attitudes found no relationship between religiosity and cheating (Michaels & Miethe, 1989; Smith, Ryan, & Digging, 1972). Sutton and Huba (1995) found that religious students had a lower threshold for considering a task to be classified as cheating and were also less likely to justify cheating behaviors.

Although demographic variables are popular and extensive in the research literature, they are, in most cases, fixed variables that lead only to shocking statistics and very general profiles of cheaters, drawing attention to the problem of academic dishonesty. They do not identify the processes underlying the decision to cheat, or lead to constructive recommendations on ways to reduce the amount of cheating that occurs in our schools and universities. However, identifying the behavioral characteristics of students who cheat can impact the ways in which cheating is detected and curbed in classrooms.

ACADEMIC CHARACTERISTICS

Students' school characteristics and the behaviors they exhibit while with their peers at school are common variables of interest in academic cheating research. The student's ability level as reported by test scores or grade point average (GPA), as well as self-reported self-efficacy, are frequent correlates of cheating behaviors. Whereas the most frequent finding is that students of lower ability are more likely to report engaging in cheating behaviors, there are also some mixed findings that demonstrate significant levels of cheating behaviors exhibited by higher ability students as well (Taylor, Pogrebin, & Dodge, 2002). Most of the literature investigates cheating at the college level; however, several studies focus on the high school level, and some investigate the differences between high school and college levels. Students involved in sororities and fraternities as well as athletics have also reported higher rates of academic cheating.

ABILITY

Newstead and colleagues (1996) conducted survey research among 943 college students in England and determined that cheating was more common among males, students of lower academic ability, younger students, and science students. Students were questioned about a list of 21 different cheating behaviors ranging from plagiarism to impersonating another student to sit for an exam. Males reported cheating more frequently than did females. They also found an interaction between gender and academic achievement (as measured by GPA) with larger differences between the sexes at lower achievement levels and nonsignificant differences in frequency of cheating at the highest achievement level. Lower achieving males reported cheating more frequently than did lower achieving females; this gender gap decreases as achievement increases.

The interaction between gender and academic achievement reported by Newstead and colleagues (1996) points to the complex relationship between ability and cheating. Researchers commonly refer to ability as a correlate of cheating behaviors, and it is generally believed that the students of lower ability are more likely to engage in cheating behaviors; however, the construct of ability is conceived in many different ways across the cheating literature. Although the most common definition of ability is grade point average, which is usually reported as inversely correlated with self-reported cheating (Baird, 1980; Diekhoff et al., 1996; Genereux & McLeod, 1995; Haines et al., 1986; Leming, 1980; Michaels & Miethe, 1989), ability has also been measured as academic aptitude (Kelly & Worrell, 1978) and task performance (Malinowski & Smith, 1985). Overwhelmingly, the relation between ability and cheating is accepted to be an inverse relation, often endorsing the assumption that the lower achiever has less pressure and lower expectation to succeed and experiences less risk when engaging in cheating.

Conversely, other researchers have suggested that students who have higher goals and experience higher pressure to succeed are more likely to cheat (Whitley, 1998). An emphasis on competition is also related to increased pressure to cheat (Anderman et al., 1998); in these situations, it is more likely that the higher achieving student would succumb to the increased pressure. Taylor and colleagues (2002) undertook a qualitative study to examine high-aptitude students and the pressure to succeed as related to academic dishonesty; they found that the pressure these high achievers experience in highly competitive environments was a primary reason many of the students interviewed stated for engaging in dishonest academic behaviors. It appears that even though ability is frequently listed as a correlate of cheating, much more is going on than a simple inverse relationship with cheating.

SUBJECT AREA

Often it is the hard sciences, business, and engineering areas that are identified as disciplines with higher incidences of cheating as compared to the fine arts and social sciences (Bowers, 1964; Davis & Ludvigson, 1995; Newstead et al., 1996). The Newstead and colleagues (1996) study cited in the preceding section also found that cheating was more common among science and technology students than in other academic disciplines including health, social sciences, and the humanities. The only reported differences between these groups were reasons for studying, which were also investigated with an open-ended response question in which answers were categorized into three groups: those avoiding the "real world", called *stopgap* students; those trying for a better job or more money; and those pursuing an education for personal development. Cheating frequency was highest with the stopgap group and lowest for the personal development group; all three groups were significantly different. Education and social work students were least likely to be classified in the stopgap group, which had the highest incidence of reported cheating. Similar findings were also reported by Newstead et al. (1996), who found that rates of cheating were higher in science and technology courses, and by Schab (1991), who found higher levels of cheating in science and math.

INSTITUTIONS AND ORGANIZATIONS

Again, in the academic realm as in the demographic variable section, we find that personal characteristics become entangled in the context. Although some may consider membership in organizations and living groups a contextual variable, it is often treated as a characteristic of the student in research. For instance, in his study of cheating at a small college campus, Dawkins (2004) notes the seeming "group mentality" of cheating whereby students who live together in dormitories and attend campus events as a group rather than as individuals are more likely to engage in cheating. It is likely that this commonality is also germane to larger campus settings, for it appears that students at larger institutions tend to cheat more than students at small private colleges (Brown & Emmett, 2001).

College students at nine different public institutions were surveyed, and fraternity or sorority membership was found to be correlated with self-reported cheating (McCabe & Bowers, 1996). A larger examination of this same data also investigated the role of athlete status and involvement in extracurricular activities in general and found them both to be correlated with cheating (McCabe & Trevino, 1997). In this study, the authors hypothesize that extracurricular activities create more demands on college students' time and therefore increase the propensity of these heavily involved students to cheat. Overall, this study argues that individual characteristics contribute far less to cheating than contextual influences; however, both sets of variables in this study only accounted for approximately 30 percent of the variance in the dependent variable of cheating. These findings were replicated by Storch & Storch (2002), who found that higher rates of cheating occurred among sorority and fraternity members than their non-Greek counterparts. Self-reports were used asking students about various forms of cheating, from copying homework to use of illegal cheat sheets on exams. Level of participation in the Greek system was positively correlated to cheating behaviors. Storch and Storch also found that athletes reported cheating more than nonathletes; however, the effect size was very small.

Academic characteristics including ability, subject area, and organizational membership have all been found to be related to students' engagement in cheating behaviors. Although these characteristics of cheaters are of particular interest in terms of gaining knowledge about who is likely to cheat, they are limited in terms of contribution to efforts to curb and control cheating.

MOTIVATION

In the past five years, a large increase has taken place in studies of cheating framed within theories of achievement motivation (see Murdock & Anderman, 2006). Given that cheating is, in fact, one way to improve one's achievement outcomes, it is not surprising that variations in achievement motivation have also been linked to dishonesty. In this chapter, we focus on cheating in relation to two indicators of achievement motivation: students' academic self-efficacy and their academic goals.

SELF-EFFICACY

Self-efficacy refers to people's task-specific beliefs in their ability to execute the actions required to bring about a desired performance accomplishment (Bandura, 1986). Students with higher levels of academic self-efficacy are more confident in their abilities to achieve their goals and persist more in the face of difficulty (Pajares, 1996). Not surprisingly, data also suggest that students who are less confident that they can master a given task are also more likely to cheat. For example, one study of middle school students reported inverse relations between cheating and self-efficacy, even after controlling for the students' personal goals and a range of classroom context variables (Murdock, Hale, & Weber, 2001). Among college students, self-efficacy predicted cheating when students' actual level of achievement was controlled (Finn & Frone, 2004).

Factors that would presumably lead students to have lower self-efficacy in a given assessment situation are also predictors of dishonest behavior. Cheating decreases as the quality and quantity of study time increases (Kerkvliet, 1994; Norton, Tilley, Newstead, & Franklyn-Stokes, 2001) and also as class attendance increases (Michaels & Miethe, 1989). Similarly, students who procrastinate are also more likely to be cheaters than are those who plan their study time appropriately (Roig & DeTommaso, 1995). Together, these studies suggest that cheating occurs more often when students are unprepared and thus probably less confident or efficacious.

Self-efficacy beliefs are theorized to develop based on four types of feedback information: performance accomplishments, vicarious experiences, modeling, and physiological arousal (Bandura, 1986). Several of these sources of efficacy information have also been linked to dishonest behavior. For example, several studies suggest that students cheat more often when they have physiological evidence of low self-efficacy including a fear of failure (Calabrese & Cochran, 1990; Michaels & Miethe, 1989), test anxiety (Malinowski & Smith, 1985), and worry over their performance (Anderman et al., 1998).

It also appears that students who cheat have fewer academic successes than those who are more honest, as evidenced by relationships between dishonesty and a priori achievement. Correlational studies at the high school and college level repeatedly find small to moderate, but significant, inverse relations between self-reported cheating and GPA (Finn & Frone, 2004; McCabe & Trevino, 1993; Michaels & Miethe, 1989; Newstead et al., 1996; Roig & DeTommaso, 1995). In addition, some field-based evidence demonstrates links between actual (versus self-reported) cheating and prior achievement. When college students were given the opportunity to cheat on six self-graded quizzes, 27 percent of the students cheated, and cheating was higher among those with lower GPAs and lower final course grades (Ward & Beck, 1990). Cheating was assessed by comparing the students' self-graded papers with scores obtained by the teaching assistants who had secretly photocopied the exams before returning them to the students to grade.

Links between prior performance and cheating behavior are not entirely clear, however. A meta-analysis based on studies of college populations found that the average effect size for this relationship is small (Whitley, 1998), and high-achieving students are also known to regularly cheat

(Taylor et al., 2002). Self-efficacy judgments may provide the explanation for these seemingly disparate findings. Recall that, although achievement is a source of self-efficacy, the two constructs do not share all of their variance and that typically, in motivational studies, self-efficacy is a better predictor of actual achievement outcomes. In a recent study described above (Finn & Frone, 2004), self-efficacy beliefs not only predicted levels of cheating beyond what was explained by GPA, but self-efficacy judgments also moderated the relations between achievement and cheating. Performance was only inversely related to cheating among students with high self-efficacy; for those with lower self-efficacy, cheating and achievement were unrelated.

GOAL THEORIES AND REASONS FOR LEARNING

If you were to survey students in any school or classroom, you would find many individual differences in their stated reasons for doing (or not doing) their academic work and, at the college level, for choosing to remain in school or pursue a particular area of study. For some students, the primary motivation is an intrinsic interest in the subject area or a desire to learn or master the material (sometimes called task or mastery goals). For others, the goals are more extrinsic to the learning itself, such as getting a good enough grade to keep a sports scholarship, to cross a requirement off a list, or to keep one's parents from punishing them. A third group of students may prioritize being seen as smart and capable (ego or performance focused). Not all of these students are equally likely to engage in academic dishonesty.

Studies of cheating that are based explicitly in achievement goal theory continually confirm that cheating occurs more often among students who are not invested in learning the material. College students who endorse learning aims on the Learning and Grade Orientation Questionnaire (LOGO) (Eison, 1981) are less likely to cheat than their peers who have lower scores on the learning scale; in contrast, a stronger endorsement of a performance orientation is positively associated with cheating (Huss et al., 1993; Weiss, Gilbert, Giordano, & Davis, 1993).

Similar relations between goals and cheating have been found among younger students as well. Two studies of middle school students that measured goals using the *Patterns of Adaptive Learning Survey* (PALS) (Midgley et al., 1995) reported inverse relations between cheating and endorsement of mastery goals within a specific class (Anderman et al., 1998; Murdock et al., 2001); in the Anderman study, having a personal extrinsic goal was also a predictor of dishonesty.

Dweck and Sorich (1999) provide evidence that even elementary students' decisions to cheat may be influenced by their goals. Fifth graders were presented with a vignette depicting a student who had tried hard on an exam, thought she had done well, and failed. Those with entity views of intelligence (e.g., see intelligence as a fixed, finite trait) were more likely to say that in similar circumstances they would "try to cheat" next time than were students with views that intelligence could be improved through learning. Prior research suggests that an incremental view of intelligence underlies the adoption of mastery goals, whereas an entity view is associated with performance beliefs.

Outside of the achievement goal theory tradition, there is other evidence that goals and cheating may go hand in hand. As Schraw and his colleagues have demonstrated, there is significant evidence that cheating increases when students' intrinsic interest is low (see Schraw, this volume). Similarly, Genereux and McLeod (1995) found that extrinsic needs such as the exam grade's effect on their long-term grades and/or their ability to garner future financial support were two of the top five reasons for cheating by Canadian community college students. Among this same group of students, cheating was much more prevalent among those who were pursuing school for extrinsic reasons, such as a better paying job, than it was for those who saw schooling as fulfilling one's personal interests.

How fixed are students' goals, and how much of what they hope to achieve is affected by the environment? Even though we have presented personal goals as characteristics of individuals, significant evidence also suggests that students' goals, and perhaps student cheating, can be influenced by their learning and family environment. Furthermore, the relationship between self-efficacy and cheating is also likely to be moderated by contextual effects of the environment, as well as interactions with peers and instructors. Although these motivational characteristics and their relationship to cheating are helpful to researchers in learning about the mechanisms of cheating, they are probably best understood and will lead to more insightful research when combined with context variables.

PERSONALITY TRAITS

Whereas achievement motives (e.g., self-efficacy, goals) appear to be clear predictors of dishonesty, other more general personality traits are also related to dishonesty. In this chapter, we focus on three characteristics that are just beginning to receive attention (impulsivity, sensation-seeking, and self-control), as well as two others that have been examined more thoroughly in relation to cheating: attitudes (including moral reasoning) and locus of control.

IMPULSIVITY AND SENSATION-SEEKING

One area that has received relatively little attention in the academic cheating literature is the relation between personality variables and academic cheating. Impulsivity and sensation-seeking represent two constructs prevalent in the personality psychology literature that may be related to academic cheating.

When individuals have a high need for sensation, they need to experience novel, exciting experiences and sensations. Tendencies toward either high need for sensation or impulsivity have been found to be related to adolescents' engagement in risky behaviors (Baer, 2002; Donohew, Zimmerman, Novak, Feist-Price, & Cupp, 2000). The need for sensation probably has evolutionary roots in that humans needed to be able to react successfully to novel stimuli in order to survive (Zuckerman, 1994). In addition, research clearly indicates that the need for sensation is rooted in human biology (Zuckerman, 1988).

Individuals who are highly impulsive tend to act without thinking across a variety of situations. Researchers of impulsivity (also commonly referred to as *impulsive decision making*) conceptualize impulsivity along a continuum, wherein individuals at one end of this continuum are considered to be highly rational, and those at the other end of the continuum are highly impulsive (Langer, Zimmerman, Warheit, & Duncan, 1996). Impulsivity is also related to engagement in risky behaviors, particularly among adolescents (Donohew et al., 2000; Martins et al., 2004).

There is some debate in the personality psychology literature regarding distinctions between impulsivity and sensation-seeking. For example, in the Five-Factor Model of Personality, impulsivity and the need for sensation are represented as subcomponents under other personality processes (Costa & McCrae, 1992; McCrae & Costa, 1990; Whiteside & Lynam, 2001). Specifically, impulsivity is categorized as part of *Neuroticism*, whereas sensation-seeking is part of *Extraversion-Introversion*. In the Alternative Five-Factor Model of Personality, impulsivity and sensation-seeking receive much attention, with *Impulsive Unsocialized Sensation-Seeking* representing one of the five factors (Zuckerman et al., 1993).

There is reason to believe that individuals who are high in need for sensation or in impulsivity would be more likely to cheat. From the perspective of impulsivity, when an individual makes decisions on the basis of impulse rather than reason, that individual may be more tempted to cheat when given the opportunity. From the perspective of need for sensation, individuals high in need for sensation may be more likely to cheat because of the risky nature of cheating. Such individuals may experience reinforcement from the "rush" associated with getting away with cheating on an academic examination or assignment.

Some limited research has been done to support these hypotheses. For example, female college students high in impulsivity were more likely to cheat than females who were lower in impulsivity in a study where students could potentially falsify self-reported performance on a task in order to earn course credit (Kelly & Worrell, 1978). A more recent study by

Anderman and Cupp (2006) examining cheating behaviors in high school health classrooms indicates positive relations between academic cheating and both impulsive decision making and sensation-seeking.

Nevertheless, studies of the relations between impulsivity, sensation-seeking, and academic cheating are largely absent from the academic cheating literature. For example, Cizek's comprehensive summary of individual differences related to academic cheating does not cite one study relating cheating to either impulsivity or the need for sensation (Cizek, 1999). The study of sensation-seeking and impulsivity as they relate to academic cheating represents a broad arena for future research. The relations of sensation-seeking and impulsivity to engagement in risky behaviors are well documented. Consequently, future studies examining the perceived risk factors associated with cheating may yield positive predictive relationships between these personality variables and cheating.

SELF-CONTROL

A construct closely related to the impulsivity construct just examined is self-control. Grasmick, Tittle, Bursik, and Arneklev (1993) found that self-control and perceived opportunity were related to cheating behaviors. The lack of self-control was posited to be a personality trait that predisposes persons to engage in deviant behaviors such as cheating. Bolin (2004) examined the role of self-control, cheating attitudes, and perceived opportunity as predictors of cheating with 799 college students. He found that attitudes toward cheating fully mediated the relationship between self-control and self-reported cheating behaviors. That is, even though self-control is not directly related to cheating behaviors, it does play a significant role in a student's decision to cheat due to the relationship between self-control and attitude towards cheating.

MORAL DEVELOPMENT AND ATTITUDES
ABOUT CHEATING

Literature that examines students' beliefs about cheating typically focuses on how cheating varies as a function of students' level of moral development or their attitudes about cheating, including the use of neutralizing strategies (see Murdock & Stephens, this volume). Across this body of work, several findings are clear. In the abstract, students report that cheating is wrong; however, their behaviors are more clearly related to their more contextualized concrete attitudes about the acceptability of cheating in a specific circumstance. We provide a brief review of moral versus attitudinal predictors of cheating below.

Studies of cheating in relation to students' morality most frequently adopted Kohlbergian (1981) notions of moral development. From this perspective, moral reasoning can be viewed as progression from the most concrete lower levels (stages 1 and 2) to the highest levels (5 and 6), where moral decisions are grounded in universal ethical principles. Although students at any level might theoretically cheat (with different reasons behind it), it is also assumed that higher levels of moral development will generally be associated with fewer instances of cheating.

Moral reasoning and cheating behavior are not, however, clearly related. Whereas correlational evidence is often consistent with the morality/behavior hypothesis, the effect sizes for these results are quite small (Whitley, 1998). Moreover, studies that have looked at morality in relation to actual cheating behavior find few, if any, effects of moral reasoning on students' actual cheating. For example, Bruggeman and Hart (1996) assessed moral reasoning among high school students using Rest's Defining Issues Test (DIT; derived from Kohlberg's theory). Students were then given two tasks to do and were asked to score their performance and report their grades. Unbeknownst to the students, the performance was also independently scored by the experimenters, allowing them to determine who "cheated" in reporting their scores. Rates of cheating were found to be unrelated to level of moral reasoning.

Among studies that do find morality-cheating links, these relations are often moderated by another variable. More specifically, two studies have found that students with higher versus lower levels of morality cheat less frequently, but only when there are no incentives to do so. In the presence of incentives, the groups behave that same way (Corcoran & Rotter, 1987; Malinowski & Smith, 1985). Surveillance also interacts with level of morality when actual cheating is measured; once again, if students are not being closely watched, students with higher levels of moral reasoning behave no differently than those whose reasoning is less mature (Leming, 1978).

In contrast to students' abstract levels of moral reasoning, students' attitudes about cheating are assessed by asking them to rate the acceptability of cheating either generally or in specific situations. More lenient attitudes toward cheating have been consistently related to cheating behavior (Whitley, 1998). For example, Anderman and colleagues (1998) surveyed 285 middle school students and found that of the students who stated that cheating was unacceptable, 21.3 percent reported having cheated in the past. For those students who indicated that cheating was sometimes acceptable, 42.7 percent reported having cheated. These same links between attitudes and behaviors have been found at the high school and college levels as well.

Variations in attitudes about cheating are often assessed based on the extent to which students neutralize cheating in a specific situation. Sykes and Matza (1957) coined the term *neutralization* to refer to strategies people use to deflect responsibility for cheating away from oneself by such tactics as denying the behavior is wrong or claiming not to have had personal control over the situation/behavior. Not only do we know that cheaters are more apt than honest students to neutralize this behavior, but we also know that students neutralize more often in those classes where they cheat the most frequently (see Murdock & Stephens, this volume).

OTHER PERSONALITY CHARACTERISTICS

In addition to the more frequently examined personality traits discussed above, other traits have been linked to cheating, including personality type and locus of control. However, results have been mixed.

In an experimental study, college students with a Type A personality described as easily aroused, competitive, and aggressive were found to engage in more cheating than students of Type B personality described as more easygoing and creative (Davis et al., 1995). These results confirmed an earlier experiment by Perry and colleagues (1990). However, two non-experimental studies completed found conflicting results: there is no effect for Type A personality behaviors (Huss et al., 1993), and Type A personality is associated with lower levels of cheating (Weiss et al., 1993). Clearly, this variable is potentially related to cheating behaviors and deserves further investigation.

Locus of control has also been investigated via experimental studies, though with more consistent findings. Three studies all concluded that persons with external locus of control are more likely to engage in cheating behaviors (Forsyth, Pope, and McMillan, 1985; Karabenick & Srull, 1978; Leming, 1980).

While demographic, academic, and motivational characteristics are more popular in the educational psychology literature, there has also been a note-worthy insurgence of personality variables into the academic cheating literature. Many additional personality variables can be studied as predictors of academic cheating in future research. Some of these include extraversion, agreeableness, conscientiousness, neuroticism, and openness.

A NOTE ON METHODOLOGY

Concerned that numbers of cheaters may be inaccurate owing to reliance on self-report techniques commonly used in the majority of cheating studies, some researchers have attempted to investigate the construct of cheating using other methods and techniques. Kerkvliet (1994) critiqued the direct questioning technique often used in college cheating studies by using two methods: (1) direct questioning about cheating, and (2) a randomized response method. He also investigated students' personal habits, socioeconomic status, and academic performance as related to cheating behaviors. The rate of cheating from the randomized response model was 42 percent as compared to only 25 percent from direct questioning.

Fraternity or sorority membership was also related to higher rates of cheating. Contrary to similar studies, men were found to be less likely to cheat than females. Undergraduate students with college-educated parents were also more likely to cheat. McCabe (2001) used focus groups to investigate students' beliefs about cheating. Thirty-two high school students participated in two-hour-long group discussion sessions. Almost all students in the study admitted to cheating. In the words of one student, "Cheating is just going to be a thing that continues. I can't say forever, but I don't know who is going to stop it or when it's going to stop" (p. 685). For more information on methodological approaches to examining cheating behaviors, see Haney and Clarke, this volume.

DISCUSSION

Who are these cheaters? They are students—male and female, intelligent and lazy, athletes and nerds—with a need to get ahead, a fear of failure, or pressure to succeed who discover cheating as an overlooked, often ignored, and technologically simple transgression. Conclusions from the multitude of cheating studies may be varied and sometimes even conflicting; however, the one conclusion we can draw from them is that there is no clear-cut profile of a student who cheats. The decision to cheat and the reasons leading up to the decision are complex and include a variety of personal and situational factors. While this chapter attempted to focus on personal characteristics, we also acknowledge that many of the characteristics reviewed and explored here do not stand alone and are part of a more complex relationship with contextual variables, which the following chapters will examine in detail. Many of the factors that predict cheating can be considered as either individual or contextual characteristics, depending on the reader's viewpoint as well as operational definition of the variable in cheating studies. For example, one's country of origin can be seen as an individual difference variable or as an assessment of culture, which is contextual. It is difficult and perhaps even meaningless to attempt to separate the characteristics of those who cheat from the contexts in which cheating occurs and other situational factors that influence their cheating behaviors. Thus, we have managed to review and comment on the most frequently referenced descriptive variables that are attached to profiles of students who cheat and have organized them into groups of demographic, academic, motivational, and personality characteristics, each with its own pattern of relationships with academic cheating behaviors and challenges of interpretation.

Demographic variables including gender, age, culture, and religion are in many studies related to cheating behaviors; however, as discussed in this chapter, the relations between these demographic variables and cheating are quite complex, as other variables confound these seemingly simple predictors of cheating. From the plethora of studies that include the variables of age or grade level and find statistically significant relationships or differences between groups, we have been unable to tease out effects that are directly attributable to the age of the student who cheats. However, rather than focusing on our inability to do so, perhaps we should be asking why we would want to do so. Studies that successfully link demographic variables to cheating behaviors in reality contribute very little to the body of academic cheating research unless we can identify the mechanisms that account for these differences and that may be amenable to intervention. The remaining chapters in this book that examine context and interplays between context and demographics are more likely to lead to more useful and informative results that might someday change how we police and put an end to cheating in classrooms.

We expect that the current declining trend in the number of demographic and characteristics studies focusing on gender, age, and grade-level differences will continue because these studies contribute little of value. We accordingly look to studies that incorporate more malleable academic and motivational characteristics. These variables, especially when combined with contextual factors, may eventually provide results that will be of greater value to school administrators, classroom teachers, and parents of students who populate our classrooms. These characteristics include such variables as ability, subject area, student self-efficacy, and educational goals. These variables, in relationship with teacher characteristics, peers, and classroom environment, should be examined concurrently in research studies in hopes of finding out how and why students choose to engage in cheating in various settings. Some research demonstrates that students are selective and choose to cheat in some classrooms/subjects but not in others (Stephens, 2004). Furthermore, we know from motivation studies that context and teacher behaviors are related to personal motivational outcomes such as self-efficacy and personal goals (Murdock et al., 2004), which are also correlated with engagement and attitudes toward cheating (Anderman et al., 1998; Finn & Frone, 2004). If these contextual and personal characteristics are examined within a single study, more valuable information can be obtained about the mechanisms of cheating, which may lead to our development of ways to deter the maladaptive behavior.

LIMITATIONS AND DIRECTIONS FOR
FUTURE RESEARCH

One of the major limitations of interpreting the vast numbers of cheating studies is the definition of cheating itself. Each study defines cheating in its own terms, and different combinations of questions and qualifications are required to label a student as a "cheater" and include them in the analyses of cheating behaviors. Furthermore, many studies include their own definition of what cheating entails, whereas others remain vague on the use of the term, allowing respondents to invoke their own definitions of right and wrong in academics. There is some concern over the reports of rampant cheating rates, and the numbers and statistics are overwhelming. Some researchers suggest that these numbers are inflated simply because of the expansion of questionnaires to include multiple practices labeled as cheating (Brown & Emmett, 2001). These authors argue that cheating did not increase at all between 1933 and 1999 and that the inflated numbers seen in research articles are a product of the number of cheating practices included in the study definition of cheating. In addition, Brown and Emmett also found that other potential explanations for the cheating increases were not upheld, including increasing sample sizes and students' willingness to overstate their engagement in academic cheating on anonymous questionnaires. These concerns about variances in the prevalence of cheating are valid and deserve a more complete investigation; future meta-analytic studies should attempt to organize studies according to the types of cheating examined as well as the context of the study and to compare results accordingly. For more information about types and methods of cheating, see Garavalia, this volume.

In terms of future research, cheating studies in our review span secondary through graduate school levels; very little research is available on younger students, and many studies report that students who cheat continue to cheat throughout their academic careers. Identifying when cheating behaviors first occur would seem to be an important missing element of the current research literature.

Also, as previously stated, there is a great need for research that incorporates these characteristics of cheaters, especially academic and motivational characteristics, with contextual factors in order for us to more fully understand why students engage in cheating behaviors. By conducting such studies, researchers will perhaps be better able to explain some of the discrepant conclusions reached in purely demographic survey studies. For example, as we have already shown here, many of the contradictory findings between cheating and age can likely be explained by confounds with other demographic as well as situational variables.

Lastly, additional personality variables need to be considered. Whereas the literature presents different conceptualizations of personality, several prominent constructs appear across a variety of theories (e.g., extraversion, conscientiousness, neuroticism). Future studies involving these and other personality-related constructs may further illuminate our understanding of the mechanisms that underlie academic cheating.

Who cheats? In this chapter we have outlined many of the individual characteristic variables in an attempt to answer this question and have found that individual characteristics cannot be completely separated from the context in which the cheating transgression occurs. Then instead of asking "who cheats?" perhaps we should ask "when and why do some students cheat?" and "under what circumstances is cheating more likely to occur?" The answers to these questions would be much more beneficial to attempts at reducing the occurrence of the negative behavior. Some of the studies examined here, which examine individual characteristics at the exclusion of context, are often inconclusive and cursory and do not provide constructive material on how to solve the problem of rampant cheating occurring in our classrooms. Further research needs to be undertaken which combines the characteristics of those who cheat with situational factors.

REFERENCES

Anderman, E. M. (2003). School effects on psychological outcomes during adolescence. *Journal of Educational Psychology, 94*, 795–809.

Anderman, E. M., & Cupp, P. K. (2006, August). Impulsivity and academic cheating. Paper presented at the annual meeting of the American Psychological Association, New Orleans, LA.

Anderman, E. M., Griesinger, T., & Westerfield, G. (1998). Motivation and cheating during early adolescence. *Journal of Educational Psychology, 90*, 84–93.

Anderman, E. M., & Midgley, C. (2004). Changes in self-reported academic cheating across the transition from middle school to high school. *Contemporary Educational Psychology, 29*, 499–517.

Baer, J. S. (2002). Understanding individual variation in college drinking. *Journal of Studies on Alcohol, 14*, 40–53.

Baird, J. S., Jr. (1980). Current trends in college cheating. *Psychology in the Schools, 17*, 515–522.

Bandura, A. (1986). *Self-efficacy: The exercise of control.* New York: W. H. Freeman.

Bolin, A. U. (2004). Self-control, perceived opportunity, and attitudes as predictors of academic dishonesty. *Journal of Psychology: Interdisciplinary and Applied, 138*, 101–114.

Bowers, W. J. (1964). *Student dishonesty and its control in college.* New York: Columbia University Press.

Brown, B. S., & Emmett, D. (2001). Explaining variations in the level of academic dishonesty in studies of college students: Some new evidence. *College Student Journal, 35*, 529–538.

Bruggeman, E. L., & Hart, K. J. (1996). Cheating, lying, and moral reasoning by religious and secular high school students. *The Journal of Educational Research, 89*, 340–344.

Calabrese, R. L., & Cochran, J. T. (1990). The relationship of alienation to cheating among a sample of American adolescents. *Journal of Research & Development in Education, 23*, 65–72.

Callahan, D. (2004). *The cheating culture: Why more Americans are doing wrong to get ahead.* New York: Harcourt.

Cizek, G. J. (1999). *Cheating on tests: How to do it, detect it, and prevent it.* Mahwah, NJ: Lawrence Erlbaum.

Corcoran, K. J., & Rotter, J. B. (1987). Morality-conscience guilt scale as a predictor of ethical behavior in acheting situation among college females. *The Journal of General Psychology, 114,* 117–123.

Costa, P. T., & McCrae, R. R. (1992). NEO-PI-R: Revised NEO Personality Inventory (NEO-PI-R). In Odessa, FL: Psychological Assessment Resources.

Davis, S. F., Grover, C. A., Becker, A. H., & McGregor, L. N. (1992). Academic dishonesty: prevalence, determinants, techniques and punishments. *Teaching of Psychology, 19,* 16–20.

Davis, S. F., & Ludvigson, H. W. (1995). Additional data on academic dishonesty and a proposal for remediation. *Teaching of Psychology, 22,* 119–121.

Davis, S. F., Noble, L. M., Zak, E. N., & Dreyer, K. K. (1992). A comparison of cheating and learning/grade orientation in American and Australian college students. *College Student Journal, 28,* 353–356.

Davis, S. F., Pierce, M. C., Yandell, L. R., Arnow, P. S., & Loree, A. (1995). Cheating in college and the Type A personality: A reevaluation. *College Student Journal, 29,* 493–497.

Dawkins, R. L. (2004). Attributes and statuses of college students associated with classroom cheating on a small-sized campus. *College Student Journal, 38,* 116–129.

Diekhoff, G. M., LaBeff, E. E., Clark, R. E., Williams, L. E., Francis, B., & Haines, V. J. (1996). College cheating: Ten years later. *Research in Higher Education, 37,* 487–502.

Donohew, L. D., Zimmerman, R. S., Novak, S. P., Feist-Price, S., & Cupp, P. (2000). Sensation-seeking, impulsive decision-making, and risky sex: Implications of individual differences for risk-taking and design of interventions. *Journal of Personality and Individual Differences, 28,* 1079–1101.

Dweck, C. S., & Sorich, L. A. (1999). Mastery-oriented thinking. In C. R. Snyder, (Ed.), *Coping* (pp. 232–251). New York: Oxford University Press.

Eison, J. A. (1981). A new instrument for assessing students' orientations towards grades and learning. *Psychological Reports, 48,* 919–924.

Evans, E. D., & Craig, D. (1990). Teacher and student perceptions of academic cheating in middle and senior high schools. *Journal of Educational Research, 84,* 44–52.

Evans, E. D., Craig, D., & Mietzal, G. (1993). Adolescents' cognitions and attributions for academic cheating: A cross-national study. *Journal of Psychology: Interdisciplinary and Applied, 127,* 585–602.

Finn, K. V., & Frone, M. R. (2004). Academic performance and cheating: Moderating role of school identification and self-efficacy. *Journal of Educational Research, 97,* 115–122.

Forsyth, D. R., Pope, W. R., & McMillan, J. H. (1985). Students' reactions after cheating: An attributional analysis. *Contemporary Educational Psychology, 10,* 72–82.

Franklyn-Stokes, A., & Newstead, S. E. (1995). Undergraduate cheating: Who does what and why? *Studies in Higher Education, 20,* 159–172.

Genereux, G. L., & McLeod, B. A. (1995). Circumstances surrounding cheating: A questionnaire study of college students. *Research in Higher Education, 36,* 687–704.

Grasmick, H. G., Tittle, C. R., Bursik, R. J., & Arneklev, B. J. (1993). Testing the core empirical implications of Gottfredson and Hirschi's general theory of crime. *Journal of Research in Crime and Delinquency, 30,* 5–29.

Haines, V. J., Diekhoff, G. M., LaBeff, E. E., & Clark, R. E. (1986). College cheating: Immaturity, lack of commitment, and the neutralizing attitude. *Research in Higher Education, 25,* 342–354.

Huss, M. T., Curnyn, J. P., Roberts, S. L., Davis, S. F., Yandell, L. & Giordano, P. (1993). Hard driven but not dishonest: Cheating and the Type A personality. *Bulletin of the Psychonomic Society, 31,* 429–430.

Jensen, L. A., Arnett, J. J., Feldman, S. S., & Cauffman, E. (2002). It's wrong, but everybody does it: Academic dishonesty among high school and college students. *Contemporary Educational Psychology, 27,* 209–228.

Karabenick, S. A., & Srull, T. K. (1978). Effects of personality and situational variation in locus of control on cheating: Determinants of the congruence effect. *Journal of Personality, 46,* 72–95.

Kelly, J. A., & Worrell, L. (1978). Personality characteristics, parent behaviors, and sex of subject in relation to cheating. *Journal of Research in Personality, 12,* 179–188.

Kerkvliet, J. (1994). Cheating by economics students: A comparison of survey results. *Journal of Economic Education, 25,* 121–133.

Kohlberg, L. (1981). *Essays on moral development.* San Francisco: Harper & Row.

Langer, L., Zimmerman, R. S., Warheit, G. J., & Duncan, R. C. (1996). An edxamination of the relationship between adolescent decision-making orientation and AIDS-related knowledge, attitudes, beliefs, behaviors, and skills. *Health Psychology, 12,* 227–234.

Leming, J. S. (1978). Cheating behavior, situational influence, and moral development. *The Journal of Educational Research, 71,* 214–217.

Leming, J. S. (1980). Cheating behavior, subject variables, and components of the internal-external scale under high and low risk conditions. *Journal of Educational Research, 74,* 83–87.

Malinowski, C. I., & Smith, C. P. (1985). Moral reasoning and moral conduct: An investigation prompted by Kohlberg's theory. *Journal of Personality and Social Psychology, 49,* 1016–1027.

Martins, S. S., Tavares, H., da Silva-Lobo, D., Galetti, A., & Gentil, V. (2004). Pathological gambling, gender, and risk-taking behaviors. *Addictive Behaviors, 29,* 1231–1235.

McCabe, D. L. (2001). Cheating: Why students do it and how we can help them stop. *American Educator, 25,* 38–43.

McCabe, D. L., & Bowers, W. J. (1996). The relation between student cheating and college fraternity or sorority membership. *NASPA Journal, 33,* 280.

McCabe, D. L., & Trevino, L. K. (1993). Academic dishonesty: Honor codes and other conRefsual influences. *Journal of Higher Education, 64,* 522–538.

McCabe, D. L., & Trevino, L. K. (1997). Individual and contextual influences on academic dishonesty: A multicampus investigation. *Research in Higher Education, 38,* 379–396.

McCabe, D. L., Trevino, L. K., & Butterfield, K. D. (2001). Cheating in academic institutions: A decade of research. *Ethics & Behavior, 11,* 219.

McCabe, D. L., Trevino, L. K., & Butterfield, K. D. (2002). Honor codes and other conRefsual influences on academic integrity: A replication and extension to modified honor code settings. *Research in Higher Education, 43,* 357–378.

McCrae, R. R., & Costa, P. T. (1990). *Personality in adulthood.* New York: Guilford.

Michaels, J. W., & Miethe, T. D. (1989). Applying theories of deviance to academic cheating. *Social Science Quarterly, 70,* 870–885.

Midgley, C., Maehr, M. L., Hicks, L., Urdan, T., Roeser, R. W., Anderman, E., et al. (1995). *Patterns of adaptive learning survey (PALS) manual.* University of Michigan, Ann Arbor.

Murdock, T. B., & Anderman, E. M. (2006). Motivational perspectives on student cheating: Toward an integrated model of academic dishonesty. *Educational Psychologist, 41,* 129–145.

Murdock, T. B., Hale, N. M., & Weber, M. J. (2001). Predictors of cheating among early adolescents: Academic and social motivations. *Contemporary Educational Psychology, 26,* 96–115.

Murdock, T. B., Miller, A., & Kohlhardt, J. (2004). Effects of classroom conRefs variables on high school students' judgments of the acceptability and likelihood of cheating. *Journal of Educational Psychology, 96,* 765–777.

Newstead, S. E., Franklyn-Stokes, A., & Armstead, P. (1996). Individual differences in student cheating. *Journal of Educational Psychology, 88,* 229–241.

Norton, L. S., Tilley, A. J., Newstead, S. E., & Franklyn-Stokes, A. (2001). The pressures of assessment in undergraduate courses and their effect on student behaviours. *Assessment & Evaluation in Higher Education, 26,* 268–284.

Pajares, F. (1996). *Review of Educational Research, 66,* 543–578.

Perry, A. R., Kane, K. M., Bemesser, K. J., & Spicker, P. T. (1990). Type A behavior, achievement striving, and cheating among college students. *Psychological Reports, 66,* 459–465.

Rettinger, D. A., & Jordan, A. E. (2005). The relations among religion, motivation, and college cheating: A natural experiment. *Ethics & Behavior, 15,* 107–129.

Robinson, E., Amburgey, R., Swank, E., & Faulker, C. (2004). Test cheating in a rural college: Studying the importance of individual and situational factors. *College Student Journal, 38,* 380–395.

Roig, M., & DeTommaso, L. (1995). Are college cheating and plagiarism related to academic procrastination? *Psychological Reports, 77,* 691–698.

Schab, F. (1991). Schooling without learning: Thirty years of cheating in high school. *Adolescence, 26,* 839–847.

Sheard, J., Markham, S., & Dick, M. (2003). Investigating differences in cheating behaviours of IT undergraduate and graduate students: The maturty and motivation factors. *Higher Education Research & Development, 22,* 91–108.

Smith, C. P., Ryan, E. R., & Diggins, D. R. (1972). Moral decision making: Cheating on examinations. *Journal of Personality, 40,* 640–660.

Storch, E. A., & Storch, J. B. (2002). Fraternities, sororities, and academic dishonesty. *College Student Journal, 36,* 247–252.

Sutton, E. M., & Huba, M. E. (1995). Undergraduate student perceptions of academic dishonesty as a function of ethnicity and religious participation. *NASPA Journal, 33,* 19–34.

Sykes, G., & Matza, D. (1957). Techniques of neutralization: A theory of delinquency. *American Sociological Review, 22,* 664–670.

Taylor, L., Pogrebin, M., & Dodge, M. (2002). Advanced placement–advanced pressures: Academic dishonesty among elite high school students. *Educational Studies, 33,* 403–421.

Ward, D. A., & Beck, W. L. (1990). Gender and dishonesty. *Journal of Social Psychology, 130,* 333–339.

Weiss, J., Gilbert, K., Giordano, P., & Davis, S. F. (1993). Academic dishonesty, type A behavior, and classroom orientation. *Bulletin of the Psychonomic Society, 31,* 101–102.

Whiteside, S. P., & Lynam, D. R. (2001). The five factor model and impulsivity: Using a structural model of personality to understand impulsivity. *Personality and Individual Differences, 30,* 669–689.

Whitley, B. E. (1998). Factors associated with cheating among college students: A review. *Research in Higher Education, 39,* 235–274.

Whitley, B. E., Jr., Nelson, A. B., & Jones, C. J. (1999). Gender differences in cheating attitudes and classroom cheating behavior: A meta-analysis. *Sex Roles, 41,* 657–680.

Zucherman, M. (1988). Behavior and biology: Research on sensation-seeking and reactions to media. In L. Donohew, H. Spher, & T. Higgins (Eds.), *Communication, social cognition, and affect* (pp. 173–194). Hillsdale, NJ: Lawrence Erlbaum.

Zucherman, M. (1994). *Behavioral expressions and biosocial bases of sensation-seeking.* Cambridge: Cambridge University Press.

Zucherman, M., Kuhlman, D. M., Joireman, J., Teta, P., & Kraft, M. (1993). A comparison of three structural models for personality: The big three, the big five, and the alternative five. *Journal of Personality and Social Psychology, 65*(4), 757–768.

3

How Do Students Cheat?

Linda Garavalia, Elizabeth Olson,
Emily Russell, and Leslie Christensen

INTRODUCTION

A recent front-page article in the *Kansas City Star* featured different types of technology that students use to cheat in the classroom (Fussell, 2005). Pocket-size MP3 players that hold large amounts of information, cell phones with text-messaging capability, and beepers that display text or numbers to the receiver are all products students are using to gain unauthorized assistance during examinations. Out-of-class work is also easily completed with the help of technology. The Internet provides instant access to ready-made papers—some at a cost but some free of charge (O'Leary, 1999). Homework questions may be readily answered by cutting-and-pasting paragraphs of text to a word processed document. Technology provides increasingly efficient methods for students to cheat on graded work, but the use of traditional nontechnology facilitated methods is also still widespread.

In this chapter, we address the question, "How do students cheat?" First we define cheating and then we discuss the effects of cheating on grades or, more specifically, the accuracy of inferences made about students on the basis of their grades. We also examine individual differences in cheating behaviors and review prior studies that focus on cheating behaviors at different grade levels. In addition, we describe what college students reported when we asked them to provide examples of how students cheat. Finally, we present implications for educators and future directions for research.

Copyright © 2007 by Academic Press, Inc.
All rights of reproduction in any form reserved.

WHAT IS CHEATING?

Developing a conceptual definition of cheating is a first step in identifying how students cheat. Evans and Craig (1990) found that people don't always agree on what constitutes academic cheating, and, as such, developing a definition is thorny. Let us begin with a dictionary definition. Cheating is "to act dishonestly or unfairly in order to win some profit or advantage" (Ehrlich, Flexner, Carruth, & Hawkins, 1980, p. 141). According to this definition, potentially dishonest actions are acceptable as long as there is no intent to gain from the act. This point of view would likely allow clever individuals to readily defend academic dishonesty. For example, in a recent university plagiarism case, the accused used the dictionary definition to defend his actions by arguing that, although he had failed to appropriately attribute quotes in a commencement address, he had not plagiarized because "there was no intention to deceive anyone" and the address "was certainly not written with publication in mind and not prepared for that purpose" (History News Network, 2005). Even so, others readily described the borrowing of words as plagiarism (Bartlett, 2005).

Cizek (2003) provides a less limiting definition. He states that cheating behaviors fall into three categories: (1) "giving, taking, or receiving information," (2), "using any prohibited materials," and (3) "capitalizing on the weaknesses of persons, procedures, or processes to gain an advantage" on academic work (p. 42). Though broad, this definition is more likely to protect the spirit of academic honesty policies more effectively than the dictionary definition. University academic honesty policies or honor codes are other good resources to be considered when formulating a good working definition of cheating. Although these policies and codes vary widely, many include cheating on graded work, plagiarism, and falsification as major cheating behaviors (e.g., Georgetown University, 2004).

Two criteria for judging academic honesty are prominent in university honor codes. First, dishonesty involves the use of unauthorized assistance. It is important to distinguish that the assistance is unauthorized because many complex assignments and problems require the use of resources, such as academic literature, graphing calculators, the Internet, and computer software, among many others. Learning objectives often focus on a student's ability to use appropriate resources to solve authentic problems. Therefore, investigating the extent to which resources are allowable is an important issue in determining academic dishonesty. The second criterion is whether the work is graded. Generally, academic dishonesty is judged to occur when consequences are associated with the work. After reviewing prior research and university policies, we defined cheating for the present chapter as the use or provision of any unauthorized materials or assistance in academic work and/or activities that compromise the assessment process (Athanasou & Olasehinde, 2002).

HOW DOES CHEATING AFFECT ASSESSMENT?

The methods that students use to cheat are most interesting because of their effect on grades or other assessment consequences. Cheating undermines the intent and process of assessment. Typically, the primary purpose of a graded test or assignment is to determine what students have learned after instruction. Cheating interferes with an evaluator's ability to make such judgments.

Reviewing the characteristics of fair and accurate assessments provides a helpful framework for considering the impact of cheating on assessment. "Good" assessments have four characteristics. They are *reliable* (e.g., Smith, 2003). In other words, good assessments include sufficient information to consistently measure student mastery of learning objectives. More importantly, good assessments are *valid* (e.g., Frey et al., 2005; Messick, 1988, 1994). Two issues related to validity are of prime concern. The first is the content of the assessment. It should reflect what was taught and should do so in a representative manner, such that the material emphasized heavily in instruction is also emphasized heavily in assessment (content-related validity). Second, the consequences of test results should be appropriate (consequential validity). A third characteristic of good assessments is *standardization of testing procedures* across students (e.g., Kane, Crooks, & Cohen, 1999). When determining whether students have mastered learning objectives, the same system should be used for all students being assessed. (*Note*: Exceptions are those students operating under individual education plans as prescribed by IDEA (Individuals with Disabilities Education Improvement Act of 2004)). Cheating hinders standardization by varying testing procedures among assessment takers, which, as a result, may unfairly advantage or disadvantage one student over another. Lastly, good assessments are *free of bias*. That is, subgroups of students do not perform differentially on the exam owing to any form of discrimination (e.g., Scheuneman & Oakland, 1998).

To some extent, cheating affects each of these four components of good assessment, but the effect on consequential validity is likely to cause the most trouble. Given that the purpose of assessment is to measure what students have learned in relation to a specific set of instruction, any behavior that interferes with that measurement diminishes the accuracy of the assessment results and reduces the interpretability of grades. Interpretability is an issue because grades are generally attached to consequences, such as being allowed to move on to the next set of instructions, advancing to the next grade level, or gaining entrance into a special program or school. Cheating detracts from the consequential validity of test results and graded assignments by contaminating the data used to make decisions (Athanasou & Olasehinde, 2002). Stated otherwise, when a student cheats on graded work, the teacher is unable to determine whether the student really knows the material, can solve the assigned problem, or can complete the given task. The student may or may not be able to do the work without unauthorized assistance. As a result, one cannot make a fair judgment about what the student has learned and what the student should do next in the learning sequence.

Another issue to consider is the level of importance assigned to the outcomes of an assessment. The stakes are higher for some assessments, and faulty judgments have more impact and greater consequences. For higher stakes testing, such as end-of-year standardized testing at the K–12 level, test results have broader effects. For such tests, cheating by teachers, school administrators, and state Department of Education officials is more problematic than student cheating. For example, in a listing of cheating incidents related to standardized testing, teachers received substantial bonuses, students were passed on to subsequent grades, and schools received accolades, higher rankings among other schools, and more local and federal funding. Cheating occurred on standardized tests by slipping students answers, directly teaching students known test items, and changing incorrect responses on examination forms (The Chronicle of Higher Education, 1999). The consequences of these actions were much greater than simply reducing a teacher's ability to judge what one student learned. With that in mind, the consequential validity of assessment results is often the primary concern when considering the effects of cheating on assessment.

DO STUDENTS DIFFER IN THE WAY THEY CHEAT?

THEORETICAL PERSPECTIVES

When students choose to cheat, how they do so is likely related to both individual and situational variables, including motivational, moral, and developmental factors. The motivational orientation of both the individual and the academic environment may trigger different types of cheating behavior. Students who are motivated primarily by a need to demonstrate high ability relative to others or to appear competent as judged by others (i.e., performance goal orientation; Ames & Archer, 1988; Anderman, Griesinger, & Westerfield, 1998; Anderman & Midgley, 2004; Murdock, Hale, & Weber, 2001) may employ cheating methods that reflect this particular goal orientation. A student with a high-performance goal orientation might cheat by copying off another student's exam, by plagiarizing a term paper, or by using unauthorized sources to complete graded work, but likely would not participate in cheating behaviors that provide help to others. A performance goal orientation could explain why a law student purposely fails to return a case to the reference desk when other students need the case to complete an assignment, why a straight-A eleventh grader fails to help other students in her study group when she herself has a good understanding of the topic, and why a bright sixth grader surreptitiously surveys the exam answers of his neighbors during testing. The emphasis in this motivational orientation is on appearing competent in relation to others.

Aside from a student's individual need to appear competent, situational goal orientation may influence which cheating behaviors are enacted or students' attitudes toward certain cheating behaviors (Anderman & Maehr, 1994; Maehr & Midgley, 1991; Murdock, Miller, & Kohlhardt, 2004). For example, when middle school students believe the school environment fosters a performance goal orientation, they are more likely to report engaging in cheating behavior in course work (Anderman, Griesinger, & Westerfield, 1998; Murdock, Hale, & Weber, 2001). Perhaps the perceived need to appear competent in relation to others is stronger than the perceived risk associated with the cheating behavior. Cheating may also be a coping mechanism for dealing with a school environment that emphasizes ability and performance (Maehr, 1991), especially as students progress through grades and as academic work becomes not only more complex, but also more competitive (e.g., Eccles & Midgley, 1989; Wigfield, Eccles, & Pintrich, 1996). Anderman et al. (1998) propose that "if the environment does not stress competition and winning at all costs, then students may have less incentive to cheat" (p. 90). We speculate that students' perceptions of the goal orientation might also influence the manner in which students choose to cheat.

Kohlberg's (1985) theory of moral development has been a prevalent theoretical framework for investigating the moral aspects of cheating. Prior research indicates an inverse relationship between moral development level and cheating behavior in male undergraduates (Malinowski & Smith, 1985). However, even students who scored high in moral development succumbed to cheating when sufficient pressure was exerted by the situation. This finding again indicates the interrelationship between individual and situational factors in determining cheating behaviors.

Eisenberg (2004) proposes a relationship between cheating behavior and the extent to which students believe cheating is a moral issue. To investigate this relationship, he asked Israeli junior high students whether they would cheat "if there was no rule in your school that forbade copying on exams" (p. 169) and identified students as either "moral" (ruled by morals) or "a-moral" (ruled by convention). He controlled for both individual differences in moral perspective and the effects of situational variables on attitudes toward cheating during exams. "A-morals" were more approving of both active (copying off another person's exam) and passive (allowing someone to copy off your exam) cheating as compared to their peers who reported cheating as a moral rather than a convention-based issue. In addition, importance of the exam, degree of supervision, and class norms were all related to students' attitudes toward cheating.

Even though Eisenberg investigated attitudes, his findings have important implications for our discussion of individual and situational factors and their relationship to cheating behaviors. First, situational factors were related to both active and passive cheating attitudes, indicating that, at least with regard to classroom exams, situational factors may impact both whether a student is a taker or a giver of unauthorized assistance. Second, when we look at students across grades and developmental levels, the real question might be whether they perceive academic cheating to be a moral issue or one that is determined by the rules of the classroom or school. In other words, methods of cheating might vary in direct relation to students' situational assessment of the moral versus conventional prohibition of the behavior rather than to developmental differences. This finding could explain similarities in cheating behaviors across grades.

Other contextual factors may affect students' choice of cheating technique. For example, McCabe, Trevino, and Butterfield (2001) reviewed the cheating literature and reported that "peer cheating behavior, peer disapproval of cheating behavior, and perceived penalties for cheating" influenced cheating behaviors in college students significantly more than "individual factors (age, gender, GPA, and participation in extracurricular activities)" (p. 222). The researchers also report that honor codes are a deterrent to cheating behaviors on campus, but more important may be the student cultural norms regarding academic dishonesty.

McCabe et al. (2001) documented the amount of serious cheating behaviors in different contexts. "A serious test cheater is defined as someone who admits to one or more instances of copying from another student on a test or exam, using unauthorized crib or cheat notes on a test or exam, or helping someone else cheat on a test or exam" (McCabe et al., 2001, p. 223). Serious cheating on written work includes "plagiarism, fabricating or falsifying a bibliography, turning in work done by someone else, and copying a few sentences of material without footnoting them" (p. 223). McCabe and colleagues found that serious cheating was more prevalent on college campuses with no honor code (McCabe et al., 2001), on larger campuses (McCabe & Trevino, 1997), and when honor codes were not adequately communicated or enforced (McCabe & Trevino, 1993). Thus, we might expect to find variation in cheating methods in these different contexts.

Aside from climate variables, the physical characteristics of the classroom, such as the seating arrangement, may also influence the student's method of cheating. Houston (1986) found that, in contrast to conventional wisdom, students are no more likely to copy off another's paper when seated in the back of the room than in the front. However, students are more likely to cheat when seated next to a study partner.

CHEATING ACROSS GRADE LEVELS

Although researchers have investigated the prevalence of cheating in elementary school, very few recent studies have examined how students cheat at the elementary school level. This limits discussion of differences in cheating methods across grade levels. A large-scale study in California enrolled 1,037 sixth graders from 45 elementary schools and 2,265 secondary school students from 105 high schools (Brandes, 1986). High school students were found to be much more likely to cheat than sixth graders. Similarly, Anderman and Midgley (2004) conducted a longitudinal study comparing self-reported academic cheating at the middle school and high school levels. Students self-reported their cheating behaviors at the middle and end of eighth grade and again at the end of ninth grade. Like Brandes (1986), the researchers reported an increase in self-reported cheating at the high school level.

Brandes also found that the methods used by high school students differed from those used by younger students. Although both younger and older students reported copying from others on tests, only older students reported using crib notes during tests. Younger students reported plagiarizing as a frequent method of cheating.

In comparing students across grade levels, it is important to consider that cheating behaviors are directly related to the demands of the academic task. In other words, the behavior is typically enacted to help the student meet important outcomes of the assignment. For example, Syer and Shore (2001) examined cheating and help-seeking behaviors among junior high and high school students completing a science-fair project. The project required students to collect and present data, and of the 27 participants, 5 admitted cheating on their project by making up data. In this case, "making up data" was a shortcut to producing a primary outcome of the project. An implication is that cheating methods will become more complex and sophisticated over the grade levels in direct relation to the increasing complexity of learning and assignment requirements. As such, we should find differences in cheating methods across grade levels (see table 3.1).

An important consideration is the potential obstacle of accurately describing cheating methods; differential understanding of what constitutes cheating can be problematic. Earlier, when we defined cheating, we pointed out how differences in definition could allow for alternate classification of behaviors as cheating or not. Likewise, Evans and Craig (1990) found differences in middle and high school students' versus teachers' understanding of cheating behaviors. Students were much less likely to identify passive behaviors, such as sharing unauthorized advance test information, as being a form of cheating than teachers. Students also were less likely than teachers to identify missing a scheduled test without a legitimate reason as being a form of cheating.

TABLE 3.1   Empirical Studies of Cheating Behaviors by Grade Level

Age range	Authors	Type of cheating explored
K–8	Brandes (1986)	(1) Copying from others on tests (2) Plagiarizing
	Syer & Shore (2001)	(1) Making up data
High school	Brandes (1986)	(1) Copying from others on tests (2) Using crib notes during tests
College	Hetherington & Feldman (1964)	(1) Individual-planned (2) Individualistic-opportunistic (3) Social-active (4) Social-passive
	Baird (1980)	(1) Quizzes (2) Tests
	Baird (1980)	(1) Obtaining test information from other students (2) Allowing others to copy work (3) Copying someone else's assignment (4) Plagiarism (5) Copying someone else's test work
	Haines, Diekhoff, LaBeff, & Clark (1986)	(1) Class assignments (2) Major exams (3) Quizzes
	Davis, Grover, Becker & McGregor (1992)	(1) Copying from others (2) Using crib notes
	Franklyn-Stokes & Newstead (1995)	(1) Allowing others to copy work (2) Doing another's coursework (3) Paraphrasing without references (4) Copying coursework with the other's knowledge
	Genereaux & McLeod (1995)	(1) Giving answers to exam questions (2) Receiving answers to exam questions
	Hollinger & Lanza-Kaduce (1996)	(1) Taking information (2) Tendering information (3) Plagiarism (4) Misrepresentation
	Newstead, Franklyn-Stokes, & Armstead (1996)	(1) Paraphrasing material from another source without acknowledging the author (2) Inventing data (3) Allowing coursework to be copied by another student (4) Fabricating references/bibliography (5) Copying material for coursework from book or other source without citing source (6) Altering data (7) Copying another student's coursework with their knowledge

(continues)

TABLE 3.1 (*continued*)

Age range	Authors	Type of cheating explored
	Norton, Tilley, Newstead, & Franklyn-Stokes (2001)	(1) Paraphrasing material from other sources without acknowledgment (2) Inventing data
	Ahlers-Schmidt & Burdsal (2004)	(1) Passive cheating
	Dawkins (2004)	(1) Copying from the Internet
	Robinson, Amburgey, Swank, & Faulkner (2004)	(1) Copying from other students' exams (2) Making tests available for others to copy (3) Receiving questions from someone who already completed the test (4) Collaborating on take-home work when not allowed

	Bennett (2005)	(1) Plagiarizing
		  (a) Copying a few sentences
		  (b) Copying several sentences
		  (c) Copying an entire paragraph
		  (d) Copying several paragraphs
		  (e) Copying an entire piece of work
		(2) Making up references
		(3) Collaborating when not allowed to do so

Examples of plagiarism were particularly confusing, with even teachers disagreeing on some of the items. For instance, uncredited paraphrasing and using the same paper for two different assignments were classified variably among the teachers. Many students (10 to 60 percent) also failed to identify uncredited paraphrasing as cheating; however, most students (50 to 75 percent) identified copying material word for word without referencing the author as cheating (Evans & Craig, 1990). These findings imply that help-seeking or resource-using behaviors may be variously classified, especially by younger students who may be less developmentally adept at making ethical distinctions. Again, describing differences in cheating methods may be limited by differences in individuals' definitions of cheating.

POSTSECONDARY CHEATING

Perhaps because of the demands of the tasks as well as the importance of performing well, investigations of cheating behaviors focus largely on the college population, though researchers continue to label and address cheating behaviors in different ways. For example, a British study reported "paraphrasing material from another source without acknowledging the original author" and "inventing data" as the most prevalent methods of cheating among a sample of college students, with more than half of the students ($N = 267$) engaging in those behaviors (Norton, Tilley, Newstead, & Franklyn-Stokes, 2001). In a larger study with 1,672 college students, Hollinger and Lanza-Kaduce (1996) factor analyzed survey responses and identified four distinct types of academic dishonesty: "taking of information, tendering of information, plagiarism, and misrepresentation" (p. 294).

Subsequently, numerous studies have investigated a broad range of potential cheating behaviors, with some focusing on the type of task on which cheating occurs and others on the different methods of cheating utilized by college students. Haines, Diekhoff, LaBeff, and Clark (1986) investigated the type of task on which cheating took place in a group of 380 undergraduates from a small state university, with results indicating that the most prevalent form of cheating occurred on class assignments (34.2 percent) followed by major exams (23.7 percent) and quizzes (22.1 percent). Obtaining similar findings in an investigation of trends in college cheating, Baird (1980) surveyed 200 college undergraduate participants about their cheating behaviors and the tasks on which students cheat by requesting that participants complete a survey that included 33 different types of cheating behaviors. General trends indicated that students had more often cheated on less important tests (i.e., quizzes and unit tests) than on important ones (i.e., midterms and finals).

Taking things a step further, Baird (1980) also investigated the numerous methods of cheating utilized by students. When looking at specific cheating behaviors, results indicated that obtaining test information from other students, allowing someone to copy work, copying someone else's assignment, plagiarizing, and copying someone's test work were the most frequently reported cheating behaviors. Similarly, Franklyn-Stokes and Newstead (1995) surveyed 128 students from nonpsychology courses, finding the most common cheating behaviors to be cheating by allowing others to copy work (72 percent), doing another's coursework (66 percent), paraphrasing without references (66 percent), and copying coursework with the other individual's knowledge (64 percent). Other studies report comparable findings, with the most typical behaviors including activities such as copying from others and using crib notes (Davis, Grover, Becker, & McGregor, 1992); copying from the Internet (Dawkins, 2004); copying from other students' exams or making tests available for others to copy, receiving questions from someone who has completed the test, and improperly collaborating on a take-home exam (Robinson, Amburgey, Swank, & Faulker, 2004); or paraphrasing material from another source without acknowledging the author, inventing data, allowing coursework to be copied by another student, fabricating references/bibliography, copying material for coursework from a book or other source without citing the source, altering data, and copying another student's coursework with their knowledge (Newstead, Franklyn-Stokes, & Armstead, 1996).

Plagiarism is a frequent focus of research on college student cheating. Bennett (2005) demonstrated the prevalence of the behavior. In a study with 327 undergraduates, students indicated on a five-point scale how frequently they copied or inserted into their work, without proper acknowledgment of published sources, the Internet, or the work of other students. Options ranged from a couple of sentences to an entire piece of work. Although 46 percent of students strongly agreed that plagiarism was "fundamentally immoral and shameful," 80 percent had copied a couple of sentences, 71 percent had copied several sentences, 46 percent admitted to copying an entire paragraph, 31 percent copied several paragraphs, and 25 percent handed in an entire piece of work that had been copied. Furthermore, 53 percent reported making up references, and 61 percent collaborated for a course when they were not authorized to do so.

In a study of 365 Canadian community college students, giving (58 percent) and receiving (49 percent) exam questions were the most commonly self-reported methods of cheating (Genereux & McLeod, 1995). The study also examined situational factors that increase or decrease the likelihood of planned and spontaneous cheating. Attitude of the instructor toward cheating, exam fairness, and other factors not directly related to the exam environment were most likely to increase cheating. The only test administration factors likely to influence cheating were instructor vigilance and spacing of students in the exam room; high vigilance and far apart spacing decrease the likelihood of spontaneous cheating. Therefore, how students cheat may depend largely on context; however, other factors such as the instructor's view on cheating, the extent to which grades affect financial support, and the relation between grades and long-term goals may also influence how students cheat. In addition, Genereux and McLeod (1995) found that students were more likely to help someone else cheat rather than cheating for their own academic gain.

Also focusing on contextual issues and taking a slightly different approach, Hetherington and Feldman (1964) attempted to isolate four different types of cheating methods: individualistic-planned, individualistic-opportunistic, social-active, and social-passive. *Individual-opportunistic cheating* was labeled as changing answers when self-grading an exam or using materials left out during an oral exam when the professor left the room. *Independent-planned cheating* was identified as using crib notes during an exam or bringing in already completed essays into an exam rather than actually writing them during the allotted exam period. Finally, *social-active cheating* was classified as copying from others, and *social-passive cheating* was allowing others to copy. Of the 78 participants, 59 percent exhibited some type of cheating activity, with 41 percent of students engaging in opportunistic-individual cheating, 27 percent in planned-independent, 16 percent in social-active, and 14 percent in social-passive.

A final method of cheating dependent on contextual factors, and more recently explored in the literature, is the occurrence of passive cheating, defined as allowing information from others to influence responses made on certain tasks (Ahlers-Schmidt & Burdsal, 2004). To investigate the occurrence of passive cheating, 49 undergraduate psychology students were assigned to an experimental group or a control group. After completing a 40-item multiple-choice test with difficult trivia questions, the control group was asked to leave the room commenting on the high frequency of "B" responses so that the experimental group waiting to take the exam next would hear. Results indicated a significant difference in the number of B responses between the experimental and control group, with the experimental having more B responses. The researchers concluded that the experimental group was influenced by those who took the exam before them and thus had engaged in passive cheating.

CHEATING REPORTED IN THE POPULAR PRESS

Methods of cheating are frequently described in the popular press as well as in the academic literature. Popular press articles have recently focused on the occurrence of technology-facilitated cheating in both K–12 and post-secondary schools. One reporter predicts that the percentage of students who admit to cheating will likely increase every year as technology becomes more sophisticated and affordable (Gilroy, 2004). For example, Personal Digital Assistants (PDAs) are popular, fairly affordable devices and can easily be used to make databases with answers that students may access during exam time (Tweedle, 2004). Also, with the rise of cellular phone sales in the United States and Canada, cellular phones have become popular devices for cheating on exams (Tweedle, 2004; Total Telecom, 2004).

Students have commonly used text messaging through cell phones to text questions and answers to other test-takers. Tweedle (2004) reported that students turn off the phone sound, type the answer, and press send. Another person in the room receives the answer. Users can send messages to people in other classrooms and record, store, and retrieve test answers. In addition, camera cell phones make possible a new version of the "cheat sheet." For example, Total Telecom (September 10, 2004) reported that students use camera phones to take pictures during exams (instead of using the classic method of looking over someone's shoulder), to take pictures of notes and study guides prior to exams to look up answers during the test, and to send photos of answers to someone else in the room.

Short Messaging Service offers technology in which messages can be sent quickly and quietly from device to device. The *Korea Times* (November 22, 2004) reported that in a large-scale cheating scheme approximately 100 high school seniors were suspected of cheating on their college entrance exams using a Short Messaging Service (SMS) for cell phones. Police reported that 40 cell phones were purchased, and students met at a motel and practiced cheating so that a large number of them could share answers in a short period of time. Allowing cell phones in the classroom during test time is clearly problematic.

Also frequently described in the popular press is the misuse of the Internet. Middle schools, high schools, and colleges all face the dilemma of encouraging students to find up-to-date information on the Internet while trying to monitor students' work for direct lifting of material from online documents. In a recent editorial regarding copying and pasting papers from the Internet, Karpf (2004) pronounced this trend an "epidemic." Unfortunately, certain web sites make it easy for students to plagiarize. For example, a simple search on Google allows access to sites that provide completed papers on selected topics. In addition, http://www.123helpme.com offers free essays but requires a payment for well-written papers, and http://www.schoolsucks.com offers essays for $30 (Thompson, 2005). The more custom-made the papers are, the more expensive they are (Gilroy, 2004). Many teachers and schools are trying to curb this practice by investing in a plagiarism detection service called Turnitin.com and by using special honor pledges (Daily Forty-Niner, 2005).

In addition to plagiarizing materials from the Internet, other forms of online cheating are becoming more common at universities. Some classes are taught completely online, utilizing course shells that allow for online storage of class materials, discussion boards, and messaging systems (i.e., Blackboard). Many students and instructors find this method to be convenient and helpful. However, Moake (2004) at Brigham Young University indicated that Blackboard has actually made it easier for students to cheat on quizzes and homework. Online quizzes mean that students may have access to outside sources or class materials when completing the quiz. The problem is that the offsite nature of web-based work reduces or eliminates an instructor's ability to monitor the use of unauthorized resources and the independent completion of work.

Given students' access to increasingly sophisticated means of cheating and the dire predictions of widespread cheating facilitated by technology, one might expect to find changes in the college students' self-reported cheating behaviors if one were to survey college students today. We investigated this possibility by surveying students on our university campus.

COLLEGE STUDENTS DESCRIBE CHEATING

Prior research has typically asked students to respond to survey items to learn about cheating behavior. We were curious about what methods college students might report without the aid of a list of behaviors. Our research question, therefore, was open-ended and broad: "How do college students cheat on graded work?" Students were also asked to report their sex and class standing (e.g., sophomore). Participants were recruited via electronic mail using a snowball technique, acquiring names of potential participants from previous participants. Overall, 67 college students ($F = 44$, $M = 23$) from the University of Missouri–Kansas City participated. Participants represented each year of class standing, including undergraduate, master's, doctoral-level, and professional degrees.

Student responses were qualitatively analyzed to identify themes and reduce the data to meaningful categories. Four major categories emerged and allowed for quantification of different aspects of the cheating behavior (See Table 3.3). The first category is the type of assignment on which students might cheat. Subcategories include (a) exams, (b) graded homework, (c) out-of-class papers, (d) graded laboratory work, and (e) graded online work or quizzes. The remaining three categories were dichotomously coded; they are technology-facilitated versus traditional, planned versus spontaneous, and independent versus collaborative methods.

Of the 67 participants, nearly 90 percent ($N = 60$) described at least one way that students could cheat on graded work. Consistent with prior research, our findings indicate that cheating occurs on numerous types of work. Cheating frequencies are reported in Table 3.2. Because many students described multiple methods of cheating, the percentages of responses within each coding category do not equal 100 percent. For example, 34 percent of students reported technology-facilitated cheating, while 84

TABLE 3.2　Cheating Frequencies: Assignments and Behaviors ($N = 67$)

	n	Percentage
Type of assignment		
Exam	32	47.8
Out-of-class paper	11	16.4
Out-of-class graded work	23	34.3
Lab work	5	7.5
Online homework/quiz	3	4.5
Characteristics of cheating behavior		
Technology-facilitated method	23	34.3
Traditional method	56	83.6
Individual	35	52.2
Collaborative	27	40.3
Spontaneous	19	28.4
Planned	34	50.7

percent described a traditional method of cheating. Many participants included more than one method of cheating. The majority of responses ($N = 83$) described cheating on exams ($n = 39$), homework ($n = 22$), or out-of-class papers ($N = 16$), with fewer students describing cheating on lab work ($N = 4$) or online homework/quizzes ($N = 2$). Research indicates that people do not always agree on whether a behavior entails cheating (Evans and Craig, 1990); thus students may have unintentionally restricted responses to more "typical" cheating behaviors.

A total of 34 percent of students described technology-facilitated cheating methods, such as the unauthorized use of graphing calculators, cell phones, the Internet, or laptop computers during exams or to complete graded work. The methods described by participants in our study provide evidence substantiating the rising concern that tech-savvy students are enlisting increasingly sophisticated means to undermine the assessment process. For example, using the Internet to plagiarize or purchase term papers was a common response and has been described as a substantial contributor to recent increases in reported cheating (McMurty, 2001; Scanlon & Neumann, 2002).

Interestingly, 84 percent of the responses did not involve technology. It may be that students are simply more aware of traditional cheating methods. Using technology requires access to and knowledge of technological innovations, and this may limit the use of those methods by students who are either uninterested in technology or who lack resources to obtain access.

Approximately 63 percent of the responses described independent cheating, meaning that cheating behavior could be accomplished without assistance. Students who look over another student's shoulder or bring a "cheat" sheet to an exam are cheating in an independent manner. Collaborative cheating methods were equally represented, with a large number (58 percent) of responses describing methods that involved at least two cooperating individuals.

Most of the cheating behaviors described by students were planned (70 percent). Some methods were rather elaborate and likely required a time commitment comparable to studying for the exam or completing the assignment honestly. Fewer cheating behaviors were spontaneous (27 percent). An example was whispering and sharing answers when a professor briefly left the room. Some testing environments provide greater opportunity for spontaneous cheating, as was mentioned in many responses. Lecture halls with hundreds of students, oblivious instructors, and multiple-choice exams were all mentioned as conditions that made cheating easier.

Correlational analyses, presented in Table 3.4, revealed an inverse relationship between class standing and the report of technology-facilitated methods of cheating, with lower-level students being more likely to report technology-facilitated cheating methods ($r = -.25$, $p < .05$). This finding is consistent with research by Underwood and Szabo (2003), who found that

TABLE 3.3 Cross-tabulated Cheating Methods and Examples of Responses

	Independent		Collaborative	
	Spontaneous	Planned	Spontaneous	Planned
Nontechnology Facilitated	18 Ex: "Looking over shoulders."	37 Ex: "Writing answers on gum wrappers, water bottles, in calculators, etc."	1 Ex: "I've witnessed students talking/whispering to each other during a test."	42 Ex: "Borrow a friend's already graded homework from previous years in the class."
Technology Facilitated	1 Ex: "Using camera cell phones to take pictures of questions and send them to people for answers."	22 Ex: "If it's a class that allows you to use a calculator, such as Math or Physics, programming equations, problems, and even definitions into the calculator is another way to cheat."	0	7 Ex: "I have seen students use Palm Pilots or some form of instant messenger to send each other answers during a test."

TABLE 3.4   Correlations of Assignments, Behavior Characteristics, Gender, and Class Standing

Variables	Individ.	Coll.	Spont.	Planned	Gender	Class
Technology-facilitated	.36**	.10	.20	.33**	.07	−.25*
Traditional	.41**	.36**	.09	.42**	−.02	−.11
Exam	.54**	.08	.45**	.31*	.04	−.08
Take-home paper	.43**	.12	.06	.29*	−.11	.19
Graded homework	.01	.40**	−.07	.25*	−.10	.04
Lab work	−.07	.21	−.01	.02	−.05	−.10
Online homework	−.05	.15	.09	.11	−.13	−.16
Other	−.13	−.11	−.17	−.19	.03	−.05

$*= p < .05$, $**= p < .01$.

UK college students reported an inverse relationship between Internet plagiarism and year of schooling. The researchers speculated that third-year college students had more invested in their education and therefore had more to lose than first-year students, thus making Internet plagiarism a riskier behavior.

Technology-facilitated methods were also moderately associated with independent methods of cheating ($r = .36, p < .01$), such as storing equations in a graphing calculator, as well as planned methods ($r = .33, p < .01$). An implication is that technology is simply a new type of confederate in cheating schemes. Independent methods of cheating were also positively related to spontaneous cheating ($r = .33, p < .01$), which included the common example of looking over someone's shoulder during an exam. Collaborative cheating was positively related to nontechnology-facilitated cheating ($r = .36, p < .01$) and cheating that was planned ($r = .51, p < .001$), indicating that students are likely to prepare in advance to cheat. Interestingly, technology-facilitated methods were not associated with collaborative methods. This finding is surprising because recent popular press articles report a growing trend toward students sharing information by using cell phones to text message answers or to send digital pictures of exam questions (Korea Times, 2004; Total Telecom, 2004; Tweedle, 2004). A possible explanation is that the students in our sample did not readily recall those methods when asked about cheating, indicating that the average student might not be as tech-savvy as the popular press reports. Planned cheating was related to more traditional methods of cheating ($r = .42, p < .001$), and spontaneous cheating was positively correlated with exams ($r = .45, p < .001$) and independent methods ($r = .33, p < .01$).

IMPLICATIONS FOR THE CLASSROOM

Cheating continues to be a problem in today's classrooms, especially at the college level (see Keith-Spiegel, Tabachnick, Whitley, & Washburn, 1998; McCabe et al., 2001; Norton et al., 2001). Cheating occurs on graded work of all kinds and ranges from sharing answers with nearby fellow examinees to instant text messaging of answers using cell phones, from copying encyclopedia material to cutting-and-pasting homework answers found on the Internet, from bringing cheat sheets on gum wrappers into exams to advance preparation of essay answers on laptop computers. Although more traditional methods of cheating, such as sharing exam answers or plagiarizing the work of others, are still common, technology-facilitated methods are being described more frequently in the popular press and academic literature.

The use of technology-facilitated methods of cheating may be related to individual differences, such as college major, financial affluence, and technological expertise. Some college majors require greater use of technology than do other majors, and we might expect to find the most sophisticated methods of cheating among those students. An important point to bear in mind is that cheating is directly related to the demands of the academic task. Therefore, if the task requires technology, then the cheating method is more likely to involve technology. Of interest is the surge of stories reported in the popular press of students using cell phones, MP3 players, and other gadgets to cheat on in-class, independent-graded work. In our study, students described more traditional methods almost twice as frequently as technology-facilitated methods, and those methods that did utilize technology involved more common, everyday uses, such as searching the Internet for answers or using a graphing calculator with programmed formulas. None of the students in our study described using text messaging or head sets attached to recorders (e.g., MP3 players) to cheat.

Our review of the literature indicates that how students cheat also depends on the characteristics of the teacher. For example, McCabe (1999) conducted focus groups with high school students and found that most students believed that their teachers were not technologically proficient and, therefore, would not be able to detect plagiarism if Internet sources were utilized. Because plagiarism using words from textbooks or books in the library was more likely to be detected, students chose a technology-facilitated method of cheating to reduce the risk of being caught.

Another teacher characteristic that affects how students cheat is the extent to which the teacher is known to handle incidents of cheating (Keith-Spiegel et al., 1998). Teachers who fail to punish cheating may contribute to the development of cheating as a normative behavior. Instructor vigilance is another factor that affects how students cheat (Genereux & McLeod, 1995). Again, students will use methods that they believe will go undetected. The less vigilant the instructor, the less sophisticated the methods of cheating need be.

Knowing how students cheat is foundational to developing and implementing strategies to counteract cheating. To combat Internet plagiarism, educators recommend assigning papers and other writing tasks that are not "classic," meaning they are not likely to be available for sale or for free on the Internet (Phillips & Horton, 2000; Renard, 2000). Detecting plagiarism may not be as difficult as one might think (Renard, 2000). Instructors should be aware of web sites that sell papers or provide discussions of assigned topics (e.g., "The Evil House of Cheat" at http:// www.cheathouse.com; http://www.schoolsucks.com; A1 Termpaper at http://www.a1-termpaper.com/). Another strategy is to type key words or phrases from the paper into a Google search and see what turns up. Researchers have also developed methods to detect cheating on scantron forms. The likelihood of students missing the exact same items on a test can be computed and provide evidence of likely collaboration (Kawai, 2005). Lastly, indicating to students that cheating is not acceptable, that punitive consequences are associated with the detection of cheating, and that learning is of primary importance, not the grade on an assignment, are additional ways to deter cheating of any kind.

SUGGESTIONS FOR FUTURE RESEARCH

Our review of the literature indicated several gaps in research on how students cheat in academic settings. Very little research is available on cheating methods in elementary schools. Identifying and describing cheating behaviors in younger students may provide greater insight into how cheating behaviors develop and manifest themselves in the early years of schooling. Similarly, longitudinal or cross-sectional studies of cheating behaviors in which students in early elementary grades through post-secondary schooling engage may pinpoint developmental differences related to cheating. Prior research indicates that cheating behaviors may be related not only to moral development (Malinowski & Smith, 1985), but also to the extent to which students view cheating as a moral violation (Eisenberg, 2004). An interesting question would be whether cheating behavior is related to other developmental differences, such as cognitive development or the mastery of various learning strategies. Identifying the types of behaviors associated with developmental factors might allow educators to instruct students in the use of more effective, noncheating strategies that utilize the same developmental strengths as the cheating behaviors.

Another issue is how different cheating methods serve various personal goals. Prior research indicates that cheating attitudes and propensity are related to the motivational goal orientation of a classroom (Anderman et al., 1998). In particular, competitive environments foster cheating behaviors to a greater extent than mastery-oriented environments. As such, students who have personal goals that involve competitive outcomes are more likely to engage in cheating behaviors. It would also be of interest to determine whether the types of cheating behaviors students actually utilize vary with goal orientation or other personal goals. For example, students who have a high need for social belonging as related to their peers might engage in cheating behaviors that help peers (e.g., passive cheating, such as providing advance information about test items) in order to gain friends, whereas students who have a high need to appear competent in relation to others would likely only engage in cheating behaviors that further their own achievement (e.g., plagiarism).

A methodological issue frequently cited by researchers is the typical use of self-report data. As one researcher stated, "A paradox of this research is that it asks students to be honest about their own dishonesty" (Newstead et al., 1996, p. 240). Future research should utilize methods that allow for the validation of self-report data. For example, in grades K–12, teachers and administrators maintain discipline records. Although these records will likely be incomplete (i.e., fail to document all cheating behaviors), they provide another mechanism for describing the types of cheating behaviors that are occurring. In addition, grade-level comparisons would be possible. Similarly, K–12 teachers could be asked to describe cheating behaviors to supplement surveys of their students. At the postsecondary level, records of disciplinary action may be more difficult to obtain, but they do exist; therefore, a similar method could be used to study cheating behaviors at the college level.

Also related to methodology is the frequent use of forced-choice or close-ended items in surveys. Future research might produce richer descriptions or uncover innovative methods by simply asking open-ended questions rather than asking students to respond to a Likert-type item or a frequency query. This issue may be particularly important in trying to identify cheating methods that utilize technology. Rapidly changing technology and easy access to innovations provide new opportunities for unauthorized use of materials or assistance. These methods may also be increasingly difficult to detect as users become more sophisticated or outpace teachers in technological expertise.

REFERENCES

Ahlers-Schmidt, C. R., & Burdsal, C. (2004). Passive cheating in back-to-back classes. *Teaching of Psychology, 31,* 108–109.
Ames, C., & Archer, J. (1988). Achievement goals in the classroom: Students' learning strategies and motivation processes. *Journal of Educational Psychology, 80,* 260–270.

Anderman, E. M., Griesinger, T., & Westerfield, G. (1998). Motivation and cheating during early adolescence. *Journal of Educational Psychology, 90,* 84–93.

Anderman, E. M., & Maehr, M. L. (1994). Motivation and schooling in the middle grades. *Review of Educational Research, 64,* 287–309.

Anderman, E. M., & Mideley, C. (2004). Changes in self-reported academic cheating across the transition from middle school to high school. *Contemporary Educational Psychology, 29,* 499–517.

Athanasou, J. A., & Olasehinde, O. (2002). Male and female differences in self-report cheating. *Practical Assessment, Research & Evaluation, 8*(5). Retrieved February 3, 2005 from http://PAREonline.net/getvn.asp?v=8&n=5.

Baird, J. S., Jr. (1980). Current trends in college cheating. *Psychology in the Schools, 17,* 515–522.

Bartlett, T. (2005). Missouri Dean appears to have plagiarized a speech by Cornel West. Retrieved June 24, 2005 from http://chronicle.com/weekly/v51/i42/42a01301.htm.

Bennett, R. (2005). Factors associated with student plagiarism in a post–1992 university. *Assessment & Evaluation in Higher Education, 30,* 137–162.

Brandes, B. (1986). Academic honesty: A special study of California students. *California State Department of Education,* p. 36, the Chronicle of Higher Education. (1999). *Colloquy.* Retrieved June 22, 2005 from http://chronicle.com/colloquy/99/cheat/27.htm.

Cizek, G. J. (2003). *Detecting and preventing classroom cheating: Promoting integrity in assessment.* Thousand Oaks, CA: Corwin Press.

Daily Forty-Niner. (2005, January 25). Teachers have aid against cheating. *Daily Forty-Niner-California State University-Long Beach.* Retrieved February 10, 2005 from http://web.lexis-nexis.com.

Davis, S. F., Grover, C. A., Becker, A. H., & McGregor, L. N. (1992). Academic dishonesty: Prevalence, determinants, techniques, and punishments. *Teaching of Psychology, 19,* 16–20.

Dawkins, R. L. (2004). Attributes and statuses of college students associated with classroom cheating on a small-sized campus. *College Student Journal, 38,* 116–129.

Eccles, J. S., & Midgley, C. (1989). Stage-environment fit: Developmentally appropriate classrooms for young adolescents. In C. Ames & R. Ames, (Eds.), *Research on motivation in education: Vol. 3. Goals and cognition.* San Diego, CA: Academic Press.

Ehrlich, E., Flexner, S. B., Carruth, G., & Hawkins, J. M. (1980). *Oxford American dictionary, Heald Colleges edition.* Oxford: Oxford University Press.

Eisenberg, J. (2004). To cheat or not to cheat: Effects of moral perspective and situational variables on students' attitudes. *Journal of Moral Education, 33*(2), 163–178.

Evans, E. D., & Craig, D. (1990). Teacher and student perceptions of academic cheating in middle and senior high schools. *Journal of Educational Research, 84*(1), 44–52.

Franklyn-Stokes, A., & Newstead, S. E. (1995). Undergraduate cheating: Who does what and why? *Studies in Higher Education, 20*(2), 159–172.

Frey, B. B, Petersen, S., Edwards, L. M, Pedrotti, J. T., & Peyton, V. (2005). Item-writing rules: Collective wisdom. *Teaching and Teacher Education, 21*(4), 357–364.

Fussell, J. A. (2005, May 13). Cheaters use new array of gadgets to get that "A". *The Kansas City Star,* pp. A1, A8.

Genereux, R. L., & McLeod, B. A. (1995). Circumstances surrounding cheating: A questionnaire study of college students. *Research in Higher Education, 36*(6), 687–704.

Georgetown University (2004). *The honor system.* Retrieved June 22, 2005 from http://www.georgetown.edu/undergrad/bulletin/regulations6.html.

Gilroy, M. (2004, December 13). Cut and paste cheating: Rampant, but amenable to faculty action. *The Hispanic Outlook in Higher Education.* Retrieved February 10, 2005 from http://web.lexis-nexis.com.

Haines, V. J., Diekhoff, G. M., LaBeff, E. E., & Clark, R. E. (1986). College cheating: Immaturity, lack of commitment, and the neutralizing attitude. *Research in Higher Education, 25,* 342–354.

Hetherington, E. M., & Feldman, S. E. (1964). College cheating as a function of subject and situational variables. *Journal of Educational Psychology, 55*(4), 212–218.

History News Network. (2005). *Bryan Le Beau's response (#62779).* Retrieved June 22, 2005 from http://hnn.us/comments/62779.html.

Hollinger, R., & Lanza-Kaduce, L. (1996). Academic dishonesty and the perceived effectiveness of countermeasures: An empirical survey of cheating at a major public university. *NASPA Journal, 33,* 292–306.

Houston, J. P. (1986). Classroom answer copying: Roles of acquaintanceship and free versus assigned seating. *Journal of Educational Psychology, 78*(3), 230–232.

Kane, M., Crooks, T., & Cohen, A. (1999). Validating measures of performance. *Educational Measurement: Issues & Practice, 18*(2), 5–17.

Karpf, M. (2004, December 6). Ignorance isn't bliss in copy-and-paste world. *The Daily Nebraskan.* Retrieved February 10, 2005 from http://web.lexis-nexis.com.

Kawai, J. (2005). New device detects cheating on multiple-choice exams. Retrieved February 4, 2005 from http://csulb.edu/~d49er/archives/2004/fall/news/volLVno60-device.shtml.

Keith-Spiegel, P., Tabachnick, B. G., Whitley, B. E., Jr., & Washburn, J. (1998). Why professors ignore cheating: Opinions of a national sample of psychology instructors. *Ethics & Behavior, 8,* 215–227.

Kohlberg, L. (1985). *Moral education: Theory and application.* Hillsdale, NJ: Lawrence Erlbaum.

Korea Times (2004, November 22). Number of students accused in cell phone scam growing. *Korea Times.* Retrieved February 10, 2005 from http://web.lexis-nexis.com.

Maehr, M. L. (1991). The "psychological environment" of the school: A focus for school leadership. In P. Thurston & P. Zodhiates (Eds.), *Advances in educational administration* (pp. 51–81). Greenwich, CT: JAI Press.

Maehr, M. L., & Midgley, C. (1991). Enhancing student motivation: A schoolwide approach. *Educational Psychologist, 26,* 399–427.

Malinowski, C. I., & Smith, C. P. (1985). Moral reasoning and moral conduct: An investigation prompted by Kohlberg's theory. *Journal of Personality & Social Psychology, 49*(4), 1016–1027.

McCabe, D., & Trevino, L. (1993). Academic dishonesty: Honor codes and other contextual influences. *Journal of Higher Education, 64,* 522–538.

McCabe, D., & Trevino, L. (1997). Individual and contextual influences on academic dishonesty: A multicampus investigation. *Research in Higher Education, 38,* 379–396.

McCabe, D. L. (1999). Academic dishonesty among high school students. *Adolescence, 34,* 681–689.

McCabe, D., Trevino, L., & Butterfield, K. (2001). Cheating in academic institutions: A decade of research, *Ethics & Behavior, 11*(3), 219–232.

McMurty, K. (2001). E-cheating: Combating a 21st century challenge. Retrieved June 22, 2005 from http://www.thejournal.com.

Messick, S. (1988). The once and future issues of validity: Assessing the meaning and consequences of measurement. In H. Wainer and H. I. Braun, (Eds.), *Test validity* (pp. 33–48). Hillsdale, NJ: Lawrence Erlbaum.

Messick, S. (1994). Foundations of validity: Meaning and consequences in psychological assessment. *European Journal of Psychological Assessment, 10*(1), 1–9.

Moake, S. (2004, September 15). Cheaters abuse flexibility of Blackboard interface at BYU. *The Daily Universe.* Retrieved February 10, 2005 from http://web.lexis-nexis.com.

Murdock, T. B., Hale, N. M., & Weber, M. J. (2001). Predictors of cheating among early adolescents: Academic and social motivations. *Contemporary Educational Psychology*, *26*, 96–115.

Murdock, T. B., Miller, A., & Kohlhardt, J. (2004). Effects of classroom context variables on high school students' judgments of acceptability and likelihood of cheating. *Journal of Educational Psychology*, *96*, 765–777.

Newstead, S. E., Franklyn-Stokes, A., & Armstead, P. (1996). Individual differences in student cheating. *Journal of Educational Psychology*, *88*, 229–241.

Norton, L., Tilley, A., Newstead, S., & Franklyn-Stokes, A. (2001). The pressures of assessment in undergraduate courses and their effect on student behaviors. *Assessment & Evaluation in Higher Education*, *26*(3), 269–284.

O'Leary, M. (1999). The Web banishes term-paper blues: Online term-paper mills. *Information Today*, *16*(3), 14–15+.

Phillips, M. R., & Horton, V. (2000). Cybercheating: Has morality evaporated in business education? *The International Journal of Educational Management*, *14*, 150–155.

Renard, L. (2000). Cut and paste 101: Plagiarism and the net. *Educational Leadership*, *57*, 38–42.

Robinson, E., Amburgey, R., Swank, E., & Faulkner, C. (2004). Test cheating in a rural college: Studying the importance of individual and situational factors. *College Student Journal*, *38*, 380–395.

Scanlon, P. M., & Neumann, D. R. (2002). Internet plagiarism among college students. *Journal of College Student Development*, *43*, 374–385.

Scheuneman, J. D., & Oakland, T. (1998). High-stakes testing in education. In Sandoval, J., Frisby, C. L. et al. (Eds.), *Test interpretation and diversity: Achieving equity in assessment* (pp. 77–103). Washington, DC: American Psychological Association.

Smith, J. K. (2003). Reconsidering reliability in classroom assessment and grading. *Educational Measurement: Issues & Practice*, *22*(4), 26–33.

Syer, C., and Shore, B. (2001). Science fairs: What are the sources of help for students and how prevalent is cheating? *School Science and Mathematics*, *101*(4), 206–220.

Thompson, L. (2005, January 16). Educators blame Internet for rise in student cheating; Easy access to information may blur rules. *The Seattle Times*. Retrieved Feburary 10, 2005 from http://web.lexis-nexis.com.

Total Telecom. (2004, September 10). Camera phones facilitate cheating on exams (September 10, 2004). *Total Telecom*. Retrieved February 10, 2005 from http://web.lexis-nexis.com.

Tweedle, E. (2004, December 14). Students using cell phones as cheating devices. *The Herald-Arkansas State University*. Retrieved February 10, 2005 from http://web.lexis-nexis.com.

Underwood, J., & Szabo, A. (2003). Academic offences and e-learning: Individual propensities in cheating. *British Journal of Educational Technology*, *34*(4), 467–477.

Wigfield, A., Eccles, J. S., & Pintrich, P. R. (1996). Development between the ages of 11 and 25. In D. C. Berliner & R. C. Calfee, (Eds.), *Handbook of educational psychology*. New York: Macmillan.

ACHIEVEMENT MOTIVATION AND CHEATING

4

INTEREST AND ACADEMIC CHEATING

GREGORY SCHRAW, LORI OLAFSON,
FRED KUCH, TRISH LEHMAN, STEPHEN
LEHMAN, AND MATTHEW T. MCCRUDDEN

This chapter explores the relationship between student interest and academic cheating. Previous research indicates that interest is an important component of student engagement and learning (Bergin, 1999; Hidi, 1990, 1995; Schraw, Flowerday, & Lehman, 2001; Schraw & Lehman, 2001). Both longstanding personal interest and spontaneous situational interest in a topic increase engagement and learning. However, no research that we know of has systematically examined the relationship between interest and academic cheating. Prior to beginning this project, we conjectured that interest would decrease cheating for a variety of reasons such as motivation, background knowledge, and a sense of personal relevance.

Cheating is common among students of all ages, including graduate and medical students (Cizek, 2003; Love & Simmons, 1998). Recent television specials suggest that cheating has reached epidemic proportions (ABC News Productions, 2004). A recent story in *People* Magazine (2005) reported that an heiress paid her roommate to assume her academic identity in order to get good grades for her! Previous studies have also reported increased prevalence. For example, Erickson and Smith (1974) found that 43 percent of students cheated when given the chance. More recently, Whitley (1998) reported that 70.4 percent of students cheat. Anderman and Midgley (2004) also reported that cheating increases as students transition from middle to high school. Some of the common explanations of cheating include lack of time, the decline of morals in society, peer

Copyright © 2007 by Academic Press, Inc.
All rights of reproduction in any form reserved.

acceptance of cheating as necessary to get good grades, the belief that cheating is too prevalent to stop, and the use of new technology such as finding or buying information on the Internet, which makes it easier to cheat. All of these explanations are likely correct to some degree.

This chapter examines the relationship between cheating and interest and is divided into five sections. The first section provides a brief review of the personal and situational interest, and how each type of interest may be related to cheating. The second summarizes the development of a survey of academic cheating used in our ongoing research at the University of Nevada, Las Vegas (UNLV). In addition, we also describe the development of a structured interview that was used to collect data specifically for this chapter. The third section presents the quantitative and qualitative results of survey and interview data with UNLV students, as well as a replication with underachieving high school students in Greeley, Colorado. The fourth section discusses implications for classroom instruction and creating a classroom culture that decreases cheating. The fifth provides suggestions for future research that expands on the role of interest and links interest to other motivational variables such as self-efficacy, goal orientations, and student attributions.

PERSONAL INTEREST, SITUATIONAL INTEREST, AND THEIR RELATIONSHIP TO CHEATING

No known studies directly relate different types of interest to cheating. However, a variety of studies have examined the relationship between cheating and other indicators similar to interests such as academic expectations, attitudes about cheating, locus of control, class attendance, and study time. For example, Szabo and Underwood (2004) examined how personal and situational factors such as student characteristics and attitudes are related to cheating. They identified 10 personal factors (e.g., self-esteem, attitudes, and desire to learn) and 11 situational factors (e.g., teacher's knowledge, task difficulty, and pressure) that affect cheating. These findings indicate that other researchers have considered personal and situational factors that affect cheating.

We begin by referring the reader to Table 4.1, which divides correlates of cheating into five categories based on Whitley's (1998) review of the cheating literature. The five categories are student characteristics, attitudes toward cheating, personality variables, situational factors, and other variables. Based on the correlations in the table, a number of conclusions seem warranted. One is that cheating decreases with age and marital stability, although cheating continues throughout the university years, and into graduate and professional schools (Love & Simmons, 1998). Cheating may

TABLE 4.1 A Summary of Significant Correlates of Cheating

Student Characteristics	Attitudes Toward Cheating	Personality Variables	Situational Factors	Other Variables
Age (−.27)	Attitude Toward Cheating (.38)	Self-Rated Honesty (−.43)	Competition (.34)	Reward for Success (.32)
Employment (−.18)	Perceived Ability to Cheat (.30)	Moral Development (−.18)	Importance of Test (.34)	Previous Victim of Cheating (−.23)
Marital Status (−.33)	Moral Obligation Not to Cheat (−.36)	Procrastination (.22)	Risk of Detection (−.36)	
GPA (−.16)		Impulsivity (.22)	Favorable Seating Arrangements (.36)	
Course Performance (−.30)		Test Anxiety (.17)	Academic Workload (.33)	
Expectation of Success (.42)		Need for Approval (.19)		
Perceived Grade Pressure (.34)		Ego Strength (−.20)		
Previous Cheating (.49)				
Quality of Study Conditions (−.38)				
Partying (.47)				

Note: Numerical values correspond to the correlation between the variable and the amount of student cheating.

decrease with age owing to greater maturity or better self-regulation of one's learning. Second, cheating decreases when individuals perform a task well, presumably because there is less need to cheat to do well, or because they feel a greater sense of academic self-efficacy (Murdoch, Hale, & Weber, 2001). Third, when success is the expected consequence of cheating, or a reward is expected for successful cheating, cheating increases. Indeed, many studies indicate that the prevalence of cheating is correlated with a student's perceived ability to cheat without detection. Fourth, if a student has a positive attitude about cheating or has cheated previously, cheating increases; this suggests that cheating begets cheating. Fifth, if students believe themselves to be effective cheaters, cheating increases. Most studies suggest that cheating serves one of two main purposes: it saves time with tasks that students perceive as unimportant, or it enables students to complete tasks they perceive to be extremely difficult and perhaps beyond their current skill level. Sixth, if students have reservations about the moral appropriateness of cheating, cheating decreases. In addition, external moral pressure not to cheat deters cheating (King and Mayhew, 2002).

Many of the variables shown in Table 4.1 may be related to personal and situational interest. *Personal interest* refers to information that is of enduring personal value, activated internally, and topic-specific (Hidi, Renninger, & Krapp, 1992; Schiefele, 1999). Personal interest also is referred to as *individual* or *topic interest*. The basis of personal interest appears to be preexisting knowledge, personal experiences, and emotions (Renninger, 1992; Tobias, 1994). Personal interest is negatively related to procrastination and positively related to internal locus of control. We predict that personal interest will be negatively related to cheating.

Personal interest can be divided into two subcategories referred to as *feeling-related* and *value-related* interest (Schiefele, 1992). Feeling-related interest occurs when an individual experiences positive affect and emotions in conjunction with a particular topic or activity. Presumably, these positive feelings provide a strong motivational incentive to engage in the activity (Anderman, Griesinger, & Westerfield, 1998). Examples of feeling-related factors include desire to learn, guilt, and attitudes. We anticipate that strong feeling-related interest would decrease cheating due to emotional involvement.

Value-related interest refers to the assignment of personal significance to a particular topic or activity. Presumably, value-related interest increases engagement because an activity or body of knowledge is judged to be important to one's long-term goals. For example, Table 4.1 shows that cheating and course performance are negatively related. One explanation is that students with higher value-related interest are more engaged in their classes and therefore do better. Schiefele (1992) concluded that feeling-related and value-related interest are strongly interrelated, even though each makes some unique contribution to perceived interest and task engagement. The extent to which they are related likely depends on the situation as well. Studies of personal interest suggest that personal interest appears to improve the quality of writing (Benton et al., 1995), deeper understanding while reading (Schiefele, 1999), and metacognition (Tobias, 1996). In the realm of cheating, a number of studies reveal that value-related interests such as moral beliefs and fairness to others are related negatively to cheating behaviors.

Situational interest refers to information that is of temporary value, environmentally activated, and context-specific (Hidi et al., 1992). Information that is sensational (e.g., newspaper headlines about a celebrity sex scandal) or relevant within a particular context (e.g., the performance of mutual funds you have recently purchased) falls into this category. Previous research suggests that situational interest is evoked spontaneously. Table 4.1 reveals that several situational variables such as competition, test importance, and favorable seating arrangements are related positively to cheating. In addition, Szabo and Underwood (2004) describe a variety of situational factors related to cheating. For example, pressure, limited time, and task difficulty increase cheating, whereas teacher competence and knowledge, student effort, and relevance of information decrease cheating. Consistent with these findings, we anticipate that factors that increase situational interest will decrease cheating.

Previous research has examined a wide variety of factors that affect situational interest. Schraw and Lehman (2001) organized these studies into three general categories: *text-based*, *task-based*, and *knowledge-based* situational interest. These categories align closely with the personal and situational factors related to cheating reviewed by Szabo and Underwood (2004). Text-based interest refers to properties of the to-be-learned information, typically a text, that affect interest. For example, coherent texts are more interesting to readers, as are texts that are informationally complete (Schraw, 1997). Similarly, the relevance of information to the student decreases cheating (Love & Simmons, 1998).

A number of text-based factors have been identified that increase situational interest, including the unexpectedness of information, character identification, activity level, structural aspects of the text such as coherence and completeness, concreteness and vividness, suspense, imagery, and relative ease of comprehension. In the realm of cheating, it is clear that situational factors such as testing and seating arrangements are related to cheating (Whitley, 1998). We anticipate that situational factors that affect interest would decrease cheating because students find the material more relevant and engaging, and are more willing to study to learn the information.

Task-based interest refers to aspects of a task or task environment that increase interest. For example, encoding instructions that change readers' goals increase interest (Hidi, 1990; Schraw & Dennison, 1994). Dewey (1913) originally suggested that interest was something that cannot be manipulated experimentally or in the classroom. However, a number of recent studies suggest that interest can be increased by changing the way readers approach a task. Based on existing research, it appears that encoding-task manipulations increase interest by changing learners' goals and strategies (Schraw & Dennison, 1994). Change-of-text manipulations increase interest by highlighting designated text segments or by making an entire text more cohesive (Hidi, 1990). Studies of academic cheating have identified a number of task factors that affect cheating, but especially *the number and difficulty of assignments* and the *amount of time to complete assignments*. Tasks that overwhelm students increase cheating. Tasks that are boring or irrelevant increase cheating as well (Szabo & Underwood, 2004; Whitley, 1998). We believe that changing task constraints related to assignments and amount of time to study will increase situational interest, which decreases cheating.

Knowledge-based interest refers to the effect that prior knowledge has on interest. Previous research suggests that prior knowledge is related both to personal (Schiefele, 1992; Tobias, 1994, 1996) and situational interest (Alexander & Murphy, 1998), even though knowledge appears to be related more strongly to personal interest (Alexander & Jetton, 1996). Love and Simmons (1998) identified the professor's knowledge as the number one external factor in inhibiting graduate students' cheating. Murdock et al. (2001) also reported less cheating when teachers were judged to be committed to student learning, knowledgeable, and competent.

Studies examining the relationship between prior knowledge and situational interest lead to the following conclusions: (1) knowledge is related to interest and learning when readers possess an adequate degree of prior knowledge or when the text provides that knowledge (Schiefele, 1999); (2) domain knowledge is related to interest (Alexander & Jetton, 1996) and interferes with learning when it is absent (Wade, Buxton, & Kelly, 1999); and (3) topic familiarity is unrelated to interest (Schraw, 1997). These findings suggest that specific topic knowledge and general domain knowledge may have different effects on interest and may help to explain the negative correlation between performance in a class and cheating. Students who possess greater domain knowledge are more likely to be interested in a course and more likely to perform well. Higher performance eliminates some or all of the pressure to cheat. In addition, studies of cheating indicate that knowledge judged to be relevant to future endeavors deters cheating (Love & Simmons, 1998).

Schraw and Lehman (2001) drew five main conclusions in their review of personal and situation interest that are relevant to academic cheating. One conclusion is that qualitatively different kinds of interest exist. Personal interest is stable, is based on previous experience, and is topic-specific, whereas situational interest is short-lived, context-dependent, and easier to manipulate. Both types of interest may affect engagement and academic behaviors concurrently. In the present case, we expect both types of interest to be related negatively to the incidence of cheating. Personal interest and values should decrease cheating in a manner analogous to the finding that personality and attitudinal variables such as morality, honesty, and ego strength are negatively related to cheating. In addition, personal interest may be related to a variety of motivational factors such as goal orientations (Anderman & Midgley, 2004) and personal confidence (Love & Simmons, 1998). Situational interest due to engaging materials, the importance of the task, and personal knowledge about the topic should decrease cheating as well.

A second conclusion is that one can distinguish among different types of situational interest, including text-, task-, and knowledge-based interest. Previous research indicates that each of these dimensions is related to learning. Szabo and Underwood (2004) identified a variety of situational factors that increase or decrease cheating that corresponds to text-, task-, and knowledge-based categories. We expect that any variable that increases situational interest would decrease cheating.

A third conclusion is that the content of a class matters. Some students may have strong personal interest in a topic prior to the class. Other students may develop situational interest because the information in the class is judged to be relevant or meaningfully connected to their lives. In addition, teachers often make classes more interesting by being knowledgeable, prepared, and excited about what they are teaching, which decreases cheating (Murdock et al., 2001) We predict that students' interest in academic content will decrease cheating. Similarly, teachers who make classes engaging by using humor or demonstrating a thorough understanding of class content, and being able to relate that content to students' lives, should decrease cheating as well. This prediction is consistent with Murdock et al. (2001), who found that academic motives alone were insufficient to explain middle school students' willingness to cheat. Social motivational variables, such as teacher competence (or lack of it), also were important predictors of cheating.

A fourth conclusion is that situational interest in class content is related positively to learning. Virtually all studies reveal that higher levels of topic interest are related to better learning. We know of no studies that report a negative correlation between interest and learning. Studies also suggest that topic interest is not static but can be increased by teacher effectiveness or highlighting the relevance of information.

A fifth conclusion is that personal and situational interest are related to more, and often deeper, learning. As we have suggested above, higher interest may increase engagement and performance in a class, which decreases the likelihood of cheating. This may explain in part the negative relationship between cheating and performance in a class reported throughout the cheating literature (Whitley, 1998).

THE PRESENT STUDY

The review of the interest literature demonstrates that, although research on academic cheating and interest has developed along separate lines, many parallels exist between the research on cheating and the research on interest. The summary of research on cheating and interest provided earlier in this chapter was based on two critical assumptions. The first is that personal interest, including feeling-related interest and value-related interest, will decrease cheating because students become more personally and affectively involved. In addition, personal interest may make information more engaging and relevant to students. Second, situational factors that affect interest will decrease cheating because students become more curious and challenged by the information. We designed a study to explicitly explore the relationship between interest and cheating by asking students on surveys and in interviews to describe how personal and situational interest impact their cheating behavior. After briefly describing the study, we outline key findings that highlight the relationship between interest and cheating.

METHODS

We collected three separate types of data. The first type consisted of survey data from 82 undergraduates who attended a variety of beginning English and mathematics classes at the University of Nevada, Las Vegas (UNLV). The complete survey is shown in Appendix A. Students provided open-ended written responses to questions 2–27. For the purposes of this study, we focused on questions 10 and 11. Question 10 addressed the role of personal interest in cheating, and question 11 concerned the role of situational interest and factors in cheating. A content analysis was conducted on the written responses. Categories were determined by using a combination of inductive and deductive approaches. In a deductive approach, researchers use a categorical scheme suggested by a theoretical perspective (Berg, 2001). Two broad categories, interest and situations that increased or decreased cheating, were derived from the existing literature on interest. Refining the categories by developing a number of subcategories was undertaken by examining the responses systematically in an inductive approach. For example, responses to question 11 included statements such as "I'm less likely to cheat when the class is regulated," "I cheat more when I'm overwhelmed," and "I cheat more when the class is high stakes." Once the categories were determined, the coding scheme was applied to the data and frequencies were calculated.

The second type of data consisted of one-on-one interviews using the scripted interview protocol shown in Appendix B with 12 UNLV undergraduates. The interviews included three stages. Stage 1 consisted of two general questions regarding the role of personal interest and situational factors in the incidence of cheating. These questions were intended to solicit students' general beliefs about the relationship between interest and cheating. Stage 2 consisted of specific probes related to situational and personal factors that might affect cheating. These factors were culled from the surveys described above. Stage 3 of the interviews consisted of a verification of specific factors that affect cheating. Students were asked to inspect this list and to indicate whether any of these factors were particularly important to their willingness to cheat, and whether the list excluded factors that affected their cheating behavior. Interview transcripts were analyzed using the selective highlighting approach (van Manen, 1990) in which each transcript was read several times and the following question was asked: "What statement(s) or phrase(s) seem particularly essential or revealing about the phenomenon or experience being described?" (p. 90). Students' statements about cheating that included discussion of personal interest and situational factors were highlighted and summarized.

The third type of data consisted of interviews from underachieving high school students in Greeley, Colorado. The purpose of these interviews was to determine whether college and high school students mentioned similar factors. Two interviews used focus groups with six and four students, respectively. In addition, there were three individual interviews. These students typically experienced reading comprehension difficulties. All were Hispanic students; most were classified as English Language Learner (ELL). Interview transcripts were analyzed using the selective highlighting approach (van Manen, 1990).

RESULTS

The results of the study are summarized according to our two general assumptions about the relationship between interest and cheating. We first describe our findings related to personal interest and relate these findings to previous research. We then summarize our findings related to situational interest.

Personal interest is negatively related to cheating. Question 10 on the survey asked, *How does your personal interest in a topic impact the likelihood of you cheating?* We created three categories based on the effect of personal interest on cheating, including it *increased cheating*, it had *no effect*, and it *decreased cheating*. Sixty-nine of the 82 participants gave codeable responses. No participants indicated that interest *increased cheating*; 17 (21 percent) indicated that interest had *no effect* on cheating; and 52 (63 percent) indicated that interest *decreased cheating*. A typical response to the *no effect* category was, "Interest doesn't matter. I won't cheat regardless." Typical responses to the *decreased cheating* category were, "If I am interested, I am less likely to cheat," "If I'm not interested in a class, it's a lot harder to motivate myself to learn," and "If you're interested, you'll *want* to learn the material and won't need to cheat." These findings are consistent with Whitley (1998) and Szabo and

Underwood (2004), who found that factors that increased students' personal engagement decreased cheating.

All 12 university undergraduate interviewees agreed that personal interest was related to cheating. Personal interest in a topic decreases cheating behavior, as one participant described: "Interest decreases cheating because you have increased background knowledge and more motivation" (Jane). The absence of personal interest, on the other hand, was seen to increase cheating: "If you're not interested in a subject say physics or some other class that you have to take then you might be more encouraged to cheat because you don't understand and you don't want to learn it" (Kristin).

Unlike the survey responses, however, participants who were interviewed also discussed how interest in a class or topic did not necessarily remain static. One respondent noted that when teachers thought she knew something she didn't already know she was apt to lose interest in the class. Another respondent described how loss of initial interest led to cheating:

> Interest starts when a question comes up and you can't answer it but you want to know. So you get interested in it and then maybe let's say you're not so good at it so you start losing interest and then maybe you can't find somebody who can help you and you lose more interest. And then finally you get to the point where you're like, "I can't understand this, I'm never going to understand this so I'm gonna give up or cheat."

A third participant described how interest in a topic changed over time: "I liked biology classes in high school and then I got here and I took them and they were okay but we got deeper into things and it wasn't as interesting as it used to be. Suddenly, I didn't really like learning all that stuff and there were a lot of occasions where it really crossed my mind, and sometimes I did, cheat because I felt like I needed the grade."

These findings are consistent with previous research on interest, which indicates that a variety of situational factors may increase or decrease interest. These factors may include coherence and relevance of information, teacher preparedness, or study conditions such as the amount of time allowed to study. Factors that increase situational interest in an activity increase learning (Schraw & Lehman, 2001). Similarly, it appears that factors that increase situational interest decrease cheating. We found no evidence that situational interest increased cheating. In addition, we found that there were a variety of ways to increase situational interest in the classroom. These included using engaging materials, clarifying the task, reducing the pressure to compete against others, and having organized teachers.

Personal factors contributed significantly to cheating among the high school students. Some students worked or participated in sports that both took away study time and imposed grade average requirements. Many students reported being lazy or bored with classes and cheated to compensate for effort. In addition, few high school students have a declared major or take a large number of classes in a preferred topic area, which may lead to more cheating than in a university. These findings closely mirror the findings reported by Whitley (1998), which are summarized in Table 4.1.

A number of personal factors were unique to these high school students. One was the burden imposed by being a second-language learner. Classes were more difficult for these students regardless of content; thus, there was greater incentive to cheat to keep up. A second factor was overcrowded classrooms. Some students reported that it was easy to cheat because they could see other students' work during a test. A third difference was financial incentive. Roughly half of the students received financial rewards from their parents for good grades. All of these students felt a greater temptation to cheat. Fourth, some students were bullied into cheating in a way that did not occur among the university students. It is difficult to determine the frequency with which "bullied cheating" takes place, but it appears to be a significant factor for some students.

The majority of university and high school students indicated that personal interest helped them stay engaged in learning. Personal interest also was linked to other aids to learning such as background knowledge, self-efficacy for the information to be learned, and greater motivation to learn. A number of feeling-based and value-based aspects of personal interest were described by students (Schiefele, 1992). Feeling-based interest focused on liking the subject matter or topic being studied, whereas value-based interest focused on whether the topic was related to one's academic major. Students are less likely to cheat when a class is within one's academic major. These findings are similar to the finding that ego strength and moral convictions are negatively related to cheating. Overall, there is clear evidence from our results that increasing personal interest in a topic decreases cheating for two reasons: (1) information appears to be more relevant to students; and (2) students are more concerned with maintaining personal integrity.

Situational interest is negatively correlated to cheating. Question 11 asked, *Are there specific situations in which you would be more or less likely to cheat?* Students provided a number of responses, which are categorized here in Table 4.2. Three main categories emerged: situations that have *no effect*, situations that *decrease cheating*, and situations that *increase cheating*. The most common response was that specific situations have *no effect* on cheating. Interest in the class or class topic, an effective teacher, and guilt *decreased cheating*. These findings are identical to those reported by Szabo and Underwood (2004). Being unprepared and encountering trouble learning the material *increased cheating*. Fear of receiving a failing grade, lack of teacher monitoring, and high-stakes exams also increased cheating. These findings are identical to those reported by Whitley (1998).

TABLE 4.2   Specific Situations That Affect Cheating

	Frequency
Specific Situations Have No Effect	21
Situations That Decrease Cheating	
Effective teacher	5
Liking the class or topic	4
Guilt	3
Fear of being caught	1
Situations That Increase Cheating	
Being unprepared; lack of study time	14
Poor or failing grade	8
Trouble learning the material; material too difficult	8
Not being watched (teacher is absent)	6
High-stakes exams	4
If others cheat	2
If students collude to cheat	2
Trivial or boring work	2
Not the student's major	1
Keep financial aid	1

Specific situations that increased or decreased cheating described in the survey data also appeared in the individual interviews. In the interviews, students provided support for the following findings: certain task-based situations increased cheating (e.g., online courses); knowledge-based situations increased cheating (e.g., not knowing the material); classroom-based aspects increased cheating (e.g., lack of space between students during an exam); and liking the class or topic decreased cheating. These findings are consistent with previous research, although few studies in the cheating literature focus primarily on online courses. Cheating may be more endemic to these classes than to face-to-face classes. Future research should focus more attention on the incidence and consequences of cyber-cheating (Szabo & Underwood, 2004).

Peer pressure was another situation that was found to potentially increase cheating. One participant described a situation where her friends were "naturally good" in a subject and they didn't have to work as hard. Brenda said that she "didn't want to be the friend who was getting a D." Sometimes, she said, other friends tried to get her to assist with cheating behavior. In the past she was more likely to provide answers, but now she tells them that she doesn't have the homework with her to avoid collaborating on cheating. Similarly, Mary noted that if your best friend is getting straight A's and "you're sitting there pulling C's and you sit right next to them you might be inclined to cheat off them." And if your best friend is sitting next to you during the test and he asks for the right answer, Jane said that she would provide the answer "because you want them to do well and that's your friend." These findings are identical to Whitley (1998), who reported that peer pressure to cheat increases cheating. This may be due to collusion, favorable seating arrangements, less fear of detection because other students support one's cheating, or the tendency of peer cheating to reduce moral pressure not to cheat.

One consistent theme from the high school student interviews was that instruction and content must be meaningful and relevant; otherwise students become bored and cheat. Like university students, the high school students were less likely to cheat if they liked the topic, experienced moderate to high self-efficacy to learn the material, and felt the information was meaningful to them and would be useful in the future. However, many students indicated that most of their classes did not meet these criteria; thus, there was a significant amount of cheating.

Instructor behaviors are another situational factor that impact the likelihood of cheating. On the surveys, five university students noted that effective teaching decreased cheating. Notably, all 12 of the interviewees discussed instructor behaviors that were seen as either increasing or decreasing student cheating. In addition to effective teaching, described in the surveys as a deterrent for cheating, interview participants described the teacher–student relationship as a factor related to cheating. Kristin, for example, noted that "If you really like the teacher and they teach really well then you're not as likely to cheat." Other aspects of a positive teacher–student relationship that were seen to decrease cheating included mutual respect and the instructor's demonstration of personal interest in the student: "She was a good teacher and I respected that. She genuinely cared about her students and whether or not we were learning the material" (Jane). On the other hand, according to Mary, if students felt like the "teacher is out to get them," or that the "teacher doesn't care about them" then they think, "I'm going to get them back by kind of undermining what they're trying to do here." These findings are consistent with Love and Simmons (1998) and Murdock et al. (2001), who found that teacher competence and enthusiasm decreased cheating owing to higher levels of student engagement.

The effect of teacher characteristics was virtually identical for university and high school students. Cheating increases when teachers are not knowledgeable, not prepared, not able to make personal contact with students, and not well organized. One significant difference with the high school students is that they are required to attend multiple classes everyday, whereas a university student may attend one three-hour class per week. As a result, there appeared to be far more boredom and lack of meaningfulness among the high school students. Teachers who left the room or ignored cheating behaviors inadvertently increased the amount of cheating in their classroom. Another difference between university and high school students is that high school students pass along information to other students about a student's relative opportunity to cheat in a given teacher's classroom. This seemed to breed cheating more than among university students at a large institution.

For university and high school students, specific situations affect cheating in similar ways. Competition, high-stakes tests, and instructor characteristics played a large role in whether or not a student cheated. Situational factors also affected the impact of personal interest. For example, even though high personal interest usually decreased cheating, having a poor instructor or financial pressure to perform well could negate the effect of personal interest. In general, as pressure to perform well increased, particularly on a single measure of academic performance, cheating increased as well.

Summary of results. Overall, high school students differed very little from university students with respect to cheating. Personal interest was very important to the high school students and uniformly decreased cheating. The same situational factors reported by university students increased (e.g., lack of vigilance, high-stakes assessments) or decreased (e.g., well-organized instruction, teacher caring) cheating. Teachers also had a large impact on cheating for better or worse. The differences between the two groups were due to the different class format between high schools and universities, the financial incentives that many of the high school students had from parents, and the occurrence of bullying.

These findings paralleled those reported previously in the cheating literature in two important ways. First, it is possible to distinguish between personal and situational factors that affect cheating. Personal factors include longstanding interest in the topic and perceived relevance. Situational factors include the quality of the learning environment, a learning environment that reduces pressure and competition, and effective instruction. Second, some of these factors increase cheating, while others decrease it. Factors that increase personal or situational interest decrease cheating. Chief among these factors are taking classes in one's academic major and preparedness.

IMPLICATIONS FOR INSTRUCTION AND CLASSROOM CULTURE

Our findings have six implications for the classroom. The first and most important is that personal and situational interests matter! Personal interest usually decreases cheating and rarely, if ever, increases it. Some situational factors such as fair, watchful instructors decrease cheating, while other factors such as competition increase it. Unfortunately, instructors have no control over the amount of personal interest a student brings to the classroom. However, as described by Mitchell (1993), creating positive situational supports might *catch* students' interest and *hold* it long enough for it to become personal interest. Bergin (1999) and Schraw et al. (2001) also suggested several strategies for increasing situational interest, including offering students meaningful choices, using well-organized texts and learning aids, and encouraging students to become active participants in learning.

A second implication is to reduce underpreparedness due to students' unfamiliarity with course content, difficulty learning the material, or procrastination (Whitley, 1998). One suggestion is to provide as much academic support as possible via a highly organized instructor, supporting materials such as web sites, peer collaboration, and frequent review sessions. A second suggestion is to conduct frequent assessments to keep students on task. Infrequent assessments allow a student to fall so far behind they are more at risk for cheating.

A third implication is that teachers must be vigilant with regard to cheating and hold students accountable when they are caught. Most students reported that instructors are not aware of cheating or turn their back on it. Poor monitoring of cheating by the instructor lowers class morale and increases cheating, in part because "everyone is doing it." Students also report being less likely to cheat if they are held accountable. Given the frequency of cheating, instructors should provide an explicit policy in their syllabus and list the penalties associated with different varieties of cheating such as plagiarizing, collaborating inappropriately with others outside the classroom, or paying others to do the work for them. Last, instructors need to follow through when cheating occurs, as the following comment from a respondent reveals:

> The instructor actually makes a big difference. I have an instructor this semester and she is very open and honest about the fact that she doesn't like cheating and some instructors you get the feeling well, they don't really care so why should I. So the instructor really sets the tone and a lot of instructors I think find it a hassle to follow through with that kind of thing.

A fourth implication is to reduce high-stakes testing as much as possible. As a rule of thumb, the higher the stakes associated with an assessment, the more likely a student will cheat! High-stakes pressure appears to decrease or negate personal interest in a class and provides a compelling reason to cheat. In contrast, using frequent low-stakes assessments provides less reason to cheat, increases day-to-day engagement, and hopefully increases interest in the course. In addition, using different forms of a test for different sections of a class decreases cheating. Many students reported increased cheating when the assessment had been given to other students previously.

A fifth implication is that instructors should make sure that every student has an opportunity to master the material. As Table 4.1 reveals, students who cannot master the material due to difficulty and who are at risk of failing or a low grade are more likely to cheat. Many times an instructor does not have the time or resources to help every student, and sometimes students who are failing will not ask for help or accept it when given. We suggest that instructors consider referring students to tutoring centers if available, assigning students to collaborative groups, engaging quality tutors who will meet with students outside of class, and alerting students to helpful web sites.

A sixth implication is that teachers should teach well! Many students reported that a boring, unorganized instructor increases cheating. In contrast, cheating decreases when an instructor is knowledgeable, well organized, makes information relevant, provides regular and fair assessments, is vigilant to inappropriate behavior such as cheating, and punishes infractions quickly and effectively. Cheating also decreases when teachers care about their students. Many of the survey responses, as well as interviews, indicated that teacher caring creates a bond with students that dramatically decreases cheating.

DIRECTIONS FOR FUTURE RESEARCH

We close with four suggestions for future research. The first is to examine the relationship between interest, cheating, and motivational variables such as self-efficacy, intrinsic motivation, goal orientations, and attributions within a classroom setting. Many participants in our research mentioned motivational variables as being an important contributor to interest and cheating. Currently, very little research addresses the relationship between interest and motivation. One exception is Anderman et al. (1998) who found that middle school students with mastery goal orientations were less likely to cheat than students with performance goal orientations. We know of no studies that examine interest, cheating, and motivation simultaneously. One question of particular importance is how motivation and interest are related and, in turn, separately and interactively affect cheating.

A second suggestion is to examine the relationship between self-regulation and cheating. Self-regulation refers to study strategies and meta-cognitive awareness of one's learning (Winne & Perry, 2000). Many students attributed their cheating to being unprepared or unable to master the material. It is unclear at this point whether their difficulty is due to the material or to their lack of self-regulation. Research on self-regulation has focused on the types of strategies students use to learn efficiently or on interventions that improve these strategies. No known research examines whether poor self-regulation increases cheating, or whether interventions that improve self-regulation decrease cheating.

A third suggestion is to examine the impact of the number and type of assessments on cheating, as well as the quality and timeliness of feedback. Our findings and previous research (Whitley, 1998) indicate that high-stakes assessments dramatically increase cheating because they carry a great deal of weight. Frequent, low-stakes assessments may decrease cheating for two reasons: there is less pressure to perform and useful formative feedback is provided, facilitating learning. We believe that research should examine several dimensions of classroom assessment, including the frequency of assessment, type of assessment (e.g., selected response versus essays), the stakes associated with the assessment, and whether the assessment provides formative feedback to students.

A fourth suggestion is to study the frequency and impact of "bullied cheating." This form of bullying is especially insidious and appears to occur often enough to represent a phenomenon in and of itself. Little is said about the role of bullying in the current cheating literature, perhaps because many of the students who are studied are at colleges and universities.

CONCLUSIONS

This chapter explored the relationship between interest and cheating. Both personal and situational interest impact cheating. High personal interest decreases cheating. Situational factors, notably teacher effectiveness, that increase interest usually decrease cheating. However, some situational factors, notably lack of surveillance and high-stakes testing, increase cheating. We proposed six strategies to decrease academic cheating and suggested others proposed by Bergin (1999) and Schraw et al. (2001). We also suggested four avenues for future research. Cheating is an all too common phenomenon in the classroom. Researchers are slowly gaining a better understanding of why students cheat and how cheating is related to other variables. However, little is known about how to deter cheating. Our findings suggest that increasing interest in a topic and providing situational constraints in the classroom may help reduce it.

REFERENCES

Alexander, P. A., & Jetton, T. L. (1996). The role of importance and interest in the processing of text. *Educational Psychology Review*, 8, 89–121.

Alexander, P. A., & Murphy, P. K. (1998). Profiling the differences in students' knowledge, interest, and strategic processing. *Journal of Educational Psychology*, 90, 435–447.

Anderman, E. M., & Midgley, C. (2004). Changes in self-reported academic cheating across the transition from middle school to high school. *Contemporary Educational Psychology*, 29, 499–517.

Anderman, E., Griesinger, T., & Westerfield, G. (1998). Motivation and cheating during early adolescence. *Journal of Educational Psychology, 60,* 84–93.

Benton, S. L., Corkill, A. J., Sharp, J. M., Downey, R. G., & Khramtsova, I. (1995). Knowledge, interest and narrative writing. *Journal of Educational Psychology, 87,* 66–79.

Berg, B. L. (2001). *Qualitative research methods for the social sciences.* Boston: Allyn & Bacon.

Bergin, D. A. (1999). Influences on classroom interest. *Educational Psychologist, 34,* 87–98.

Cizek, G. C. (2003). *Detecting and preventing classroom cheating.* Thousand Oaks, CA: Corwin Press.

Dewey, J. (1913). *Interest and effort in education.* Boston: Riverside Press.

Erickson, M. L., & Smith, W. B. (1974). On the relationship between self-reported and actual deviance: An empirical test. *Humboldt Journal of Social Relations, 1*(2), 106–113.

Heyman, J. D. (2005, January 24). Pssst . . . what's the answer? *People* Magazine, 108–111.

Hidi, S. (1990). Interest and its contribution as a mental resource for learning. *Review of Educational Research, 60,* 549–572.

Hidi, S. (1995). A reexamination of the role of attention in learning. *Educational Psychology Review, 7,* 323–350.

Hidi, S., Renninger, K. A., & Krapp, A. (1992). The present state of interest research. In A. Renninger, S. Hidi, and A. Krapp (Eds.), *The role of interest in learning and development* (pp. 433–446). Hillsdale, NJ: Lawrence Erlbaum.

King, P. M., & Mayhew, M. J. (2002). Moral judgment development in higher education: Insights from the Defining Issues Test. *Journal of Moral Education, 31,* 247–270.

Love, P. G., & Simmons, J. (1998). Factors influencing cheating and plagiarism among graduate students in a college of education. *College Student Journal, 32,* 539–551.

Mitchell, M. (1993). Situational interest: Its multifaceted structure in the secondary school mathematics classroom. *Journal of Educational Psychology, 85,* 424–436.

Murdock, T. B., Hale, N. M., & Weber, M. J. (2001). Predictors of cheating among early adolescents: Academic and social motivations. *Contemporary Educational Psychology, 26,* 96–115.

Primetime Live. (2004, April 29). *Cheating Crisis in America's Schools.* New York: ABC News Productions.

Renninger, K. A. (1992). Individual interest and development: Implications for theory and practice. In K. A. Renninger, S. Hidi, and A. Krapp (Eds.), *The role of interest in learning and development* (pp. 361–395). Hillsdale, NJ: Lawrence Erlbaum.

Schiefele, U. (1992). Topic interest and levels of text comprehension. In A. Renninger, S. Hidi, and A. Krapp (Eds.), *The role of interest in learning and development* (pp. 151–182). Hillsdale, NJ: Lawrence Erlbaum.

Schiefele, U. (1999). Interest and learning from text. *Scientific Studies of Reading, 3,* 257–280.

Schraw, G. (1997). Situational interest in literary text. *Contemporary Educational Psychology, 22,* 436–456.

Schraw, G., & Dennison, R. S. (1994). The effect of reader purpose on interest and recall. *Journal of Reading Behavior, 26,* 1–18.

Schraw, G., Flowerday, T., & Lehman, S. (2001). Promoting situational interest in the classroom. *Educational Psychology Review, 13,* 211–224.

Schraw, G., & Lehman, S. (2001). Situational interest: A review of the literature and directions for future research. *Educational Psychology Review, 13,* 23–52.

Szabo, A., & Underwood, J. (2004). Cybercheats: Is information and communication technology fuelling academic dishonesty? *Learning and Teaching in Higher Education, 5,* 180–199.

Tobias, S. (1994). Interest, prior knowledge, and learning. *Review of Educational Research, 64,* 37–54.

Tobias, S. (1996). Interest and metacognitive word knowledge. *Journal of Educational Psychology, 87*, 399–405.

Van Manen, M. (1990). *Researching lived experience: Human science for an action sensitive pedagogy.* London, ON: Althouse Press.

Wade, S. E., Buxton, W. M., & Kelly, M. (1999). Using think-alouds to examine reader-text interest. *Reading Research Quarterly, 34*, 194–216.

Whitley, B. E. (1998). Factors associated with cheating among college students: A review. *Research in Higher Education, 39*(3), 235–274.

Winne, P., & Perry, N. (2000). Measuring self-regulated learning. In M. Boekaerts, P. Pintrich, and M. Zeidner (Eds.), *Handbook of self-regulation* (pp. 531–566). San Diego, CA: Academic Press.

APPENDIX A

STUDENT SURVEY: ACADEMIC DISHONESTY AND CHEATING

The Educational Psychology department at UNLV is attempting to learn more about academic dishonesty or cheating. We would like to know what you think about some of the various facets of the topic of academic integrity. Your knowledge and honest opinions will help. Your responses will be confidential and will not be used to identify or single out anyone. For the purposes of this survey, the terms "academic dishonesty" and "cheating" and "plagiarism" all mean the same thing. Answer all parts of a question. Please try to provide a thoughtful response for each item. If you have time, go back and add details.

Part I: Demographic Information

Gender; ethnicity; age; state residency; full or part-time status; family income; years of school completed beyond high school; academic major.

Part II: Instances

1. Based on your experience, list instances of cheating that you have seen, done, or heard about at UNLV. (Cheating doesn't just occur during exams.) List at least five instances in some detail. Make the last one be the most extreme or elaborate instance of academic dishonesty you've ever heard of.

Part III: Issues

2. Have you observed other students cheating at UNLV during the past semester? If so, about how many times?
3. What kind of cheating seems to occur most often?

4. What particular kinds of cheating are associated with specific departments (for example, English, Engineering, Math, Psychology, etc.)?

5. Are there harmful effects to a person's character when they cheat? (For example, getting into the habit of cheating might make it easier to lie about other things.) Describe what these effects might be and whom they would mostly affect.

6. What are your feelings about cheating?

7. Under what circumstances, if any, is it OK to cheat? Why?

8. Have you ever been a victim of cheating? Did a cheater use your work or your ideas in any way in order to get a better grade? If so, please describe the circumstances.

9. Did you cheat during this past semester? If so, about how many times? Would you say that is more, less, or the same as other students?

10. How does your personal interest in a topic impact the likelihood of you cheating?

11. Are there specific situations in which you would be more or less likely to cheat?

12. Some universities have a policy that states that students who *help* others cheat are just as guilty as those who do the cheating. (An example would be giving your answers to another student during an exam.) Are those who help others to cheat just as guilty? Why or why not?

13. Some universities have a policy that requires students who *witness* cheating to report it. Do you think that this is a good policy? How would you respond if this policy was implemented at UNLV—sort of a "code" among students?

14. Are you aware of UNLV's policy for handling instances of cheating? If so, where did you learn about it?

15. What are some penalties for cheating at UNLV? Please list several and explain the circumstances when each might be appropriate.

16. How do instructors respond to instances of cheating?

17. List at least five detailed reasons people cheat.

18. Do you think cheaters are bad people? Why or why not?

19. Do you think there is a specific type of person who is a cheater? What are some characteristics of cheaters?

20. What steps does the university take to discourage cheating? And what steps do departments take to discourage cheating? For each step you list, describe how effective you think that step is.

21. What other things could be done to discourage students from cheating?

22. Is it cheating if during a test a student looks at their notes just to get the right spelling of a term (and on that test, spelling counts)? Why or why not?

23. Some students say they cheat because a course is *not important* to them (like an engineering student who has to take a literature elective). Please describe your feelings about this.

24. Some students say they cheat because the course is *very important* to them (like a course in their major in which they must get at least a B). Please describe your feelings about this.

25. Everyone who cheats on a midterm gets an A. You choose not to cheat and get a C. Please describe your feelings about this and describe what you might do, if anything.

26. During an exam, a professor sees someone cheating and looks the other way. How do you feel about this?

Part IV: Suggestions

27. If it were up to you, how would you find out more about cheating at UNLV?

APPENDIX B

AN INITIAL EXPLORATION OF ACADEMIC DISHONESTY (CHEATING) AT UNLV INTERVIEW PROMPTS

Introduction

We are interested in learning more about academic dishonesty and its relationship to interest and to specific situations in which academic dishonesty might be more or less likely to occur. By *academic dishonesty* we mean all forms of cheating (on tests or other academic assignments) and plagiarism (representing the words or ideas of another as one's own).

Part 1: General Questions

1. How does your personal interest in a topic impact the likelihood of you cheating?
2. Are there specific situations in which you would be more likely to cheat? Less likely to cheat?

Part 2: Questions Related to the Sub-Categories
Situational Aspects

1. Does the kind of assignment or task given by the instructor influence the likelihood of cheating?
2. If the content of the course is not relevant to a student's program might they be more likely to cheat?
3. Are students more likely to cheat if they don't know or understood the material?
4. When students feel pressured to get good grades are they more likely to cheat?
5. Are there any situations related the physical space of the classroom that might lead to cheating?
6. Are there things about an instructor or instructor behaviors in the classroom that encourage cheating?

7. Do certain personal or social situations of students increase the likelihood of cheating?
8. Do financial considerations play a role in cheating?

Personal Interest Aspects

1. When students like the subject, are they less likely to cheat?
2. When students are interested in the topic, are they less likely to cheat?
3. When students like the instructor, are they less likely to cheat?

Part 3: List

Provide students with the list of sub-categories (attached). Ask them to provide feedback (e.g., adding additional categories and/or specific instances.)

Students are more likely to cheat during:

1. Group projects
2. In-class assignments
3. On-line courses
4. Tests
5. Take-home exams
6. Classes that are unimportant

Students are more likely to cheat when:

1. They don't know the material
2. In math and engineering because the answers are specific
3. School is hard
4. A class is way too difficult and a good grade is needed
5. Professors expect too much
6. They are pressured to perform at a high level
7. Grades affect the ability to get into a program or stay in a program
8. There isn't enough space between students during exams
9. They know they can't get caught
10. ID's aren't checked during finals for large classes
11. Everyone in class helps each other by getting a copy of previous exams
12. They don't have time to study
13. They have procrastinated
14. They are stressed
15. The teacher helps you cheat by giving answers to a standardized test
16. The teacher isn't paying attention
17. The professor is bad

18. Professors don't change the exams/essays from semester to semester
19. There are pressures from work
20. They are overloaded with tests
21. Deadlines are approaching
22. They are lazy
23. They aren't smart enough to pass without cheating
24. They have low confidence
25. They don't go to classes
26. They don't like the subject
27. They think that cheating is inevitable and that everyone does it

Students are less likely to cheat when:

1. Classes are straightforward and easy
2. Tests are easy
3. Professors are more observant
4. In-class reviews are provided
5. They are interested in the topic
6. They are good at the subject
7. The classes are part of their major

5

THE EFFECTS OF
PERSONAL, CLASSROOM,
AND SCHOOL GOAL
STRUCTURES ON
ACADEMIC CHEATING

ERIC M. ANDERMAN

Academic cheating does not occur without a reason. Students generally don't cheat "for the fun of it"; rather, cheating often is motivated by specific individual difference variables, by contextual factors, or by interactions between individual differences and social contexts.

A number of demographic and individual difference variables have been examined as precursors to academic cheating, including gender, age, grade level, ability, subject area, personality, and academic status (Cizek, 1999; Nathanson, Paulhus, & Williams, 2006; Newstead, Franklyn-Stokes, & Armstead, 1996). Contextual variables have not received much attention in the literature, although some studies have examined the effects of some social contexts, such as the country in which the student resides (Evans, Craig, & Mietzel, 1991), whether or not institutions use an honor code (McCabe & Trevino, 1993), or whether the student attends a public or private school (Calabrese & Cochran, 1990). The present chapter offers a theoretically based examination of the effects of these variables and contexts on engagement in academic cheating behaviors.

Specifically, the present chapter examines the relations of achievement goal orientation theory to academic cheating. Although individual differences are considered, the chapter focuses on motivational influences that are reflected by classroom social contexts. These contexts are created through a cross-fertilization of instructional practices, school policies,

Copyright © 2007 by Academic Press, Inc.
All rights of reproduction in any form reserved.

classroom interactions, student characteristics, and teacher characteristics. The overall purpose of the present chapter is to examine students' motives for cheating, utilizing a currently popular theoretical framework from the field of achievement motivation: goal orientation theory.

ACADEMIC SOCIAL CONTEXTS AND CHEATING

A recent qualitative study by McCabe, Trevino, and Battlefield (1999) used focus groups to examine why high school students engage in academic cheating. Whereas the study was not rooted in motivation theory, the students' comments about cheating clearly focused on motivational variables that are rooted in the social contexts of classrooms. These comments reflected the need to cheat in order to get good grades and highlighted the fact that cheating may not be needed in environments that are truly focused on mastery, effort, and improvement. For example, one student noted that "If the teacher just takes the time to make sure you actually learn the information, I think that would decrease a lot of cheating" (McCabe, 1999, p. 686), whereas another student stated that "If ... cheating is going to get you the grade, then that's the way to do it" (McCabe et al., 1999, p. 683).

It is particularly important to consider the contexts in which learning occurs (Anderman & Anderman, 2000; Turner, 2001). A contextual approach to the study of motivation and academic cheating may be useful, since academic contexts often are malleable. Thus an intervention aimed at changing the social environment of a classroom may be easier to implement (and therefore to affect a larger group of students) than interventions aimed at changing the characteristics of individual students. The study of social contexts and their relations to academic cheating takes on a heightened sense of urgency, given the current rise in the prevalence of technology and online assessment (Haney & Clarke, this volume).

CLASSROOM FACTORS THAT AFFECT CHEATING

Many studies have emerged over the past few decades that have examined classroom-level instructional practices and their relations to academic cheating. Unfortunately, most of these studies have not been grounded in theory. Nevertheless, some important findings have emerged.

In his comprehensive review of the literature on academic cheating, Cizek (1999) notes that students are less likely to cheat when classes are small, when classroom conditions are conducive to effective learning, when academic work and assessments are clear and relevant, and when teachers take steps to prevent academic cheating. Other variables that often are implemented at the classroom level that often are related to lower levels of cheating include the use of honor codes (McCabe & Trevino, 2002), the use of coercive classroom-control techniques (e.g., Houser, 1982), and the belief that the student will not be caught (see Whitley, 1998, for a review).

In addition, teachers can affect the prevalence of cheating through the ways they encourage or discourage students from engaging in effective study habits. For example, in his meta-analysis on academic cheating, Whitley (1998) demonstrated that students who study under poor conditions are more likely to cheat than are students who study in more favorable conditions. Although instructors do not always have the ability to control students' study habits, they can encourage effective study strategies.

Goal orientation theory is a research-driven theoretical perspective that offers a comprehensive theory-based framework for the study of academic cheating. Goal orientation theory is particularly useful in terms of examining the relations between student motivation and cheating. In the following sections, after reviewing the basic assumptions of the theory, its relations to academic cheating are examined.

A GOAL ORIENTATION THEORY PERSPECTIVE ON ACADEMIC CHEATING

Currently, a number of prominent theoretical perspectives are being used in the study of academic motivation, including attribution theory (Weiner, 1985, 1986), self-efficacy theory (Bandura, 1997; Pajares, 1996), self-determination theory (Deci & Ryan, 1985), flow theory (Csikszentmihalyi, 1996), and expectancy-value theory (Eccles & Wigfield, 1995; Wigfield & Eccles, 2002). However, the current motivation theory that has been most often utilized in the study of academic cheating has been goal orientation theory. In the present chapter, goal orientation theory is used as the theoretical framework for examining cheating behaviors and attitudes because goal orientations are both somewhat malleable and context-specific (Anderman & Maehr, 1994; Maehr & Midgley, 1996; Meece, Anderman, & Anderman, 2006; Urdan, 1997).

Goal orientation theory has become an important basis for much motivational research during the past 25 years. Goal theorists are interested in the general purposes that students offer for engagement in academic tasks. In terms of academic cheating, goal orientation theory is a particularly useful perspective, since the theory incorporates both individual difference characteristics and contextual variables as important components.

Briefly, goal theorists often study two general classes of achievement goals. These goals are referred to by different names in the literature, but broadly the two classes are known as *mastery goals* and *performance goals*. When a student endorses mastery goals (also called *learning goals*), the student is interested in truly mastering a task. Mastery-oriented students generally use themselves as points of comparison: they are interested in how much they have improved at a task (compared to their previous performance) rather than how their ability at a task compares to others' abilities at the same task. Mastery goal-oriented students are interested in improving and generally focus on effort, trying hard, and truly learning from their mistakes (Dweck & Leggett, 1988; Meece et al., 2006; Meece, Blumenfeld, & Hoyle, 1988; Midgley, 2002; Pintrich, 2000; Urdan, 1997).

In general, when students are performance-goal oriented, they are concerned with how their abilities compare with those of other students. There has been more debate in the literature about the nature of performance goals than mastery goals. Current research supports a distinction between two types of performance goals: *performance-approach* and *performance-avoid* goals (Elliot & Church, 1997; Elliot & Harackiewicz, 1996; Middleton & Midgley, 1997). A student who endorses a performance-approach goal is primarily interested in demonstrating that his or her ability is superior to that of other students. In contrast, a student who endorses a performance-avoid goal is primarily interested in not appearing incompetent or "stupid" compared to peers. In addition, other researchers have also identified *extrinsic goals* as another class of performance goal (Anderman, Griesinger, & Westerfield, 1998; Anderman & Johnston, 1998; Pintrich & Garcia, 1991; Wolters, Yu, & Pintrich, 1996). A student who endorses an extrinsic goal engages in an academic task in order to obtain some type of extrinsic incentive (e.g., a good grade or money).

CLASSROOM GOAL STRUCTURES

In considering the importance of social contexts in the study of academic motivation, researchers have also assessed the effects of perceived classroom goal structures on a variety of outcomes. A classroom goal structure represents the perceptions that individual students have concerning the goals that are stressed within their classrooms (Ames, 1992; Ames & Ames, 1984; Kaplan, Middleton, Urdan, & Midgley, 2002; Midgley, 2002). These goal "stresses" are communicated to students through various means, most notably teachers' instructional practices. Goal structures are subjective in nature—they may differ for two students in the same classroom (Kaplan et al., 2002).

In general, two types of goal structures have been identified: mastery goal structures and performance goal structures. When students perceive the presence of a mastery goal structure in a classroom, the students believe that the classroom (i.e., the teacher and the instructional methods) stresses improvement, effort, task-mastery, and self-comparisons. When students perceive the presence of a performance goal structure in a classroom, the students believe that the classroom stresses social comparisons, grades, and competition (Ames, 1992; Midgley, 2002). Some studies also have identified a classroom extrinsic goal structure (i.e., classrooms that are perceived as stressing the receipt of certain extrinsic incentives as a reward for performance or completion of work) (Anderman et al., 1998). Students' interpretations of goal structures are their own perceptions, which may or may not map on to specific observable practices and behaviors in the classroom (Urdan, Kneisel, & Mason, 1999).

EFFECTS OF PERSONAL GOALS AND PERCEIVED GOAL STRUCTURES ON ACADEMIC OUTCOMES

The personal achievement goals that students adopt and students' perceptions of the goals that are stressed in their classrooms are related to important educational outcomes. Research generally suggests that students' personal goals emanate from the goals that are perceived as being emphasized in classrooms and schools (Ames & Archer, 1988; Anderman et al., 2001; Anderman & Maehr, 1994; Meece et al., 2006; Roeser, Midgley, & Urdan, 1996). Some research has examined the relations between students' personal goals and perceived classroom goal structures and other variables.

Personal Goal Orientations and Learning

In terms of personal goal orientations, the adoption of mastery goals is related for the most part to adaptive educational outcomes, including persistence at classroom tasks (Miller et al., 1996), the use of deep-level cognitive processing strategies and more intense cognitive engagement (Meece et al., 1988; Nolen, 1988; Wolters, 2004), and lower levels of academic self-handicapping (Midgley & Urdan, 2001). Interestingly, research to date has not documented a direct relation between academic achievement and mastery goals (Ames & Archer, 1988; Barron & Harackiewicz, 2001; Elliot & Church, 1997; Skaalvik, 1997).

The relations of performance goals to academic outcomes have been inconsistent, primarily because performance-approach and -avoid goals often were confounded in earlier measures of achievement goals. Results generally are mixed for performance-approach goals, with some studies finding positive results (Elliot, McGregor, & Gable, 1999; Pintrich, 2000), some producing null results (Wolters, 2004), and others indicating that the adoption of performance-approach goals is related to maladaptive outcomes such as the avoidance of help-seeking (Ryan, Hicks, & Midgley, 1997). In contrast, most researchers agree that performance-avoid goals are related to maladaptive educational outcomes (Elliot et al., 1999; Middleton & Midgley, 1997).

Classroom Goal Structures and Learning

Students' perceptions of classroom goal structures also are related to learning outcomes. However, it is important to recall that students' perceptions of classroom goal structures also are predictive of students' personal goals (Anderman et al., 2001; Wolters, 2004). Consequently, for many classrooms, disentangling the relations between goal structures, personal goals, and learning outcomes is not a simple task.

In general, the results for classroom goal structures parallel the results for personal goal orientations: perceptions of mastery goal structures are for the most part related to beneficial educational outcomes (e.g., use of effective cognitive strategies, positive affect, etc.), whereas perceptions of performance goal structures are related to maladaptive outcomes (e.g., the use of ineffective strategies, negative affect) (Ames & Archer, 1988; Kaplan & Maehr, 1999; Kaplan et al., 2002; Turner et al., 2002; Urdan, Ryan, Anderman, & Gheen, 2002).

Similar to the research on personal achievement goals, research examining the relations between perceptions of classroom goal structures and academic achievement is somewhat mixed. Most studies suggest that perceptions of a classroom performance goal structure are related inversely to grades, whereas perceptions of a classroom mastery goal structure are unrelated to grades (Anderman & Wolters, 2006), although some research (e.g., Midgley & Urdan, 2001) does suggest a positive relation between perceptions of a classroom mastery goal structure and achievement.

GOAL ORIENTATION THEORY AND ACADEMIC CHEATING

Several studies reported in recent years have examined the relations between achievement goals and academic cheating. These studies have examined a variety of variables related to cheating (e.g., self-reported cheating behaviors, attitudes about cheating). Whereas some studies have focused on personal achievement goals, others have concentrated on perceived classroom goal structures, or even on perceived schoolwide goal structures. In addition, whereas most of these studies have focused on college student populations, some more recent studies have examined cheating attitudes and behaviors in adolescent populations.

PERSONAL GOALS AND ACADEMIC CHEATING

Most studies have focused on the relations between students' personal goals (i.e., their adoption of either mastery goals, performance goals, or multiple goals) and indices of academic cheating. The findings from these studies are remarkably similar.

Newstead and his colleagues (1996) examined cheating in college-aged students. These researchers were among the first to suggest that academic cheating might be related to students' motivational goals. Specifically, they noted that

> Taken together, the results point strongly to the importance of motivation, and especially the distinction between performance and learning goals, in the explanation of cheating behavior. All of the individual differences studied lend themselves readily to explanation in these terms. In contrast, the evidence for moral judgment as an explanation of the observed differences is rather weak. (p. 239)

Although Newstead et al.'s research did not directly examine the relations between achievement goals and cheating, some of their results did suggest the existence of a relation. For example, they reported that males were more likely to cheat in order to increase their grades than were females.

The basic argument is that when students are mastery-oriented, they are unlikely to engage in academic cheating behaviors because cheating will not serve a useful purpose. If the goal of an academic task is to truly master the material, and to truly improve one's own learning, then an academic shortcut such as cheating will not lead to the attainment of one's goal. Indeed, if the goal is to learn, then cheating will not enhance the student's capacity as a learner. In fact, cheating could actually hinder a mastery or learning goal because the development of self-efficacy might be hindered (Anderman et al., 1998).

In contrast, when students endorse performance goals, they are either interested in demonstrating their ability relative to others (a performance-approach goal) or avoiding appearing "dumb" or incompetent (a performance-avoid goal). In such situations, engaging in academic cheating behaviors might help the learner accomplish his or her goals. As noted by Cizek (1999), a performance-oriented student "participates in class discussions, completes assignments, and puts forth maximal effort on tests, term papers, and so forth, in order to obtain the best prize—a high grade" (p. 101). For example, if a student is enrolled in a physics class, and the student is performance-approach oriented, then the student would want to demonstrate that she is more competent at physics than are her classmates. Consequently, the student might consider cheating as instrumental in attaining this goal, since cheating could allow the student to more easily reach the goal. In contrast, another student enrolled in the same physics class might be performance-avoid oriented; this student would not want to appear as a poor physics student to his peers. This student might therefore see cheating as a viable alternative since it could help him to reach his goal of not appearing "dumb" compared to his peers in the physics class.

Some researchers have directly examined the relations of students' personal achievement goals and students' perceptions of classroom goal structures to academic cheating. Anderman, Griesinger, and Westerfield (1998) examined the relations between personal goal orientations (mastery and extrinsic goals) and two measures of academic cheating (self-reported cheating behaviors and beliefs about the acceptability of cheating) in a sample of middle school students. Self-reported cheating behaviors and beliefs in the acceptability of cheating were related positively to personal extrinsic goals and negatively to personal mastery goals. After controlling for other variables, including perceived classroom goal structures, perceived school goal structures, worry about school, self-handicapping, and deep-level cognitive strategy usage, results indicated that students were more likely to report that they believed that cheating was acceptable when they reported holding personal extrinsic goals in science. Interestingly, personal extrinsic goals were unrelated to self-reported cheating behaviors. In addition, personal mastery goals were unrelated to either self-reported cheating behaviors or beliefs about the acceptability of cheating.

In a subsequent study, Murdock, Hale, and Weber (2001) extended the Anderman et al. (1998) study by examining the same relations, but also adding an array of classroom-social variables to the analyses. Since academic honesty is generally considered a fundamental philosophical principle in most schools, Murdock and her colleagues argued that students will be more likely to adopt the schools' tenets if the students have good relationships with their teachers. Consequently, in addition to examining the relations between personal goal orientations, classroom goal structures, and cheating, Murdock et al. also examined the relations of social variables in the classroom to cheating.

Using a sample of nearly 500 seventh and eighth grade students, Murdock and her colleagues found that students were more likely to report that they cheated when they held personal extrinsic goals. Interestingly, perceptions of a democratic participation structure were related *positively* to student cheating. Classroom goal structures were unrelated to cheating, once classroom social variables were controlled.

Using the LOGO II (Eison, 1981), Weiss, Gilbert, Giordano, and Davis (1993) examined the relations of what they refer to as a "learning orientation" and a "grade orientation" to academic cheating. Results indicated that in a sample of college-aged students, endorsement of a learning orientation was related to lower levels of academic dishonesty, whereas endorsement of a grade orientation was related to higher levels of dishonesty. In another study also using the LOGO II, Wryobeck and Whitley (1999) found that college students who reported high levels of learning orientation were more likely to recommend a severe punishment for a cheater than were students with lower self-reports of learning orientation. In addition, respondents who were high on the measure of grade orientation indicated that they would be more likely to cheat than would those lower in grade orientation.

Other studies have assessed constructs similar to personal achievement goals and have yielded similar results. For example, Genereux and McCleod (1995) found that college students reported that grades and future financial considerations were variables that justified cheating for some students. Results of an early study by Hill and Kochendorfer (1969) indicate that students tend to cheat more when they are aware of their peers' scores compared to when they are unaware of peers' scores; such findings suggest that contexts that promote comparisons between students (i.e., a focus on performance goals) may be conducive to increased cheating. In addition, results of a study by Dweck and Sorich (1999) suggest that when faced with academic difficulties, incremental theorists (i.e., students who believe in the malleability of intelligence, who are likely to be mastery oriented) are more likely to report that they will exert more effort, whereas entity theorists (i.e., students who believe that intelligence is a fixed entity, who are likely to be performance oriented) report that they might choose to cheat in order to succeed (see also Dweck & Leggett, 1988).

In summary, these studies taken together suggest that when students are focused on mastery and improvement, they may be less inclined to engage in cheating behaviors. In contrast, when students are focused on getting high grades, demonstrating their abilities relative to others, or avoiding appearing incompetent, they may be more likely to choose cheating as a viable behavior.

CLASSROOM GOAL STRUCTURES AND ACADEMIC CHEATING

The stressors that students perceive from their classroom teachers (as well as from parents and other external sources) to get good grades and to do well in school are profound. One student recently made the following statement to CNN: "I believe cheating is not wrong. People expect us to attend 7 classes a day, keep a 4.0 GPA, not go crazy and turn in all of our work the next day. What are we supposed to do, fail?" (Slobogin, 2002). Students often perceive these stressors as reflections of classroom goal structures. Indeed, it is possible to argue that goal structures convey messages to students that ultimately lead them to engage in cheating behaviors. As noted by McCabe, Trevino, and Butterfield (2001), when students perceive that the ultimate goal of learning is to get good grades, they are more likely to see cheating as an acceptable, justifiable behavior.

Only limited research has been done on the relations between classroom goal structures and cheating. However, the few studies that have been completed yield similar results. In general, perceptions of a performance goal structure in classrooms are related to increased reports of cheating, whereas perceptions of a mastery goal structure are related to decreased reports of cheating, even after controlling for a host of other variables.

Anderman et al. (1998) examined the relations between middle school students' perceptions of classroom mastery and extrinsic goal structures and both (a) cheating behaviors and (b) students' beliefs in the acceptability of cheating. Using logistic regression analyses that controlled for personal goals and perceptions of the school goal structure, the researchers showed that perceptions of a focus on extrinsic goals in science classrooms were significantly related to both cheating behaviors and beliefs in the acceptability of cheating (with greater perceptions of an extrinsic goal structure being related to more occurrences of academic cheating, and a stronger endorsement of cheating as acceptable). In terms of cheating behaviors, when both personal and extrinsic goals were examined simultaneously, only perceptions of the classroom extrinsic goal structure were related to cheating.

Murdock, Hale, and Weber (2001) also examined the relationship of classroom goal structures to students' self-reports of cheating. Interestingly, in their study, which also controlled for students' perceptions of teacher commitment, participation structures, school belonging, and respect for teachers, perceptions of classroom goal structures were not related significantly to academic cheating, whereas personal extrinsic goals were still related to cheating, even after controlling for other variables.

In an experimental study using vignettes, Murdock, Miller, and Kohlhardt (2004) manipulated descriptions of classroom goal structures. Results indicated that perceptions of a performance goal structure in hypothetical classrooms were related to indications that cheating was both more justifiable and more likely to occur than in classrooms where a mastery goal structure was prominent.

Anderman and Midgley (2004) conducted a longitudinal study of changes in students' reports of cheating in mathematics and changes in classroom goal structure. Data were collected at three time periods (fall of eighth grade, spring of eighth grade, and spring of ninth grade); students were in middle school during the eighth grade and in high school during the ninth grade. Thus the study examined changes across the high school transition. Students' perceptions of classroom goal structures were measured at all three time points. From a longitudinal perspective, students' reports of cheating increased over time. When measures of the perceived mastery- and performance-goal structures were included as time-varying covariates in the statistical models, results indicated that perceptions of a classroom mastery goal structure were related negatively to academic cheating, whereas perceptions of a classroom performance goal structure were related positively to academic cheating.

Followup analyses indicated that perceptions of the classroom goal structure were related to changes in self-reported cheating across the transition from middle school to high school. In terms of perceptions of a mastery goal structure, academic cheating increased when students moved from math classrooms with a high perceived mastery goal structure (in middle school) to math classrooms with a low-mastery goal structure (in high school). In addition, cheating decreased when students moved from low-mastery middle school classrooms into high-mastery high school class-rooms. When perceptions of performance goal structures were examined, results indicated that students who moved from math classrooms with a low-performance goal structure (in middle school) to classrooms with higher-performance goal structures (in high school), cheating increased. No significant changes in cheating were observed for transition from highly performance-oriented classrooms to less performance-oriented classrooms.

Integrating Personal Goals and Classroom Goal Structures as Predictors of Academic Cheating

In a study of motivation and cheating with college students, Jordan (2001) used both measures of personal mastery and extrinsic goals and classroom goal structures (or "course" goals in his study). Using a some-what different approach, Jordan factor analyzed the four goal orientation scales, along with several other measures. Personal and "course" mastery formed one index, whereas personal and "course" extrinsic goals formed a second index.

Those two indices were used to predict engagement in cheating behav-iors. Results indicated that the combined mastery scale was inversely related to academic cheating, whereas the combined extrinsic scale was related positively to cheating, after controlling for attitudes, norms, and knowledge. In addition, when these indices were averaged across courses in which students had reported cheating, students who had reported cheat-ing reported lower levels of mastery orientation than did noncheaters, whereas the students who reported cheating also reported higher extrinsic motivation scores than did the noncheaters. Interestingly, these results suggest that students' goal orientations and cheating behaviors differ across courses (Jordan, 2001). Similar results have been obtained in other studies (e.g., Murdock, Miller, & Anderman, 2005).

Related Studies

Some other research has examined variables related closely to classroom goal structures, in terms of how these variables relate to academic cheating. For example, McCabe, Trevino, and Butterfield (1999, 2001) examined how instructors can manage cheating in classroom settings. Many of the recom-mendations emphasize moving away from the use of performance-oriented instructional practices toward the use of mastery-oriented instructional practices. For example, some of the suggestions include "focus on learning, not on grades," and "when possible, reduce pressure by not grading students on a strict curve" (McCabe et al., 2001, p. 229).

Taken together, the results of these studies suggest that students' perceptions of the goals that are stressed in classrooms are predictive of academic cheating. When students perceive that their classrooms are focused on ability, competition, and high grades, students often are more likely to report engagement in cheating behaviors, compared with students in classrooms that are more focused on mastery and improvement.

SCHOOL GOAL STRUCTURE AND
ACADEMIC CHEATING

Some research indicates that students perceive an overall school "culture" or school goal structure, which is an entity that is distinct from individual classroom goal structures (Anderman & Maehr, 1994; Maehr & Fyans, 1989; Maehr & Midgley, 1991, 1996). The specific policies and practices implemented at a schoolwide level influence the perceptions of these practices (Maehr & Buck, 1992). Thus a school that emphasizes grades and test scores (e.g., by hanging up an honor roll at the door of the school, by printing the names of their best students in the newspaper, by always having conversations about assessments) is likely to be perceived as a performance-oriented school. In contrast, a school that stresses self-improvement, intrinsic learning, and task mastery is likely to be perceived as a mastery-oriented school (Anderman & Maehr, 1994; Maehr & Anderman, 1993).

Maehr (1991) found that the relations of students' perceptions of the school culture to motivation increases as students progress from elementary school to middle school and then to high school. This is particularly important in discussions of academic cheating because research indicates that cheating increases as students progress through schooling, particularly as they transition from middle school into high school (Anderman & Midgley, 2004). Thus the potential relationship between perceptions of the school culture and cheating is likely to be enhanced as students move from elementary school into middle school and then on to high school. This outcome is logical because assessment and testing are particularly prevalent at the high school level, and because the results of such assessments often have higher stakes for the students themselves after the transition.

Anderman and colleagues' 1998 study of academic cheating in science classrooms did include measures of students' perceptions of the school culture. Specifically, they examined perceptions of both a school performance goal structure and a school mastery goal structure. After controlling for personal goals and perceived classroom goal structures, perceptions of a school performance goal structure were related positively to increased self-reports of cheating, whereas perceptions of a school mastery goal structure were related inversely to beliefs in the acceptability of cheating (Anderman et al., 1998). Interestingly, a perceived performance goal structure predicted actual cheating behaviors, whereas a perceived mastery goal structure predicted beliefs about the acceptability of cheating. Although this research was exploratory, it does suggest that a schoolwide focus on performance goals may be related to actual engagement in cheating behaviors. Such a focus may be what pushes students over the edge, causing them to cross the line and cheat.

Related Studies Examining Schoolwide Perceptions and Academic Cheating

Numerous other variables are related to academic cheating (Cizek, 1999), but it is worth specifically mentioning a few additional studies that emanate from motivation theory, which demonstrate that students' perceptions of psychological phenomena at the school level are related to outcomes such as academic cheating. As will be argued in the final section of this chapter, phenomena of this nature are important because they can often be changed, as schools adopt various reform efforts.

For example, although Murdock and her colleagues did not specifically examine perceived school goal structures in terms of goal theory, they did examine several schoolwide variables, including schoolwide teacher respect and perceived school belonging. Results indicated that cheating was less prevalent when students perceived a schoolwide perception of teacher respect (Murdock et al., 2001).

In addition, a recent study by Finn and Frone (2004) examined cheating behaviors in a sample of 16- to 19-year-old high school and college students. This study examined the relations of students' identification with school and cheating. Whereas this study did not specifically examine perceived schoolwide goal structures, it did study perceptions of a related schoolwide psychological variable, lending support to the notion that other schoolwide variables are related to academic cheating. In this study, identification with school is operationalized in terms of "the extent to which students feel a sense of valuing and belonging in school" (Finn and Frone, 2004, p. 118). Results indicated that low-achieving students in particular were more likely to report engaging in academic cheating behaviors when they felt little identification with school.

Other studies also indicate that schoolwide policies such as the use of honor codes may be related to less frequent cheating in certain educational contexts (e.g., McCabe & Trevino, 1993), although the strength of such effects is inconsistent (e.g., Baldwin, Daugherty, Rowley, & Schwarz, 1996).

DISCUSSION

Cheating is prevalent and rampant in today's schools (Cizek, 1999; Schab, 1991). Although many students admit that cheating is wrong, the same students often report that they engage in academic cheating behaviors anyway (Anderman et al., 1998). Even preservice teachers have reported that they would be willing to cheat in the administration of standardized examinations to students, if the students would benefit (Wellhousen & Martin, 1995).

The present chapter reviewed literature examining the relationships of students' personal goal orientations, students' perceptions of classroom and schoolwide goal structures, and academic cheating. In general, results indicate that a focus on mastery goals (or a perception of a mastery goal structure) is related to less engagement in academic dishonesty, whereas a focus on performance and extrinsic goals (or perceptions of performance or extrinsic goal structures) is related to greater academic dishonesty.

Nevertheless, the extensive correlational evidence for the relations between achievement goals and cheating does not guarantee that manipulation of such goals and goal structures will lead to decreased cheating. One area ripe for future exploration is the study of *interventions* based in goal orientation theory, which are aimed at preventing cheating. Researchers often have made recommendations regarding the prevention of academic cheating. Indeed, practices such as the use of honor codes may help to prevent cheating (McCabe & Trevino, 1993). In his comprehensive book on academic cheating, Cizek (1999) recommends various practices that can be implemented to help prevent cheating (e.g., not using grades punitively, addressing cheating and plagiarism in advance, and ensuring test security). Nevertheless, virtually no controlled studies have been conducted on the effectiveness of cheating prevention programs, using either a goal orientation theory approach or any other theoretical approach. Since many recommendations for preventing academic cheating center on helping students to become better learners and avoid focusing on competition and grades (e.g., Cizek, 1999), carefully controlled studies based in motivation theory that are aimed at controlling and preventing cheating may be the next logical step for researchers in this area.

Some research does suggest that manipulating students' goals and students' perceptions of goal structures can lead to changes in student outcomes. Using a quasi-experimental design, some researchers have shown that teachers in both elementary and middle grades schools can effectively change instructional practices to be supportive of the adoption of mastery rather than performance goals (e.g., Anderman, Maehr, & Midgley, 1999; Maehr & Midgley, 1996). In a study of students in a middle school which focused on enhancing mastery goals and gave less stress to performance goals, Anderman et al. (1999) found that students who transitioned from traditional elementary schools into this middle school (compared to another middle school) were less likely to report endorsing performance and extrinsic goals after the transition than students who moved into the other, more performance-oriented school. Although academic cheating was not examined in that study, students who moved into the more mastery-oriented school may have cheated less than did students who moved into the more performance-oriented middle school, since grades and competition were not stressed in the former school. Similar work by Ames and her colleagues on changing the practices of teachers in order to affect students' goal orientations leads to similar conclusions (e.g., Ames, 1990).

Another area of particular importance for future research will be the relationship between goals/goal structures and cheating within the context of online and distance education courses. Whereas the contexts of traditional classrooms and schools can be altered in order to change students' goals or students' perceptions of goal structures, it is not necessarily as simple to make these changes in online and distance education courses. Specifically, the "context" is no longer the traditional classroom. Thus any generalizations drawn from this literature may be tenuous when applied to these newer technologies.

CONCLUSION

The National Research Council Institute of Medicine recently issued a comprehensive report called *Engaging Schools* (National Research Council Institute of Medicine, 2004) that summarized much of the extant literature on student learning and engagement. The authors of the report concluded with several recommendations, including the following:

> The committee recommends that preservice teacher preparation programs provide high school teachers deep content knowledge and a range of pedagogical strategies and understandings about adolescents and how they learn, and that schools and districts provide practicing teachers with opportunities to work with colleagues and to continue to develop their skills. (p. 4)

The literature reviewed in the present chapter, as well as in the other chapters in this volume, strongly suggests that the instructional practices used by teachers in their classrooms can have an impact on academic cheating. When teachers structure their classrooms around grades, competition, and ability differences, students will be able to justify cheating behaviors. Indeed, in schools where cheating is the norm, students who feel a sense of belonging in that school may internalize cheating as an acceptable behavior in that environment (Anderman, Freeman, & Mueller, this volume).

Teacher education programs should focus on academic cheating. All too often, preservice teachers are not forced to face the reality that academic cheating is prevalent. Research from a goal orientation theory perspective strongly suggests that a critical examination of academic cheating among preservice teachers may be an effective means of preventing cheating in the future.

Of course, individuals will often argue that the prevalence of cheating in society in general will make it difficult to prevent cheating from occurring in school settings (Callahan, 2004). However, although schools represent our larger society, they also are unique, with their own unique philosophies, values, and cultures. Educators can and should try to make a "dent" in these behaviors. Goal orientation theory offers a tested theoretical framework for critically examining instructional practices that may foster increased cheating behaviors in children, adolescents, and adults.

REFERENCES

Ames, C. (1990, April). The relationship of achievement goals to student motivation in classroom settings. Paper presented at the Annual Meeting of the American Educational Research Association, Boston, MA.

Ames, C. (1992). Classrooms: Goals, structures, and student motivation. *Journal of Educational Psychology, 84*(3), 261–271.

Ames, C., & Ames, R. E. (1984). Goal structures and motivation. *Elementary School Journal, 85*(1), 39–52.

Ames, C., & Archer, J. (1988). Achievement goals in the classroom: Students' learning strategies and motivation processes. *Journal of Educational Psychology, 80*(3), 260–267.

Anderman, E. M., Maehr, M. L., & Midgley, C. (1999). Declining motivation after the transition to middle school: Schools can make a difference. *Journal of Research and Development in Education, 32*, 131–147.

Anderman, E. M., & Anderman, L. H. (2000). The role of social context in educational psychology: Substantive and methodological issues. *Educational Psychologist, 35*.

Anderman, E. M., Eccles, J. S., Yoon, K. S., Roeser, R. W., Wigfield, A., & Blumenfeld, P. (2001). Learning to value math and reading: Individual differences and classroom effects. *Contemporary Educational Psychology, 26*, 76–95.

Anderman, E. M., Griesinger, T., & Westerfield, G. (1998). Motivation and cheating during early adolescence. *Journal of Educational Psychology, 90*, 84–93.

Anderman, E. M., & Johnston, J. (1998). Television news in the classroom: What are adolescents learning? *Journal of Adolescent Research, 13*(1), 73–100.

Anderman, E. M., & Maehr, M. L. (1994). Motivation and schooling in the middle grades. *Review of Educational Research, 64*(2), 287–309.

Anderman, E. M., & Midgley, C. (2004). Changes in self-reported academic cheating across the transition from middle school to high school. *Contemporary Educational Psychology, 29*, 499–517.

Anderman, E. M., & Wolters, C. (2006). Goals, values, and affect: Influences on student motivation. In P. Alexander & P. Winne (Eds.), *Handbook of educational psychology* (Vol. 2). (pp. 369–389). Mahwah, NJ: Lawrence Erlbaum Associates.

Baldwin, D. C., Daugherty, S. R., Rowley, B. D., & Schwarz, M. D. (1996). Cheating in medical school: A survey of second-year students at 31 schools. *Academic Medicine, 71*(3), 267–273.

Bandura, A. (1997). *Self-efficacy: The exercise of control.* New York: W. H. Freeman.

Barron, K. E., & Harackiewicz, J. M. (2001). Achievement goals and optimal motivation: Testing multiple goal models. *Journal of Personality and Social Psychology, 80*(5), 706–722.

Calabrese, R. L., & Cochran, J. T. (1990). The relationship of alienation to cheating among a sample of American adolescents. *Journal of Research and Development in Education, 23*, 65–72.

Callahan, D. (2004). *The cheating culture: Why more Americans are doing wrong to get ahead.* San Diego, CA: Harcourt.

Cizek, G. J. (1999). *Cheating on tests: How to do it, detect it, and prevent it.* Mahwah, NJ: Lawrence Erlbaum.

Csikszentmihalyi, M. (1996). *Creativity: Flow and the psychology of discovery and invention.* New York: HarperCollins.

Deci, E., & Ryan, M. (1985). Intrinsic motivation and self-determination in human behavior. New York: Plenum.

Dweck, C. S., & Leggett, E. L. (1988). A social-cognitive approach to motivation and personality. *Psychological Review, 95*(2), 256–273.

Dweck, C. S., & Sorich, L. A. (1999). Mastery-oriented thinking. In C. R. Snyder (Ed.), *Coping* (pp. 232–251). New York: Oxford University Press.

Eccles, J. S., & Wigfield, A. (1995). In the mind of the actor: The structure of adolescents' achievement task values and expectancy-related beliefs. *Personality and Social Psychology Bulletin, 21*(3), 215–225.

Eison, J. A. (1981). A new instrument for assessing students' orientations towards grades and learning. *Psychological Reports, 48*, 919–924.

Elliot, A. J., & Church, M. A. (1997). A hierarchical model of approach and avoidance achievement motivation. *Journal of Personality and Social Psychology, 72*(1), 218–232.

Elliot, A. J., & Harackiewicz, J. M. (1996). Approach and avoidance achievement goals and intrinsic motivation: A mediational analysis. *Journal of Personality and Social Psychology, 70*(3), 461–475.

Elliot, J., McGregor, H. A., & Gable, S. (1999). Achievement goals, study strategies, and exam performance: A mediational analysis. *Journal of Educational Psychology, 91*(3), 549–563.

Evans, E. D., Craig, D., & Mietzel, G. (1991). Adolescents' cognitions and attributions for academic cheating: A cross-national study. Paper presented at the Society for Research in Child Development, Seattle, WA.

Finn, K. V., & Frone, M. R. (2004). Academic performance and cheating: Moderating role of school identification and self-efficacy. *Journal of Educational Research, 97*, 115–122.

Genereux, R. L., & McCleod, B. A. (1995). Circumstances surrounding cheating: A questionnaire study of college students. *Research in Higher Education, 36*, 687–704.

Haney, W., & Clarke, M. (in press). In E. M. Anderman & T. B. Murdock (Eds.), *Psychological perspectives on academic cheating.* San Diego, CA: Elsevier.

Hill, J. P., & Kochendorfer, R. A. (1969). Knowledge of peer success and risk of detention as determinants of cheating. *Developmental Psychology, 1*, 231–238.

Houser, B. B. (1982). Student cheating and attitude: A function of classroom control technique. *Contemporary Educational Psychology, 7*, 113–123.

Jordan, A. E. (2001). College student cheating: The role of motivation, perceived norms, attitudes, and knowledge of institutional policy. *Ethics & Behavior, 11*(3), 233–247.

Kaplan, A., & Maehr, M. L. (1999). Achievement goals and student well-being. *Contemporary Educational Psychology, 24*, 330–358.

Kaplan, A., Middleton, M. J., Urdan, T., & Midgley, C. (2002). Achievement goals and goal structures. In C. Midgley (Ed.), *Goals, goal structures, and patterns of adaptive learning* (pp. 21–53). Mahwah, NJ: Lawrence Erlbaum.

Maehr, M. L., & Anderman, E. M. (1993). Reinventing schools for early adolescents: Emphasizing task goals. *Elementary School Journal, 93*(5), 593–610.

Maehr, M. L., & Buck, R. (1992). Transforming school culture. In M. Sashkin & H. J. Walberg (Eds.), *Educational leadership and school culture* (pp. 40–57). Berkeley, CA: McCutchan.

Maehr, M. L., & Fyans, L. J., Jr. (1989). School culture, motivtion, and achievement. In M. L. Maehr & C. Ames (Eds.), *Advances in motivation and achievement: Motivation enhancing environments* (Vol. 6, pp. 215–247). Greenwich, CT: JAI.

Maehr, M. L., & Midgley, C. (1991). Enhancing student motivation: A schoolwide approach. *Educational Psychologist, 26,* 399–427.

Maehr, M. L., & Midgley, C. (1996). *Transforming school cultures.* Boulder, CO: Westview Press.

McCabe, D. L., & Trevino, L. K. (1993). Academic dishonesty: Honor codes and other contextual influences. *Journal of Higher Education, 64,* 522–538.

McCabe, D. L., Trevino, L. K., & Butterfield, K. D. (1999). Academic integrity in honor-code and non-honor code environments: A qualitative investigation. *Journal of Higher Education, 70,* 211–234.

McCabe, D. L., Trevino, L. K., & Butterfield, K. D. (2001). Cheating in academic institutions: A decade of research. *Ethics & Behavior, 11*(3), 219–232.

Meece, J., Anderman, E. M., & Anderman, L. H. (2006). Educational psychology: Structures and goals of educational settings. *Annual Review of Psychology, 56.* Stanford, CA: Annual Reviews.

Meece, J. L., Blumenfeld, P. C., & Hoyle, R. H. (1988). Students' goal orientations and cognitive engagement in classroom activities. *Journal of Educational Psychology, 80*(4), 514–523.

Middleton, M. J., & Midgley, C. (1997). Avoiding the demonstration of lack of ability: An underexplored aspect of goal theory. *Journal of Educational Psychology, 89*(4), 710–718.

Midgley, C. (Ed.). (2002). *Goals, goal structures, and patterns of adaptive learning.* Mahwah, NJ: Lawrence Erlbaum.

Midgley, C., & Urdan, T. (2001). Academic self-handicapping and achievement goals: A further examination. *Contemporary Educational Psychology, 26*(1), 61–75.

Miller, R. B., Greene, B. A., Montalvo, G. P., Ravindran, B., & Nichols, J. D. (1996). Engagement in academic work: The role of learning goals, future consequences, pleasing others, and perceived ability. *Contemporary Educational Psychology, 21*(4), 388–422.

Murdock, T. B., Hale, N. M., & Weber, M. J. (2001). Predictors of cheating among early adolescents: Academic and social motivations. *Contemporary Educational Psychology, 26*(1), 96–115.

Murdock, T. B., Miller, A., & Anderman, E. M. (2005). Is cheating a function of classroom context? A reanalysis of two studies using HLM. Paper presented at the Annual Meeting of the American Educational Research Association, Montreal, Canada.

Nathanson, C., Paulhus, D. L., & Williams, K. M. (2006). Predictors of a behavioral measure of scholasting cheating: Personality and competence but not demographics. *Contemporary Educational Psychology, 31,* 97–122.

National Research Council Institute of Medicine. (2004). Engaging schools: Fostering high school students' motivation to learn. Washington, DC: The National Academies Press.

Newstead, S. E., Franklyn-Stokes, A., & Armstead, P. (1996). Individual differences in student cheating. *Journal of Educational Psychology, 88,* 229–241.

Nolen, S. B. (1988). Reasons for studying: Motivational orientations and study strategies. *Cognition and Instruction, 5,* 269–287.

Pajares, F. (1996). Self-efficacy beliefs in academic settings. *Review of Educational Research, 66*(4), 543–578.

Pintrich, P. R. (2000). An achievement goal theory perspective on issues in motivation terminology, theory, and research. *Contemporary Educational Psychology, 25*(1), 92–104.

Pintrich, P. R. (2000). Multiple goals, multiple pathways: The role of goal orientation in learning and achievement. *Journal of Educational Psychology, 92*(3), 544–555.

Pintrich, P. R., & Garcia, T. (1991). Student goal orientation and self-regulation in the college classroom. In M. L. Maehr & P. R. Pintrich (Eds.), *Advances in motivation and achievement* (Vol. 7, pp. 371–402). Greenwich, CT: JAI Press.

Roeser, R. W., Midgley, C., & Urdan, T. (1996). Perceptions of the school psychological environment and early adolescents' psychological and behavioral functioning in school: The mediating role of goals and belonging. *Journal of Educational Psychology, 88,* 408–422.

Ryan, A. M., Hicks, L., & Midgley, C. (1997). Social goals, academic goals, and avoiding seeking help in the classroom. *Journal of Early Adolescence, 17*(2), 152–171.

Schab, F. (1991). Schooling without learning: Thirty years of cheating in high school. *Adolescence, 26,* 839–847.

Skaalvik, E. M. (1997). Self-enhancing and self-defeating ego orientation: Relations with task and avoidance orientation, achievement, self-perceptions, and anxiety. *Journal of Educational Psychology, 89,* 71–81.

Slobogin, K. (2002). Survey: Many students say cheating's OK. Retrieved April 5, 2002, from http://archives.CNN.com/2002/fyi/teachers.ednews/04/05/highschool.cheating/

Turner, J. C. (2001). Using context to enrich and challenge our understanding of motivational theory. In S. Volet & S. Jarvela (Eds.), *Motivation in learning contexts: Theoretical advances and methodological implications* (pp. 85–104). Amsterdam: Pergamon.

Turner, J. C., Midgley, C., Meyer, D. K., Gheen, M., Anderman, E. M., Kang, Y., et al. (2002). The classroom environment and students' reports of avoidance strategies in mathematics: A multimethod study. *Journal of Educational Psychology, 94*(1), 88–106.

Urdan, T. (1997). Achievement goal theory: Past results, future directions. In M. L. Maehr & P. R. Pintrich (Eds.), *Advances in motivation and achievement* (Vol. 10, pp. 99–141). Greenwich, CT: JAI Press.

Urdan, T., Kneisel, L., & Mason, V. (1999). The effect of particular instructional practices on student motivation. In T. Urdan (Ed.), *Advances in motivation and achievement* (Vol. 11, pp. 123–158). Stamford, CT: JAI.

Urdan, T., Ryan, A. M., Anderman, E. M., & Gheen, M. (2002). Goals, goal structures, and avoidance behaviors. In C. Midgley (Ed.), *Goals, goal structures, and patterns of adaptive learning* (pp. 55–83). Mahwah, NJ: Lawrence Erlbaum.

Weiner, B. (1985). An attribution theory of achievement motivation and emotion. *Psychological Review, 92,* 548–573.

Weiner, B. (1986). *An attributional theory of motivation and emotion.* New York: Springer-Verlag.

Weiss, J., Gilbert, K., Giordano, P., & Davis, S. F. (1993). Academic dishonesty, type A behavior, and classroom orientation. *Bulletin of the Psychonomic Society, 31*(2), 101–102.

Wellhousen, K., & Martin, N. K. (1995). Preservice teachers and standardized test administration: Their behavioural predictions regarding cheating. *Research in the Schools, 2*(2), 47–49.

Whitley, B. E. (1998). Factors associated with cheating among college students: A review. *Research in Higher Education, 39,* 235–270.

Wigfield, A., & Eccles, J. S. (2002). The development of competence beliefs, expectancies for success, and achievement values from childhood through adolescence. In A. Wigfield & J. S. Eccles (Eds.), *Development of achievement motivation. A volume in the educational psychology series* (pp. 91–120). San Diego, CA: Academic Press.

Wolters, C. A. (2004). Advancing achievement goal theory: Using goal structures and goal orientations to predict students' motivation, cognition, and achievement. *Journal of Educational Psychology, 96*, 236–250.

Wolters, C. A., Yu, S. L., & Pintrich, P. R. (1996). The relation between goal orientation and students' motivational beliefs and self-regulated learning. *Learning and Individual Differences, 8*, 211–238.

Wryobeck, J. M., & Whitley, B. E. J. (1999). Educational value orientaion and peer perceptions of cheaters. *Ethics and Behavior, 9*, 231–242.

6

UNDER PRESSURE AND UNDERENGAGED: MOTIVATIONAL PROFILES AND ACADEMIC CHEATING IN HIGH SCHOOL

JASON M. STEPHENS AND
HUNTER GEHLBACH

*There's a lot of pressure and you don't have enough time
to get A's in every single class. So, if someone else did it,
they cheat on their homework because it's faster, you
know . . . and you get the homework done. You get full
credit.* Where does that pressure come from? *I think it is
mostly from their parents. Pressure to get into college
since college applications are getting harder. Yes, it's
getting harder to get into schools.*

(Jin, tenth grade male honors student)

Although it's been a long time since the publication of Hartshorne and
May's (1928) oft-cited *Studies in Deceit*, many still wish to locate the
problem of academic cheating and other morally problematic behaviors
within the individual. They view cheating as a function of poor judgment,
weak will, or lack of character. The thesis and popular success of William
J. Bennett's (1993) *The Book of Virtues* seems emblematic of how this
(incomplete) psychological view of human behavior still resonates deeply
with many Americans. The case of Jin,[1] the high school student quoted
above, offers the sobering view that to *know* the good is not necessarily to

---

[1] The names used in this chapter are pseudonyms.

Copyright © 2007 by Academic Press, Inc.
All rights of reproduction in any form reserved.

do the good (see Blasi, 1980, for a review of empirical research on relations between moral cognition and moral action). Jin believes cheating to be morally wrong, and yet he cheats—frequently.

Unfortunately, Jin is not alone. Cheating among high school adolescents has become commonplace. Steinberg (1996), for example, found that two-thirds of students reported cheating on a test, and nearly 9 out of 10 indicated that they had copied someone else's homework. At the same time, most students, like Jin, believe that cheating is wrong or unacceptable, but they report doing it anyway (Anderman, Griesinger, & Westerfield, 1998; Jordan, 2001; Stephens, 2004b). This research and the frequency of stories like Jin's lead to many questions: Why do so many students cheat, even when they think it is wrong? Are they pursuing certain goals that are associated with increased cheating? Do they cheat to cope with pressure for grades or to avoid doing work they find unengaging?

Research on students' motivation to learn constitutes an important approach to understanding why students might cheat. Over the past two decades, achievement goal theory has become a prominent approach to examining how students are motivated to achieve in school and has informed how educators can foster interesting, supportive classroom environments (Ames, 1992; Midgley, 2002). As described in more detail below, goal theorists posit the existence of two types of achievement goals: mastery goals that orient individuals toward *developing* their ability and performance goals that focus individuals on demonstrating their ability. Recent research has connected goal theory approaches with beliefs and behaviors related to academic cheating (Anderman, Griesinger, & Westerfield, 1998; Anderman & Midgley, 2004; Jordan, 2001; Murdock, Hale, & Weber, 2001; Murdock, Miller, & Kohlhardt, 2004; Rettinger, Jordan, & Peschiera, 2004).

This chapter extends this work at the intersection of goal theory and cheating. We examine which types of goals are most closely associated with personal beliefs about cheating (i.e., beliefs about which behaviors constitute "cheating" as well as beliefs about the acceptability of cheating under certain circumstances), perceptions of cheating in the environment (specifically what the norms are regarding cheating among peers), and engagement in cheating behaviors (assignment and test cheating as well as plagiarism). In addition, we investigate how high school students with distinct profiles of personal goal orientations and perceptions of classroom goal structures differ with respect to these same outcomes. In doing so, the present chapter sheds light on the complex psychological, social, and situational problem that is academic cheating. Specifically, we explore the possibility that students who cheat most frequently are "under pressure" to perform (i.e., get high grades) and "underengaged" in learning content (i.e., mastering the subject matter).

CHEATING BELIEFS: DEFINITIONS
AND JUSTIFICATIONS

We begin by reviewing selected theoretical and empirical work on students' beliefs and behaviors related to cheating. Specifically, we focus on two distinct types of student beliefs related to cheating: beliefs about what behaviors constitute cheating and beliefs about the acceptability of cheating. These beliefs have the potential to impact cheating behavior in distinct ways. For example, if students define working together on an assignment as collaboration, they are more likely to do so than students who view working together as cheating. Similarly, students who tend to see some situations as justifying cheating are more likely to cheat than students who have a less context-dependent view of cheating.

DEFINITIONS: WHAT CONSTITUTES "CHEATING"?

There are over 100 empirical studies of academic cheating (for comprehensive reviews, see Bushway & Nash, 1977; Whitley, 1998). Only a handful of these studies (e.g., Barnett & Dalton, 1981; Livosky & Tauber, 1994; Newstead, Franklyn-Stokes, & Armstead, 1996; Stephens, 2004a) have investigated students' beliefs concerning which behaviors constitute "cheating." More importantly for our purposes, none of these studies has explored the relations between achievement goals and definitions of cheating. However, students' conceptions of cheating may vary depending on their personal goal orientations or perceptions of the classroom goal structure. For example, perhaps copying homework or asking someone for answers during an exam is not regarded as "cheating" among students who are primarily concerned with getting high grades and/or avoiding studying material that they are not especially interested in learning.

The existing research on students' beliefs about what behaviors constitute cheating indicates that when students define a behavior as "cheating," they are less likely to report seeing others doing it, and they are less likely to report engaging in that behavior themselves. For example, Newstead et al. (1996) used a multistudy design to explore cheating beliefs and behaviors among undergraduates in Great Britain. They found an inverse relationship between the self-reported behaviors of students in Study 1 and the extent to which different students in Study 2 believed those behaviors constituted cheating behavior. Thus, the more certain students are that a behavior is cheating, the less likely they are to report engaging in that behavior.

In this chapter, we extend the work of Newstead et al. (1996) by examining the relationship between achievement goals and cheating beliefs and behaviors. In particular, we investigate whether students who are "under pressure" and "underengaged" are less likely to likely to identify a set of academic behaviors as "cheating" and more likely to report engaging in them (as compared to students who feel less pressure to perform and are highly engaged in learning).

JUSTIFICATIONS: WHEN IS CHEATING ACCEPTABLE?

As the case of Jin reminds us, although many students believe cheating to be wrong, they report doing it anyway. For these students, situational pressures or other priorities seem to matter more in determining their behavior than abstract or absolute judgments about the rightness or wrongness of cheating. In their classic theory of delinquency, Sykes and Matza (1957) identified several types of "neutralization techniques" that adolescents used to avoid or reduce self-recrimination for engaging in criminal or immoral behavior. These neutralization techniques, or moral disengagement mechanisms (Bandura, 1986), reduce or even negate one's responsibility by attributing one's conduct to others or to situational contingencies. In doing so, the behavior in question (be it cheating or something more serious) becomes more acceptable or even morally justified. Jin, for example, appears to neutralize his personal responsibility for cheating by implying that he cheats because (1) others do it, (2) there is too little time to get the A's he needs without doing so, and (3) his parents put pressure on him.

Several studies have demonstrated a strong positive association between cheating and a "neutralizing attitude" among college undergraduates (e.g., Diekhoff et al., 1996; Haines, Diekhoff, LaBeff, & Clark, 1986; LaBeff, Clark, Haines, & Diekhoff, 1990; McCabe, 1992; Michaels & Miethe, 1989). Of the five neutralization techniques described by Sykes and Matza (1957), "denial of responsibility" was the most prevalent among undergraduates: 61 percent of students who reported cheating rationalized their cheating by blaming others and/or some aspect of the situational context (McCabe, 1992). Similarly, Evans and Craig (1990b) found that high school students were more likely to blame their cheating on teachers than teachers were to claim responsibility for student cheating. This displacement of responsibility to the teacher was most pronounced among college-bound and high-achieving students (Evans & Craig, 1990a).

More recently, Jensen and her colleagues (Jensen, Arnett, Feldman, & Cauffman, 2002) evaluated high school and college students' beliefs about the acceptability of cheating under different circumstances. Specifically, participants were presented with 19 different cheating vignettes, each depicting a different motive for looking at a classmate's answers on a math exam. Participants were then asked to rate how acceptable the behavior would be under the given circumstance. As expected, Jensen et al. found a significant positive relationship between students' acceptability beliefs and their self-reported engagement in cheating behavior.

In sum, the acceptability or justifiability of cheating is strongly related to engagement in cheating behavior. In other words, the situation matters, and students regard cheating as more acceptable under certain situations (e.g., high stakes, poor instruction). The empirical illustration described later in this chapter explores the relations between neutralizing beliefs and achievement goals. In particular, we investigate whether students who feel "under pressure" for grades are more likely to neutralize or justify cheating behavior than students who are highly "engaged" in learning (perhaps because they displace the responsibility for their cheating to the teacher or somebody else who is pressuring them).

ENVIRONMENTAL PERCEPTIONS: PEER ATTITUDES AND BEHAVIORS

There is a rich history, particularly in social psychology, on the powerful effect peers can have on each other's norms, beliefs, and behaviors (Asch, 1951; Festinger, 1954; Newcomb, 1943; Sherif, 1936). Several empirical studies have investigated the relations of peer-related variables to academic dishonesty among college students (e.g., Bowers, 1964; Graham, Monday, O'Brien, & Steffen, 1994; Jordan, 2001; Lanza-Kaduce & Klug, 1986; McCabe & Trevino, 1997; McCabe & Trevino, 1993; McCabe, Trevino, & Butterfield, 2001).

McCabe and Trevino (1997), for example, found peer-related contextual factors to be the strongest predictors of cheating in their multicampus investigation of individual and contextual influences related to academic dishonesty. Specifically, peer disapproval of cheating and perceptions of peer cheating behavior were the two most influential contextual factors associated with self-reported cheating: students who perceived that their peers disapproved of academic dishonesty were less likely to cheat, while those who perceived higher levels of cheating among their peers were more likely to report cheating. Similarly, Jordan (2001) found a significant correlation between college students' perceived social norms and their self-reported cheating. Students who cheated reported not only higher estimates of the percentage of students who cheated at their college than noncheaters (31.2 percent vs. 20.6 percent, respectively), but also significantly higher rates of having seen someone else cheat (70.8 percent vs. 40.5 percent, respectively).

Although there is no empirical research relating peer norms to academic cheating in high school, several studies have shown that peers affect other forms of deviant behavior during this period of adolescence. A review of recent literature revealed numerous studies that showed peer influences on a variety of behaviors, such as cheating in sports (Guivernau & Duda, 2002), cigarette smoking (Alexander, Piazza, Mekos, & Valente, 2001), alcohol-related driving offenses (Shope, Raghunathan, & Patil, 2003) and substance abuse (Urberg, Luo, Pilgrim, & Degirmencioglu, 2001). Thus, it is likely that high school students' perceptions of their peers' attitudinal and behavioral norms also relate to cheating. We also believe that there may be a significant relationship between peer norms and achievement goals. In particular, we investigate whether students who feel "under pressure" for grades are more likely to perceive their peers as less disapproving of cheating and more involved in various forms of cheating than students who are highly "engaged" in learning.

CHEATING BEHAVIOR: TYPES AND PREVALENCE

Academic dishonesty is a multifaceted and pervasive phenomenon. By most accounts in the literature, most students seem to be doing it and are doing it in more than one way (for comprehensive reviews of the types and prevalence of cheating in high school and college, see Cizek, 1999; Whitley, 1998, respectively). In his series of large-scale studies, Steinberg (1996) found that during the past academic year, nearly 70 percent of high school students reported cheating on a test and nearly 90 percent indicated that they had copied someone else's homework. More recently, in McCabe's (2005) national survey of over 18,000 high school students, over 70 percent of students reported one or more instance of test cheating and over 60 percent admitted to some form of plagiarism.

Research examining the relation between cheating and academic ability (often indexed by grade point average, or GPA) has shown a small negative association in both high school (Josephson Institute of Ethics, 2002) and college (e.g., McCabe & Trevino, 1997; Newstead, Franklyn-Stokes, & Armstead, 1996). The Annual Survey of Who's Who among American High School Students (1998), however, illustrates that cheating is not a problem simply among low-achieving students. In their twenty-ninth Annual Survey, 80 percent of America's highest achieving students admitted to having cheated (the highest percentage in the history of the survey).

In sum, cheating is a pervasive problem in high schools and affects both high- and low-achieving students. In the empirical illustration presented in this chapter, we investigated the prevalence of three types of academic dishonesty—assignment cheating, test cheating, and plagiarism—among students at two distinct high schools, one average achieving and the other very high achieving. We thought that students with goals to get good grades or to outperform others might be more likely to cheat than those who were more focused on learning the material.

ACHIEVEMENT GOALS AND
ACADEMIC CHEATING

Today, one of the most well-established approaches to understanding students' academic motivation is achievement goal theory (e.g., Weiner, 1990). This theoretical framework is well suited to constructing potential explanations of students' cheating beliefs, environmental perceptions, and behaviors. As described by Midgley (2002), achievement goal theory posits the existence of two types of achievement goals: "to develop ability (variously labeled a mastery goal, learning goal, or task goal) and the goal to demonstrate ability or to avoid the demonstration of lack of ability (variously labeled a performance goal, ego goal, or ability goal)" (p. xi). Mastery goals orient individuals toward developing their knowledge, learning new skills, and using self-referenced evaluation criteria. Performance goals, on the other hand, focus individuals on appearing to be smart relative to others, displaying skills, avoiding the appearance of inability, and using norm-referenced evaluation criteria. While mastery goals have been consistently related to adaptive "patterns of learning" (e.g., higher levels of academic efficacy and achievement), performance approach goals have been associated with mixed cognitive, affective, and behavioral outcomes (Ames, 1992; Harackiewicz et al., 2002; Midgley, Kaplan, & Middleton, 2001).[2]

Recently, numerous studies using goal theory have furthered our understanding of academic cheating. Studies at this intersection of goal theory and cheating have evolved over the past decade and fall into four basic types: interindividual differences (Anderman, Griesinger, & Westerfield, 1998; Murdock, Hale, & Weber, 2001), intra- and interindividual differences (Jordan, 2001; Stephens & Roeser, 2003), longitudinal (Anderman & Midgley, 2004), and experimental (Murdock, Miller, & Kohlhardt, 2004; Rettinger, Jordan, & Peschiera, 2004).

INTERINDIVIDUAL DIFFERENCES IN MOTIVATION
AND CHEATING

This first type of study focuses on how individuals differ in their cheating patterns. These types of studies are useful in identifying which student characteristics (particularly which types of goals) are associated with cheating.

In their study of middle school adolescents, Anderman, Griesinger, and Westerfield (1998) predicted students' cheating beliefs and behavior in

---

[2] Only performance approach goals, not performance-avoidance goals, have been found to be adaptive. Thus, this chapter restricts its examination of performance goals to approach goals.

science class. Anderman et al. found a significant, positive relationship between students' beliefs about the acceptability of cheating and their self-reported cheating behavior. Their results also showed significant positive main effects for personal and classroom-level extrinsic orientations[3] on students' beliefs about the acceptability of cheating. In addition, they noted that classroom extrinsic and school performance goals were positively associated with cheating behavior. Mastery goals at all levels—personal, classroom, and school—were significantly negatively correlated with both cheating beliefs and behaviors. However, these correlations were modest ($r = -.13$ to $r = -.27$).

Murdock, Hale, and Weber (2001) extended the work of Anderman et al. (1998) by examining the relation between seventh and eighth grade students' academic *and* social motivation and cheating. Mastery and extrinsic goals at the personal and classroom levels were significantly predictive of being a cheater. However, only personal extrinsic goals were significant in a final regression model. Specifically, students who cheated were significantly more likely to endorse personal extrinsic goals. Other unique predictors of cheating included year in school, academic self-efficacy, teacher commitment, classroom participation structure, and schoolwide teacher respect. Specifically, eighth graders and students with lower academic self-efficacy were more likely to be "cheaters." Students who perceived their classroom teachers as less committed and the teachers in the school as less well respected were also more likely to cheat.

INTRA- AND INTERINDIVIDUAL DIFFERENCES IN MOTIVATION AND CHEATING

Another set of studies supplement findings of differences between students with an examination of how a group of students change their beliefs, perceptions, and behaviors across different settings. These studies also help clarify which factors relate to individual differences in cheating and enlarge our understanding of which situational factors relate to cheating.

For example, in studying college students, Jordan (2001) assessed the personal mastery and extrinsic goal orientations of students who had cheated in two distinct contexts: courses in which they had cheated versus not cheated. Results of this intraindividual component of the study indicated that cheaters were less mastery oriented and more extrinsically motivated in courses in which they had cheated compared to courses in which

[3] This personal extrinsic scale differs from the performance goal orientation described above. It measures the extent to which students do their schoolwork for extrinsic, non-ego-involved reasons, such as to get good grades, "because it's required" or to avoid getting into trouble for not doing it.

they had not cheated. Additional analyses revealed that cheaters were less mastery oriented and more extrinsically motivated than noncheaters. Finally, Jordan found that cheaters were significantly more likely than noncheaters to report having seen other students cheat and found higher levels of cheating among those who reported observing others cheat.

Stephens and Roeser (2003) examined cheating among adolescents at two high schools. Results indicated that compared to the class in which they cheated least often (if at all), students reported significantly lower personal mastery goals than in the class in which they cheated most often (or would be most likely to cheat in). For one of the schools, students also reported significantly lower personal performance goals in the latter class. Students at both schools perceived the classroom goal structures of the class that they cheated most often in as less focused on mastery goals and more focused on performance goals. Thus, although the classroom-level results of this study are consistent with existing studies, the finding that personal performance goals were associated with less cheating appears to be incongruous with Jordan's (2001) findings.

LONGITUDINAL AND EXPERIMENTAL STUDIES OF MOTIVATION AND CHEATING

Although these correlational studies provide important information regarding individual differences and environmental factors, these designs do not allow researchers insights into certain changes that might be occurring over time. Similarly, they do not provide insights into which variables might be *causing* changes in cheating beliefs, perceptions of the environment, or behaviors.

In order to address these issues, Anderman and Midgley (2004) examined changes in cheating behavior across the transition from middle school to high school. Their longitudinal analyses revealed that self-reported levels of cheating in math classes increased significantly for two groups of students: those who moved from an eighth grade class with high mastery goals to a ninth grade class with low mastery goals and those who moved from an eighth grade class with low performance goals to a ninth grade class with high performance goals. Also noteworthy was the significant decrease in self-reported cheating behavior among those students who moved from an eighth grade class with low mastery goals to a ninth grade class with high mastery goals.

To further disentangle the relations between the context variables on cheating beliefs and behaviors, Murdock, Miller, and Kohlhardt (2004) used hypothetical vignettes to isolate the effects of classroom goal structures and teacher pedagogy on students' beliefs about the acceptability and likelihood of cheating. Consistent with previous research, they found that students believed cheating to be more justifiable (as well as more likely) when the classroom in their hypothetical vignettes was portrayed as focused on grades and the teacher as a poor instructor. Murdock et al. also assessed students' beliefs about the morality of cheating (i.e., an absolute, as opposed to context-dependent, belief about the rightness or wrongness of cheating). Not surprisingly, students' beliefs about the morality of cheating were less influenced by contextual factors and less strongly related to the perceived likelihood of cheating.

Similarly, Rettinger, Jordan, and Peschiera (2004) used hypothetical vignettes to assess undergraduates' beliefs about the likelihood of cheating. In this experiment, however, it was the personal goal orientation and academic competence of the vignette's protagonist that were manipulated, and not the classroom goal structure and teacher pedagogical competence. Consistent with previous findings, students rated protagonists portrayed as mastery oriented and highly competent as less likely to cheat. Thus, these experiments provide evidence of variables that cause changes in students' beliefs about the acceptability and probability of cheating behavior. Because of the obvious ethical problems, no studies have examined factors that cause cheating behaviors.

Taken together, the methodologies of these different studies indicate that mastery goals (at the personal and classroom levels) are unrelated or negatively associated with cheating. Conversely, extrinsic or performance goals (at the personal and classroom levels) are either unrelated or positively associated with such behavior. Although the general pattern is relatively clear, it is hard to understand why some of these studies manifest significant relationships between goals and cheating while others do not. One potential explanation for some of these differences is that important differences exist between extrinsic goals and performance goals.

These methodologies do not address the question of what happens to students who have certain configurations of goals. In other words, how do students who are personally quite mastery oriented and perceive a minimal performance press from the classroom environment compare to students who are low in personal mastery goals but perceive a strong press from the environment to achieve certain performance goals?

In the remainder of the chapter, we provide an empirical illustration that accomplishes two ends. First, we examine how the different personal goals and classroom goal structures relate to different cheating outcomes. Second, we examine students whose profiles consist of different configurations of these personal goals and classroom goal structures. Specifically, regression analyses are used to ascertain which of the four achievement goals (mastery and performance goals at the personal and classroom levels) significantly predict students' cheating beliefs, perceptions of peer norms related to cheating, and cheating behaviors. Using the significant achievement goal predictors from the regression results to identify students with unique goal profiles, we then conduct an extreme

Perceived Classroom
Performance Goal Structure

		Low	High
Personal Mastery	Low	Unpressured but Underengaged	Under Pressure and Underengaged
Goal Orientation	High	Engaged and Unpressured	Engaged but Under Pressure

FIGURE 6.1   Motivational profiles based on personal mastery goal orientation and perceived classroom goal structure.

group analysis of four groups of students with distinct achievement goal profiles. A conceptual representation of these four motivational profiles is presented in Figure 6.1.

We expected personal mastery goals and classroom performance goals to be the most powerful and consistent predictors of the attitudinal, perceptual, and behavioral outcomes. In particular, we expected personal mastery goals (i.e., engagement in learning course material) to be a negative predictor of students' cheating outcomes. In other words, students who set goals to master material would have no reason to cheat; cheating would not help them achieve their desired end of acquiring new knowledge and/or skills. This effect seems likely to be less robust if others are encouraging those students to engage in and master material (i.e., classroom mastery goal structures). We also expected classroom performance goals (i.e., the press for high grades) to be a positive predictor of their cheating. When parents or teachers encourage students to demonstrate their ability, students may feel that they are being pressured into adopting certain goals. They may respond to this pressure by cheating. For personal performance goals, this pressure is likely to be diminished given that the goals would be set by the students themselves, who would likely set their goals at a level that is relatively comfortable for them.

In addition to these main effects for personal mastery and classroom performance goals, we expected to see a significant interaction between these two goals, believing that those students who are "under pressure" (i.e., perceive high classroom performance goals) and "underengaged" (i.e., hold low personal mastery goals) would be especially likely to cheat. This effect should be particularly robust in comparison to those students who are "engaged" (i.e., hold high personal mastery goals) and "unpressured" (i.e., perceive low classroom performance goals).

METHOD

To test these predictions, sophomores and juniors attending two high schools located in California participated. The sample of 337 students was as follows: 54 percent female; 51 percent sophomores; and 62 percent white, 31 percent Asian, and 7 percent Hispanic. The demographic and academic characteristics of the two high schools were distinct in several ways: East High ($n = 113$) was a high-achieving high school with a predominantly Asian American student body, whereas South High ($n = 224$) was an average-achieving school with a predominantly Caucasian student body.

Students completed a questionnaire during a 90-minute block period (of an English or literature class) in February or March of 2002. Employing a technique adapted from Adams (1960), the first author visited each classroom two weeks before administering the survey. During a 15- to 20-minute interactive presentation, students were informed of the nature of the research and confidentiality procedures, and were encouraged *not* to participate if they felt they could not respond honestly. On the day data were collected, between 80 and 90 percent of students present in each classroom had returned the parental consent and student assent forms and completed the survey. Most students completed the questionnaire in 50 to 70 minutes.

The questionnaire consisted of items that measured students' personal goals and their perceptions of the classroom goal structures. In addition, three types of outcomes were assessed: cheating beliefs, perception of peer cheating norms, and actual cheating behavior. The *cheating beliefs* outcomes assessed whether students believed that different manifestations of assignment cheating, plagiarism, and test cheating constituted cheating. An additional cheating belief assessed neutralization—the extent to which students attributed blame for cheating to social or situational circumstances. *Perceptions of peer cheating norms* included variables assessing students' perceptions of whether peers engaged in assignment cheating, plagiarism, and test cheating. In addition to these behavioral norms, one variable assessed peer attitudinal norms (i.e., the extent to which peers disapproved of cheating). In the third group of outcomes, we examined students' *actual cheating behavior* through their reports of engaging in assignment cheating, plagiarism, and test cheating.

The questionnaire included original items as well as measures adapted from Midgely et al.'s (2000) Patterns of Adaptive Learning Survey, McCabe's (2001) Survey of Academic Behaviors, and Diekhoff et al.'s (1996) neutralization scale. In addition, students indicated their gender, ethnicity, and grade level (tenth or eleventh). The scale items and their reliabilities are presented in the Appendix.

To simplify the interpretation of results, all scales were transformed into scales that ranged from 0 to 1. Putting all the scales on the same metric facilitates comparisons between scales. In addition, these unstandardized beta coefficients in regression equations can be more easily interpreted (e.g., a coefficient of .25 indicates that a unit change in an independent variable is associated with a change of 25 percentage points in the outcome variable).

RESULTS

Results are divided into two main sections. The first consists of variable-centered analyses—analyses that examined how well specific variables predicted our cheating outcomes in comparison to other variables. We used this technique to investigate which types of personal goals and/or goal structures most closely relate to cheating beliefs and cheating behaviors. Specifically, these analyses test the assumption that personal mastery goals and perceptions of classroom performance goal structures are most likely to relate to cheating outcomes.

The second section, describing person-centered analyses, focuses on the individuals in the study rather than the variables (see Peck & Roeser, 2003, for more discussion of person-centered and variable-centered analyses). By looking at types of people who have different configurations of goals, we can learn more about how actual groups of students might differ on cheating outcomes. Specifically, we wanted to know more about students who are both "under pressure" (i.e., perceive their classrooms to be high in performance goal structures) and "underengaged" (i.e., are low in personal mastery goals). We compared students with this particular goal profile that appears to put them at risk for cheating with students who have the opposite profile (i.e., students who are high in personal mastery goals and who perceive their classroom goal structures as being less performance oriented). To create these groups we first took the extreme thirds for each variable (i.e., the thirds of the population that were most and least personally mastery oriented, and the thirds that were highest and lowest in their perceptions of classroom performance goal structures). Next we cross-tabulated the two variables to see which students remained in each cell. For example, these analyses retained students who were in the highest third in personal mastery goals and in the lowest third in their perceptions of the classroom performance goal structure. By contrast, students who were high on one variable but only in the middle of the distribution on the other variable were dropped from these analyses.

Means and standard deviations of the variables are presented in Table 6.1. The table also presents the bivariate correlations between the two grouping measures (personal mastery goal orientation and classroom performance goal structures) and each of the remaining measures. Two observations from this table are noteworthy. First, although the personal and classroom mastery scales appear similar (their means are similar and they correlate highly), the means of the personal and classroom performance scales are quite different and the variables correlate only moderately. Thus,

TABLE 6.1   Means, Standard Deviations, and Pearson r Correlations with Personal Mastery Goal Orientations and Perceptions of Classroom Performance Approach Goal Structures

| | | | Correlations with | |
| | | | Personal mastery | Classroom performance |
	Mean	SD		
Goals				
(1) Personal mastery goal	.66	.21	—	
(2) Classroom performance goal	.72	.17	.17**	—
(3) Personal performance goal	.39	.23	.26**	.35**
(4) Classroom mastery goal	.68	.17	.72**	.22**
Cheating Beliefs				
(4) Assignment cheating	.53	.31	.03	.02
(5) Plagiarism	.76	.27	.10	−.08
(6) Test cheating	.82	.21	.13*	−.06
(7) Neutralization	.43	.20	−.33**	−.00
Perceptions of Peer Norms				
(8) Peer assignment cheating	.89	.16	−.15**	.08
(9) Peer plagiarism	.36	.25	−.17**	.13*
(10) Peer test cheating	.56	.26	−.25**	.12*
(11) Peer disapproval of cheating	.32	.24	.35**	−.03
Cheating Behaviors				
(12) Assignment cheating	.76	.26	−.26**	.03
(13) Plagiarism	.27	.26	−.21**	.13*
(14) Test cheating	.50	.30	−.34**	.15**

Note: $N = 325–337$. All scales have been converted to a 0–1 metric.
*$p ≤ .05$ **$p ≤ .01$.

it seems plausible that personal performance goal orientations may function differently than classroom performance goal structures. Second, there appears to be a pattern in the correlations whereby personal mastery goals were significantly and negatively associated with most cheating outcomes. In contrast, classroom performance goal structures were significantly and positively related to four outcomes. The tendency for personal mastery goal orientations to be negatively related to cheating outcomes (and positively related to peer disapproval of cheating) and for classroom performance goal structures to have positive associations with cheating aligned with our expectations.

VARIABLE-CENTERED ANALYSES

The regression analyses in Table 6.2 help to further explore the trends observed in the first table. Specifically, they allow us to control for other variables that could potentially relate to cheating (e.g., sex, ethnicity, school,

TABLE 6.2 Regression Results for Cheating Beliefs, Perceptions of Peer Norms, and Cheating Behavior: Unstandardized Betas

	Cheating Beliefs				Perceptions of Peer Norms				Cheating Behavior		
	Assign.	Plag.	Test	Neutralization	Assign.	Plag.	Test	Peer disapproval	Assign.	Plag.	Test
Constant	0.43**	0.57**	0.78**	0.66**	0.79**	0.32**	0.38**	0.18*	0.85**	0.27**	0.45**
Demographic Controls											
Sex (0 = female; 1 = male)	0.08	0.08*	0.02	-0.04	0.03	-0.04	0.00	0.06*	0.04	-0.05	-0.05
Grade (0 = 10th, 1 = 11th)	-0.02	-0.02	0.01	0.01	0.00	0.11**	0.03	0.02	0.00	0.03	0.04
High School (0 = East; 1 = South)	0.13	0.13**	-0.02	-0.03	0.08**	-0.06	0.18**	-0.09*	0.03	-0.05	0.14**
Asian (0 = other; 1 = Asian)	0.04	0.04	0.05	-0.02	0.03	0.02	0.00	0.05	-0.09*	-0.02	0.01
Hispanic (0 = other; 1 = Hispanic)	-0.05	-0.05	0.03	0.02	0.01	0.07	0.06	-0.01	0.03	0.08	0.03
Goals											
Personal Mastery	0.15	0.15	0.09	-0.25**	-0.11	-0.25*	-0.26**	0.33**	-0.26*	-0.28**	-0.42**
Personal Performance	0.00	0.00	-0.04	0.14**	0.03	0.08	-0.03	0.11	0.01	-0.03	0.02
Classroom Mastery	-0.02	-0.02	0.09	-0.14	-0.03	0.10	-0.07	0.05	-0.13	0.11	-0.11
Classroom Performance	-0.04	-0.04	-0.13	0.02	0.14*	0.15	0.37**	-0.20*	0.20*	0.23*	0.42**
Interaction											
Personal Mastery X Classroom Performance	-0.35	-0.35	-0.33	0.36	0.13	0.83*	0.63	-0.25	-0.32	0.20	-0.05
N	307	308	308	309	308	310	302	308	306	303	305
F	1.02	2.54**	1.49	5.33**	2.46**	4.95**	7.46**	8.32**	3.78**	2.97**	7.87**
Total Adjusted R^2	.00	.05	.02	.12	.05	.11	.18	.19	.08	.06	.18

Note: Assign. = Assignment cheating; Plag = Plagiarism; Test = Test cheating.

and grade level); simultaneously examine the relationship between all four types of goals and cheating; and determine how much variance the demographic factors and goals explain in these cheating outcomes. The demographic measure indicating which high school students attended is the only variable consistently related to students' cheating definitional and acceptability beliefs, perceptions of peer cheating norms, and self-reported cheating behavior. Sex, race, and grade are rarely related to the cheating outcomes, and the associations that were significant were weak in magnitude.

Because all the goal variables (except for the interaction term) were converted to the same 0 to 1 scale, their relative magnitudes can be compared within each equation. None of the goal orientations manifest any significant relationship to cheating beliefs with the exception of neutralization. Students' personal mastery goals are negatively related, and students' personal performance goals are positively related to their tendency to make attributions that neutralize the responsibility for cheating. For the remaining outcomes (perceptions of peer norms and cheating behaviors), a clear pattern emerges. Personal performance goals and classroom mastery goals are unrelated to these outcomes. However, personal mastery goals and classroom performance goal structures are significantly related to six out of seven of these outcomes, though in opposite directions. In other words, personal mastery goals are negatively related to perceptions of peer cheating (though positively related to peer disapproval of cheating) and cheating behavior. Classroom performance goal structures are positively related to perceptions of peer cheating (though negatively related to peer disapproval of cheating) and cheating behavior.

The interaction of "personal mastery X classroom performance" reached significance only once (for the observation of plagiarism outcome). Because several significance tests were conducted, it seemed more sensible to investigate this relationship further rather than jumping to conclusions from a single result. The need to further explore this result provided an additional rationale for conducting the person-centered analyses.

In spite of the consistency of the relationships between personal mastery and classroom performance goals with cheating outcomes, these results should be interpreted with some caution due to the modest amount of variance that is explained in each. Within the cheating beliefs outcomes, the variables in the regression equation explain more than 10 percent variance for only the neutralization outcome; all the other equations predicting cheating beliefs leave over 90 percent of the variance unexplained in the outcomes. For perceptions of peer cheating norms, the equations explained little variance in assignment cheating, and modest amounts of variance in plagiarism, test cheating, and peer disapproval of cheating. In terms of actual cheating behavior, the equations explained small but significant amounts of variance in assignment cheating (8 percent) and plagiarism (6 percent) and about 18 percent of the variance in test cheating.

In sum, students' personal mastery goals and their perceptions of the classroom performance goal structures are consistently related to the cheating outcomes examined in this chapter. However, the combination of these goals and the demographic characteristics examined here explain modest amounts of variance in these student cheating outcomes.

PERSON-CENTERED ANALYSES

The strong trend of associations from the personal mastery goal orientations and the classroom performance goal structures made the examination of a subgroup of students who were particularly high or low on these variables of particular theoretical interest. Specifically, we were interested in examining how students with an "under pressure and underengaged" goal profile compared to students who were "engaged and unpressured."

After creating the groups of students who were high mastery/high performance, high mastery/low performance, low mastery/high performance, and low mastery/low performance (as discussed earlier), we described these student groupings in two ways. First, we examined their mean levels of personal mastery goal orientation and classroom performance goal structures. As Figure 6.2 indicates, the levels of each type of goal are significantly different for both the "under pressure and underengaged" and "engaged and unpressured" students ($t_{(72)} = 19.13$; $p < .01$ for personal

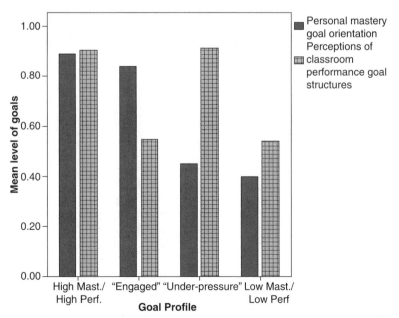

FIGURE 6.2   Mean level of goal orientation for each of four student goal profiles.

mastery; and $t_{(66.7)} = 18.63$; $p < .01$ for classroom performance).[4] Second, the clusters were described in terms of how different demographic characteristics were distributed across groups. There were no significant differences in the distribution of gender ($\chi^2_{(3, N=180)} = 4.60$, $p = .20$), ethnicity ($\chi^2_{(6, N=180)} = 12.02$, $p = .06$), or grade ($\chi^2_{(3, N=180)} = 2.08$, $p = .56$). However, there was a significant overrepresentation of students from East High School in the "under pressure and underengaged" group ($\chi^2_{(3, N=180)} = 12.39$, $p < .01$). Because of this difference and because of the importance of school in the regression results, analyses examined the impact of the school as well as goal profile.

The data were analyzed with a multivariate analysis of variance (MANOVA) design, with motivational goal profile and school as the two between-subjects factors. As seen in Table 6.3, there were effects of students' motivational goal profile and of school for each set of cheating outcomes. A motivational goal profile by school interaction emerged for cheating beliefs only. The univariate statistics are presented in Table 6.4 for motivational goal configuration only and are described here. The table illustrates results for all four groups of students with different profiles, although the description of the results focuses just on comparing the "under pressure and underengaged" students to their "engaged and unpressured" counterparts. Significant school effects emerged for plagiarism beliefs, perceptions of peer assignment and test cheating, peer disapproval of cheating, and self-reported test cheating. In addition, significant interactions emerged for beliefs about assignment cheating and test cheating.

TABLE 6.3 Omnibus MANOVA Results: F-values and η^2 for Goal Profile, School and their Interaction

| | Main and Interaction Effects | | | | | |
| | Goal Profile | | School | | Interaction | |
Outcome	F	η^2	F	η^2	F	η^2
Cheating Beliefs	2.11*	.05	3.60**	.08	2.11*	.05
Perceptions of Peer Norms	3.19**	.07	10.78**	.21	.76	.02
Cheating Behaviors	3.38**	.07	5.95**	.10	.30	.01

Note: F-values are Wilks' Lambda.
*$p \leq .05$ **$p \leq .01$.

[4] The t-test for the classroom performance goal structures violated the assumption of the homogeneity of variances; the corrected test is what caused the df to be unusual.

TABLE 6.4 Mean Group Comparisons of Personal Cheating Beliefs, Perceptions of Peer Norms, and Cheating Behaviors by Four Motivational Goal Profiles

Outcome	Motivational Profile				Univariate $F_{(3,158)}$	η^2
	High mast/ high perf (n = 53–56)	High mast/ low perf (n = 40–43)	Low mast/ high perf (n = 30)	Low mast/ low perf (n = 46–50)		
Cheating Beliefs						
Assignment cheating	.51	.54	.59	.45	.87	.02
Plagiarism	.80	.83	.74	.72	.97	.02
Test cheating	.83	.85	.83	.76	.93	.02
Neutralize responsibility	.37[ab]	.32[a]	.48[bc]	.50[c]	6.60**	.10
Perceptions of Peer Norms						
Assignment cheating	.87	.86	.90	.91	.64	.01
Plagiarism	.36[ab]	.25[a]	.41[b]	.42[b]	3.06*	.05
Test cheating	.53[b]	.39[a]	.63[b]	.61[b]	8.05**	.13
Peer disapproval	.37[ab]	.47[a]	.25[bc]	.22[c]	8.56**	.14
Cheating Behaviors						
Assignment cheating	.68	.73	.81	.82	1.91	.03
Plagiarism	.22[ab]	.17[a]	.33[b]	.30[ab]	2.35[†]	.04
Test cheating	.42[ab]	.32[a]	.65[c]	.57[bc]	10.52**	.16

Note: Tukey Honestly Significant Differences were used to test between group differences. Groups with different superscripts for a particular variable are significantly different from one another at the $p \leq .05$ level.

Mast = Personal Mastery Goal Orientation, Perf = Perceived Classroom Performance Approach Goal Structure.

$^\dagger p \leq .10$, $^* p \leq .05$, $^{**} p \leq .01$.

However, between-school differences are described elsewhere (see Stephens and Roeser, 2003); thus, these results are not included in Table 6.4, nor are they the focus of the discussion.

Personal Beliefs about Cheating

Within the cheating beliefs outcomes, students only differed in their neutralization scores. As indicated in Table 6.4, students in the under pressure and underengaged group were significantly more likely than students in the engaged and unpressured group to make attributions absolving students of responsibility in different types of cheating. The effect size indicates that the group means are a tenth of a standard deviation away from one another.

Perceptions of Peer Norms

In terms of perceptions of peer engagement in cheating behavior, significant differences emerged between groups for plagiarism and test cheating. Students in the under pressure and underengaged group reported seeing significantly more plagiarism and test cheating than did engaged and unpressured students. The effect size was larger for test cheating (.13) than for plagiarism (.05). With respect to peer disapproval of cheating, students in the under pressure and underengaged group were significantly less likely to perceive that their peers disapproved of cheating as compared to their more engaged and less pressured classmates (with an effect size of .14).

Cheating Behaviors

Significant differences emerged on self-reported engagement for plagiarism and test cheating. The under pressure and underengaged students reported plagiarizing and cheating on tests significantly more than their engaged and unpressured peers. The effect sizes were much larger for test cheating (.16) than for plagiarism (.04).

In sum, the under pressure and underengaged students were much more likely to perceive the school climate as a place where cheating occurred regularly and to actually engage in cheating behaviors than their engaged and unpressured classmates. These groups did not differ on most cheating beliefs or on their observations of, or engagement in, assignment cheating. Effect sizes were much larger for test cheating than for plagiarism.

DISCUSSION

At the outset of the chapter, we noted that Jin, like so many students, believes that cheating is morally wrong. Yet, he does it anyway. We then posited that such belief–behavior incongruity might be explained by examining students' personal goals and their perceptions of classroom goal structures. As expected, personal mastery goals and classroom performance goals were the strongest and most consistent predictors of the attitudinal, perceptual, and behavioral outcomes in this chapter. In particular, personal mastery goal orientation negatively predicted students' tendency to neutralize responsibility for cheating and the extent to which they observed and engaged in various types of cheating behavior. Conversely, students' perception of classroom performance goals was a positive predictor of the extent to which they observed and engaged in various types of cheating.

In addition to these main effects for personal mastery and classroom performance goals, we expected to see a significant interaction between these two goals, believing that those students who are *under pressure* (i.e., perceive high classroom performance goals) and *underengaged* (i.e., hold low personal mastery goals) would be especially likely to cheat. Regression analyses revealed only one such significant interaction (for perceptions of peer plagiarism). Results from person-centered analyses indicated that students with this under pressure and underengaged goal profile differed significantly from students who perceived less pressure in their classes and reported being more engaged in mastering material on many of the cheating outcomes. In other words, although there was little evidence that this particular combination of goals was disproportionately worse than other combinations, it appears that the additive effects of being under pressure and underengaged are associated with the observation of and engagement in cheating behavior. Incidentally, Jin is a student with this motivational goal profile.

Before discussing these results in greater detail, it is important to note that neither the regression nor person-centered analyses revealed significant relationships between these goals and students' beliefs about which behaviors constituted cheating. It seemed plausible that students who were particularly performance oriented might be more instrumental in their approach to school and, consequently, would adopt a belief system that defined cheating more narrowly. This belief system would allow them a wider range of instrumental actions to accomplish their goals without feeling guilty for having violated school rules or perhaps even their moral beliefs. The data did not support this possibility.

VARIABLE-CENTERED ANALYSES

The findings of this chapter are consistent with previous studies that have shown personal mastery goals to be either unrelated to or negatively associated with cheating (Anderman, Griesinger, & Westerfield, 1998; Haines, Diekhoff, LaBeff, & Clark, 1986; Jordan, 2001; Murdock, Hale, & Weber, 2001; Rettinger, Jordan, & Peschiera, 2004) and perceived classroom performance goals to be positively associated with such behavior

(Anderman & Midgley, 2004; Murdock & Miller, 2003). These results are congruent with common sense: students who seek to master course material would have no reason to cheat. Cheating would undermine their desired end of acquiring new knowledge and/or skills. However, goals that are imposed from external sources (e.g., parents or teachers) that press students to demonstrate their ability are likely to put an unpleasant amount of pressure on students.

Ryan and Deci's (2000) work on self-determination theory offers an important way to understand the results from the empirical illustration in this chapter. According to these scholars, it is important to distinguish whether students' motivation comes from within (intrinsic) or from other people or from situational pressures (extrinsic). When students' motivation comes more from the intrinsic end of the continuum, they have more autonomy and an increased capacity to determine their own fate (hence "self-determination theory"). In our empirical example, this theory may function in the following way.

Students with high personal mastery goals in a certain domain are more internally motivated to engage in mastering material in that domain. Learning in this domain is something they are choosing to do. In other words, because they have chosen to learn the content of a certain domain, they are determining their own fate and their autonomy is supported. Cheating would undermine this learning goal. Students who perceive a press from others or from their environment to master a certain content or domain are in a different situation. They may also choose to master that domain of content. However, they may be less interested in engaging that particular domain of content because they did not get to choose that particular content. Thus, in those cases where students have adopted mastery goals (perhaps as a result of being in classrooms with high mastery goals), cheating would be an unreasonable response. However, in those cases where students are pushed to master a certain domain of material that does not interest them, cheating might serve as a way to appear to have achieved the goals that others wanted them to achieve. Our data's consistent negative relationships between cheating and personal mastery goals and lack of a relationship with classroom mastery goal structures are congruent with this self-determination explanation.

There are similarly important differences between personal and classroom performance goal structures. Students who set personal performance goals are determining their own standards of comparison (i.e., they can choose to try and outperform their friends, the other students in their same track, or the smartest kid in the class). Not only does this choice give them a substantial amount of autonomy, but it also allows them the luxury of changing the comparison set. For example, a student might begin by trying to outperform her best friend. If her best friend is destined to be the class valedictorian, she will regularly fail to measure up. After some time, she might decide to outperform her twin brother instead, believing that he makes for a more sensible (and favorable!) comparison. In this manner, she can exert substantial control over whether she is successful in meeting her goals. If the same student finds herself pressured to achieve extrinsic goals or takes classes with teachers promoting performance goal structures, her situation could be quite different. In this case, she would not choose her goals, her standards of success, or her comparison sets. Instead, her parents would urge her to get certain grades, or her teacher might ask her why she cannot do as well as others on her tests. If she cannot live up to those standards or be competitive with those other students, there is nothing that she can do to change the yardstick by which she is being measured. She lacks the autonomy to determine whether she is successful in meeting her goals in this case. She may perceive that her only option for coping with the pressure is to cheat. As noted previously, the correlation between personal performance goals and classroom performance goal structures is only .35, indicating that being in a performance-oriented classroom is not particularly indicative of being performance oriented.

In sum, self-determination theory provides a compelling explanation for the results of the empirical example. Furthermore, it helps to explain some of the previously inconsistent findings. For example, the extrinsic goals of several of the previously discussed studies (e.g., Anderman, Griesinger, & Westerfield, 1998; Jordan, 2001; Murdock, Hale, & Weber, 2001) may have pressured students in a way that the personal performance goals of Stephens and Roeser's (2003) study did not. Thus, an important possibility to consider may be that students who are given the autonomy to choose their own goals are less likely to cheat than students who are pressured to achieve extrinsic goals.

PERSON-CENTERED ANALYSES

The person-centered analyses provided additional evidence that the students who are under pressure and underengaged are particularly likely to observe and engage in academic cheating. However, these analyses do more than just confirm the general trends from the regression analyses. In today's high-stakes testing climate, with the pressure of the No Child Left Behind legislation, this "under pressure and underengaged" profile may become increasingly prevalent in our schools. If teachers continue to be assessed and even paid based on the performance of their students, they are more and more likely to teach to standardized tests.[5] Because standardized testing (both the taking of tests and studying for them) is often not motivating for students (Paris, Lawton, Turner, & Roth, 1991), they are

---

[5] Recent research has also shown that these teachers are also more and more likely to cheat, or have their students cheat, on these high-stakes tests (Jacob & Levitt, 2004).

likely to become less engaged in the content that is presented to them. This shift toward focusing on standardized tests will likely be accompanied by students feeling more pressure from teachers whose careers are likely to be impacted by student performances on these tests. The combination of less interesting material (e.g., a focus on breadth over depth) and more teacher-centered pedagogy (e.g., "drill and kill"), with more pressure on students to perform at a high level, could set the stage for a substantial increase in the number of under pressure and underengaged goal profiles.

It is important to remember that this chapter examines only correlates of cheating rather than causes. However, it does seem plausible that a rise in the numbers of students who were under pressure and underengaged could lead to a rise of cheating in schools. Based on the specific goal profiles that were linked with cheating in this chapter, we suspect that students' self-determination in choosing the goals that they pursue plays an important role in mediating their goals and their cheating. More clarity on these issues of causality and the role of self-determination would be particularly useful to both researchers and educators.

FUTURE DIRECTIONS

One goal for future studies will be to overcome some of the limitations of the empirical example given in the present chapter. First, the chapter relied exclusively on self-report data. Future studies should augment students' self-reports with teacher reports and observations where possible. Second, the empirical illustration in this chapter used personal mastery goals as a proxy for student engagement and classroom performance goal structure as a proxy for pressure. Future research that can more directly assess students' engagement in learning and the extent to which they feel pressured for grades might provide a more accurate and powerful picture of these phenomena and their relationships to cheating beliefs and behaviors.

In addition to improving these issues, future research will need to ascertain whether a causal connection exists between either of these goals and cheating. Several questions will be of particular interest. For instance, if teachers press students to demonstrate their ability, will rates of cheating increase? If teachers can foster classroom climates that encourage students to adopt mastery goal orientations, will incidents of cheating decline? Laboratory studies that manipulate environments to be more or less interesting and more or less pressure-filled could offer helpful insights as to the causal connections between goals and cheating.

A second factor to be examined in future research is to further explore the potential for self-determination theory in order to explain the relationship between goals and cheating. Experiments that allow participants to choose their own goals (in one condition) or to assign goals to participants (in the other condition) might provide particularly important insights into the role of autonomy in cheating.

Finally, in this chapter, the regression equations explained relatively modest amounts of variance in the cheating outcomes, and the person-centered analyses resulted in similarly modest effect sizes. Thus, examining a wider range of independent variables to provide a better sense of the factors that are influential in whether students cheat will be an important goal for future research. Other fundamental motivators such as students' perceptions of their own competence and their sense of relatedness (see Connell & Wellborn, 1991) may play an important role in whether students cheat. For example, students may be more likely to cheat when tasks are too hard for them or when they do not feel an emotional connection to the teacher. Situational variables such as the amount of time pressure students are under, the existence of an honor code, perceived costs of getting caught, and parental engagement in students' schooling experiences will also be important to examine in future work.

IMPLICATIONS AND CONCLUSION

Even without causal evidence, this chapter can still provide ideas for teachers. The results suggest that teachers should help foster students' engagement in learning for its own sake. Past research on interest (e.g., Renninger, Hidi, & Krapp, 1992) makes it clear that little harm will come from increasing students' engagement in coursework and a great many benefits may arise. It seems safe to encourage teachers to help students adopt a mastery-oriented approach to learning. Our data also indicate that performance goals at the classroom level are likely to be maladaptive as far as cheating outcomes are concerned.

Finally, there may be critical differences between students assuming a performance goal orientation of their own free volition and being placed in a performance-oriented classroom. Thus, at this point in time, the safest recommendation appears to be to advise teachers against promoting performance goals in their classes. Classroom formats and activities where students establish self-referenced goals, regularly assess their own progress against their previous efforts, and are provided with choice may help make classrooms more engaging and less pressured.

REFERENCES

Adams, H. R. (1960). Cheating: Situation or problem. *Clearing House, 35*, 233–235.
Alexander, C., Piazza, M., Mekos, D., & Valente, T. (2001). Peers, schools, and adolescent cigarette smoking. *Journal of Adolescent Health, 29*(1), 22–30.

Ames, C. (1992). Classrooms: Goals, structures, and student motivation. *Journal of Educational Psychology, 84*(3), 261–271.

Anderman, E., Griesinger, T., & Westerfield, G. (1998). Motivation and cheating during early adolescence. *Journal of Educational Psychology, 90*(1), 84–93.

Anderman, E., & Midgley, C. (2004). Changes in self-reported academic cheating across the transition from middle school to high school. *Contemporary Educational Psychology, 29,* 499–517.

Asch, S. E. (1951). Effects of group pressure upon the modification and distortion of judgments. In H. Guetzkow (Ed.), *Groups, leadership and men: Research in human relations* (pp. 177–190). Oxford, England: Carnegie Press.

Bandura, A. (1986). *Social foundations of thought and action: A social cognitive theory.* Englewood Cliffs, NJ: Prentice-Hall.

Barnett, D. C., & Dalton, J. C. (1981). Why college students cheat. *Journal of College Student Personnel, 22,* 545–551.

Bennett, W. J. (Ed.). (1993). *The book of virtues: A treasury of great moral stories.* New York: Simon & Schuster.

Blasi, A. (1980). Bridging moral cognition and moral action: A critical review of the literature. *Psychological Bulletin, 88,* 1–45.

Bowers, W. J. (1964). *Student dishonesty and its control in college.* New York: Columbia Bureau of Applied Research.

Bushway, A., & Nash, W. R. (1977). School cheating behavior. *Review of Educational Research, 47*(4), 623–632.

Cizek, G. J. (1999). *Cheating on tests: How to do it, detect it, and prevent it.* Mahwah, NJ: Lawrence Erlbaum.

Connell, J. P., & Wellborn, J. G. (1991). Competence, autonomy, and relatedness: A motivational analysis of self-system processes. In M. R. Gunnar & L. A. Sroufe (Eds.), *Self processes and development* (Vol. 23, pp. 43–77). Hillsdale, NJ: Lawrence Erlbaum.

Diekhoff, G. M., LaBeff, E. E., Clark, R. E., Williams, L. E., Francis, B., & Haines, V. J. (1996). College cheating: Ten years later. *Research in Higher Education, 37*(4), 487–502.

Evans, E. D., & Craig, D. (1990a). Adolescent cognitions for academic cheating as a function of grade level and achievement status. *Journal of Adolescent Research, 5*(3), 325–345.

Evans, E. D., & Craig, D. (1990b). Teacher and student perceptions of academic cheating in middle and senior high schools. *Journal of Educational Research, 84*(1), 44–52.

Festinger, L. (1954). A theory of social comparison processes. *Human Relations, 7,* 117–140.

Graham, M. A., Monday, J., O'Brien, K., & Steffen, S. (1994). Cheating at small colleges: An examination of student and faculty attitudes and behaviors. *Journal of College Student Development, 35*(4), 255–260.

Guivernau, M., & Duda, J. L. (2002). Moral atmosphere and athletic aggressive tendencies in young soccer players. *Journal of Moral Education, 31*(1), 67–85.

Haines, V. J., Diekoff, G. M., LaBeff, E. E., & Clark, R. E. (1986). College cheating: Immaturity, lack of commitment, and the neutralizing attitude. *Research in Higher Education, 25*(4), 342–354.

Harackiewicz, J. M., Barron, K. E., Pintrich, P. R., Elliot, A. J., & Thrash, T. M. (2002). Revision of achievement goal theory: Necessary and illuminating. *Journal of Educational Psychology, 94*(3), 638–645.

Hartshorne, H., & May, M. A. (1928). *Studies in deceit.* New York: Macmillan.

Jacob, B. A., & Levitt, S. D. (2004). To catch a cheat. *Education Next,* Winter, 69–75.

Jensen, L. A., Arnett, J. J., Feldman, S., & Cauffman, E. (2002). It's wrong, but everybody does it: Academic dishonesty among high school and college students. *Contemporary Educational Psychology, 27*(2), 209–228.

Jordan, A. E. (2001). College student cheating: The role of motivation, perceived norms, attitudes, and knowledge of institutional policy. *Ethics & Behavior, 11*(3), 233–247.

Josephson Institute of Ethics. (2002). Report Card 2002: The ethics of American youth. Retrieved April 1, 2003, from http://www.josephsoninstitute.org/Survey2002/survey2002-pressrelease.htm.

LaBeff, E. E., Clark, R. E., Haines, V. J., & Diekhoff, G. M. (1990). Situational ethics and college student cheating. *Sociological Inquiry, 60*(2), 190–198.

Lanza-Kaduce, L., & Klug, M. (1986). Learning to cheat: The interaction of moral-development and social learning theories. *Deviant Behavior, 7*(3), 243–259.

Livosky, M., & Tauber, R. T. (1994). Views of cheating among college students and faculty. *Psychology in the Schools, 31*, 72–82.

McCabe, D. (1992). The influence of situational ethics on cheating among college students. *Sociological Inquiry, 62*(3), 365–374.

McCabe, D. (2005). New CAI Research. Retrieved July 29, 2005, from http://www.academicintegrity.org/cai_research.asp.

McCabe, D., & Trevino, L. K. (1997). Individual and contextual influences on academic dishonesty: A multicampus investigation. *Research in Higher Education, 38*(3), 379–396.

McCabe, D. L., & Trevino, L. K. (1993). Academic dishonesty: Honor codes and other contextual influences. *Journal of Higher Education, 64*(5), 522–538.

McCabe, D. L., Trevino, L. K., & Butterfield, K. D. (2001). Dishonesty in academic environments: The influence of peer reporting requirements. *Journal of Higher Education, 72*(1), 29–45.

Michaels, J. W., & Miethe, T. D. (1989). Applying theories of deviance to academic cheating. *Social Science Quarterly, 70*(4), 870–885.

Midgley, C., Maehr, M. L., Hruda, L. Z., Anderman, L., Freeman, K. E., Gheen, M., Kaplan, A., Kumar, R., Middleton, M. J., Nelson, J., Roeser, R., and Urdan, T. (2000). Manual for the Patterns of Adaptive Learning Scales (PALS). Ann Arbor, MI: University of Michigan. Retrived on January 15, 2002 from http://www.umich.edu/~pals/manuals.html.

Midgley, C. (Ed.). (2002). *Goals, goal structures, and patterns of adaptive learning.* Mahwah, NJ: Lawrence Erlbaum.

Midgley, C., Kaplan, A., & Middleton, M. (2001). Performance-approach goals: Good for what, for whom, under what circumstances, and at what cost? *Journal of Educational Psychology, 93*(1), 77–86.

Murdock, T. B., Hale, N. M., & Weber, M. J. (2001). Predictors of cheating among early adolescents: Academic and social motivations. *Contemporary Educational Psychology, 96*(1), 96–115.

Murdock, T. B., & Miller, A. (2003). Teachers as sources of middle school students' motivational identity: Variable-centered and person-centered analytic approaches. *Elementary School Journal, 103*(4), 383–399.

Murdock, T. B., Miller, A., & Kohlhardt, J. (2004). Effects of classroom context variables on high school students' judgments of the acceptibility and likelihood of cheating. *Journal of Educational Psychology, 96*(4), 765–777.

Newcomb, T. M. (1943). *Personality and social change; attitude formation in a student community.* Fort Worth, TX: Dryden Press.

Newstead, S. E., Franklyn-Stokes, A., & Armstead, P. (1996). Individual differences in student cheating. *Journal of Educational Psychology, 88*(2), 229–241.

Paris, S. G., Lawton, T. A., Turner, J. C., & Roth, J. L. (1991). A developmental perspective on standardized achievement testing. *Educational Researcher, 20*(5), 12–20.

Peck, S. C., & Roeser, R. W. (Eds.). (2003). *Person-centered approaches to studying development in context: New directions for child and adolescent development* (Vol. 101). San Francisco: Jossey-Bass.

Renninger, K. A., Hidi, S., & Krapp, A. (Eds.). (1992). *The role of interest in learning and development*. Hillsdale, NJ: Lawrence Erlbaum.

Rettinger, D. A., Jordan, A. E., & Peschiera, F. (2004). Evaluating the motivation of students to cheat: A vignette experiment. *Research in Higher Education, 45*(8), 873–890.

Ryan, R. M., & Deci, E. L. (2000). Self-determination theory and the facilitation of intrinsic motivation, social development, and well-being. *American Psychologist, 55*(1), 68–78.

Sherif, M. (1936). *The psychology of social norms*. Oxford, England: Harper.

Shope, J. T., Raghunathan, T. E., & Patil, S. M. (2003). Examining trajectories of adolescent risk factors as predictors of subsequent high-risk driving behavior. *Journal of Adolescent Health, 32*(3), 214–224.

Steinberg, L. (1996). *Beyond the classroom: Why school reform has failed and what parents need to do*. New York: Simon & Schuster.

Stephens, J. M. (2004a, April 15). Beyond reasoning: The role of moral identities, socio-moral regulation and social context in academic cheating among high school adolescents. Paper presented at the Annual Meeting of the American Educational Research Association, San Diego, CA.

Stephens, J. M. (2004b). Just cheating? Motivation, morality and academic (mis)conduct among adolescents. Unpublished Ph.D. Dissertation, Stanford University, Stanford, CA.

Stephens, J. M., & Roeser, R. W. (2003, April 25). Quantity of motivation and qualities of classrooms: A person-centered comparative analysis of cheating in high school. Paper presented at the Annual Meeting of the American Educational Research Association, Chicago, IL.

Sykes, G. M., & Matza, D. (1957). Techniques of neutralization: A theory of delinquency. *American Sociological Review, 22*, 664–670.

Urberg, K. A., Luo, Q., Pilgrim, C., & Degirmencioglu, S. M. (2001). A two-stage model of peer influence in adolescent substance use: Individual and relationship-specific differences in susceptibility to influence. *Addictive Behaviors, 28*(7), 1243–1256.

Weiner, B. (1990). History of motivational research in education. *Journal of Educational Psychology, 82*(4), 616–622.

Whitley, B. E., Jr. (1998). Factors associated with cheating among college students: A review. *Research in Higher Education, 39*(3), 235–274.

Who's Who among American High School Students. (1998). Cheating and succeeding: Record numbers of top high school students take ethical shortcut: 29th annual survey of high achievers. Retrieved March 30, 2003, from http://www.eci-whoswho.com/highschool/annualsurveys/29.shtml.

APPENDIX

SCALES AND
RELIABILITIES

Achievement Goals

Scale	Item	Reliability (α) (N = 337)
Personal Mastery Goal Orientation	It's important to me that I learn a lot of new concepts in this class this year. One of my goals in this class is to learn as much as I can. It's important to me that I thoroughly understand my work in this class. It's important to me that I improve my skills in this class this year. One of my goals is to master a lot of new skills in this class this year.	.89 (LC[6]) .93 (MC[7])
Personal Performance Approach Orientation	It's important to me that other students in this class think that I am good at my class work. One of my goals is to show others that I am good at my class work. One of my goals is to show others that work in this class is easy for me. One of my goals is to look smart in comparison to the other students in this class. It's important to me that I look smart compared to others in this class.	.88 (LC) .93 (MC)
Classroom Mastery Goal Structure	Working hard to do your best counts a lot in this class. In this class, really understanding the material is the main goal. In this class, it's okay to make mistakes as long as you are learning. In this class, learning and understanding new ideas and concepts is very important.	.72 (LC) .75 (MC)
Classroom Performance Approach Goal Structure	In this class getting good grades is the main goal. In this class, it's important to get high scores on tests. In this class, getting right answers is very important.	.63 (LC) .68 (MC)

[6] LC connotes the class in which the student reports cheating in least often or would be least likely to cheat in.
[7] MC connotes the class in which the student reports cheating in most often or would be most likely to cheat in. The scores used in the study are an average of "least cheat" and "most cheat."

Personal Beliefs about Cheating

"Do you think this behavior is cheating?" (Yes, No, Not Sure)

Homework Cheating	.63
Letting another student copy your homework.	
Turning in work you copied from another student.	
Working on an assignment with other students when the teacher asked for individual work.	
Plagiarism	.73
Copying a few sentences from a site on the Internet without footnoting them in a paper or assignment you submitted.	
Copying a few sentences from a book, magazine, etc. without footnoting them in a paper or assignment you submitted.	
Copying almost word for word from a book, magazine, or other source and turned it in as your own work.	
Turning in a paper that was obtained in large part from a term paper "mill" or web site.	
Test Cheating	.56
Getting questions or answers from someone who had already taken a test in an earlier class period.	
Helping someone else cheat on a test.	
Copying from another student during a test or exam.	
Using unpermitted crib notes or cheat sheets during a test.	

"Mark should *NOT* be blamed for cheating if . . ."

Neutralization of Responsibility	.89
the course material is too hard	
he's in danger of losing his scholarship	
he doesn't have the time to study	
the teacher doesn't seem to care	
the teacher acts like his/her course is the only one	
his cheating isn't hurting anyone	
everyone else in the room seems to be doing it	
people sitting around him made no attempt to cover their papers	
a friend asked him to help him/her cheat	
the instructor left the room	
the course is required	

Perceptions of Peer Norms

Scale	Item	Reliability (α) ($N = 337$)
Peer Assignment Cheating	How frequently do you see or hear about other students copying someone else's homework?	.76
	How frequently do you see or hear about your friends copying someone else's homework?	
Peer Plagiarism	How frequently do you see or hear about other students engaging in plagiarism on written assignments?	.83
	How frequently do you see or hear about your friends engaging in plagiarism on written assignments?	
Peer Test Cheating	How frequently do you see or hear about other students cheating on tests or exams?	.83
	How frequently do you see or hear about your friends cheating on tests or exams?	
Peer Disapproval of Cheating	I would be disappointed if I found out that a friend of mine cheated on a test.	.75
	My friends would be disappointed in me if they found out I had cheated on a test.	
	Among my friends, minor forms of cheating aren't considered a big deal.*	
	* Negatively keyed item	

Cheating Behaviors

"During this school year, how often have you engaged in the following behaviors?" (Never, Once, More than once)

Homework Cheating	.61
Let another student copy your homework.	
Turned in work you copied from another student.	
Worked on an assignment with other students when the teacher asked for individual work.	
Plagiarism	.72
Copied a few sentences from a site on the Internet without footnoting them in a paper or assignment you submitted.	
Copied a few sentences from a book, magazine, etc. without footnoting them in a paper or assignment you submitted.	
Copied almost word for word from a book, magazine, or other source and turned it in as your own work.	
Turned in a paper that was obtained in large part from a term paper "mill" or web site.	
Test Cheating	.70
Got questions or answers from someone who had already taken a test in an earlier class period.	
Helped someone else cheat on a test.	
Copied from another student during a test or exam.	
Used unpermitted crib notes or cheat sheets during a test.	

7

APPLYING DECISION THEORY TO ACADEMIC INTEGRITY DECISIONS

DAVID A. RETTINGER

ABSTRACT

This chapter explores academic integrity decisions from the perspective of behavioral decision theory. Decision principles of economic utility, the psychology of decision making, emotions, neuroeconomics, and evolutionary psychology are all explained and applied to students' decisions to cheat or plagiarize. Making these connections highlights the process by which a student comes to commit academic integrity code violations as well as providing a means of exploring that process experimentally. An understanding of the integrity decision process will help educators guide students toward more ethical and ultimately beneficial choices.

We are all painfully aware that students at all levels are choosing to cheat and plagiarize in increasing and startling numbers. The research presented in this book and elsewhere paints a compelling picture of the causes and correlates of cheating: student factors like motivation and moral development; situational factors like the chances of being caught and the behavior of peers; long-term influences like students' relationships with their teachers; and the motivational structures of classes and honor codes. These factors combine to represent the state-of-the-art view of academic integrity, but they do not address the decision process that a student undertakes when the opportunity to cheat arises. A process-oriented approach is useful to integrate the contributing factors from the integrity literature with the generic understanding of decisions from the judgment and decision-making literature.

Copyright © 2007 by Academic Press, Inc.
All rights of reproduction in any form reserved.

The field of judgment and decision making has origins in both psychology and economics. This economic focus has historically led to an emphasis on the manner in which choices and judgments are made and the relationship of those choices to the ideal or rational decision. A good portion of the literature is devoted to competing descriptive theories of the judgment and choice processes, with the emphasis squarely placed on the concept of maximizing the value that one gains as a result of one's choice. This is predicated on the assumption that decision makers seek to find the option that is of the most value to them, usually by means of considering one's options deliberatively. From that basis, the field has expanded to include theories that emphasize emotions, decision shortcuts, and social constraints, among many other advancements. This chapter outlines principles of decision theory and applies them to the field of academic integrity in order to improve understanding of how these factors lead to cheating by individual students and to inspire future empirical research on the decision process involved in academic integrity decisions.

It is widely believed (see Schwarz, 2000, for an example) that both deliberative and emotional processes are involved in making decisions. Within this chapter we will explore how these two processes, separately and together, can help us understand academic integrity decisions. There are many competing theories within the field of judgment and decision making, and it is certainly not possible to do justice to them all in this context. We begin with deliberative decisions, focusing on basic decision principles derived from economics. Subjective expected utility theory (von Neumann & Morgenstern, 1947) and prospect theory (Kahneman & Tversky's, 1984) were built to explain how people make decisions under risky circumstances. Prospect theory can be used *post hoc* to explain certain features of the cheating literature. Following that, emotion-based approaches to decision making will be used to explain why so many students are willing to take serious risks with their academic standing by cheating or plagiarizing. Schwarz's (1990) three aspects of emotional decision making—the effects of current moods, the effects of prior emotional experiences, and the effects of anticipated emotions—will be discussed, as will Damasio's (1994) somatic marker hypothesis, which views decision making as a biological and emotional process. Finally, the deliberative and emotional viewpoints will be integrated, making use of more recent research on the evolutionary (Cosmides, 1989) and biological (Sanfey et al., 2003) bases of judgment processes. This work seems to indicate that a biologically based notion of fairness takes place in decision contexts.

As we examine this multifaceted approach to integrity decisions, an overall model will emerge. A basic diagram of that model is included as Figure 7.1. As this chapter progresses, it will attempt to explicate the various elements of the model. First, we will explore deliberative decision theories, including utility theory, which is the basis against which the others are

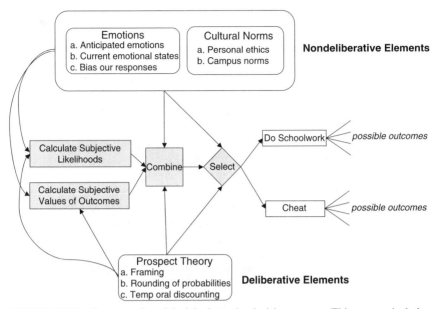

FIGURE 7.1   A proposed model of the integrity decision process. This process includes both deliberative (e.g., Prospect Theory) and nondeliberative (e.g., emotionally driven decisions) elements.

judged. We will then discuss prospect theory, a descendant of utility theory that has important implications for integrity decisions. Special attention will be paid to the decision biases that these models predict and to how those biases play out in integrity decisions. These deliberative theories will then be contrasted with nondeliberative conceptions of the decision process, namely emotion-based decision making, specialized models of integrity decisions, evolutionary decision theories, and brain-based understanding of the decision process. The focus of the chapter then turns to the application of theories of emotional and deliberative decision processes to students' academic integrity decisions. Neutralizing attitudes, the importance of intrinsic and extrinsic motivations toward courses, and the role of enforcement and penalties will all be discussed in light of decision theory. The conclusion discusses how to use this information to improve the integrity culture in our schools and describes a course for future research.

DECISION THEORIES

DELIBERATIVE THEORIES

Utility theory, the major component of the judgment and decision-making literature, takes its origins from the rational decision principles laid out by economists (von Neumann & Morgenstern, 1947). This theory and those following (Kahneman & Tversky, 1984; Savage, 1954) are intended to explain how we choose among options in cases where we are unsure of the outcomes. For example, we may hold or sell a stock, not knowing what its price will be in the future. According to these theories, a decision maker evaluates each choice on the basis of the value (utility) of each possible outcome of that choice. In the case of academic integrity decisions, the choices are roughly whether or not to cheat/plagiarize— answer a question oneself, skip it, and so on. The outcomes in both cases are measured in a handful of ways, including the number of points received for the exam question, possible sanctions if caught, and any personal value the student places on learning and self-reliance. In order to make a deci- sion, the student must weigh the likelihood of each possible result of each option (to cheat or not) and balance the values of the possible results for each choice with each outcome's likelihood. For example, if a student believes that there is a 1 percent chance of being caught cheating, then the attendant severe penalties and their associated negative value to the student will be weighed lightly against the positive value and the much more likely possibility of using these dishonest means to answer correctly without being detected. In this context, given the "right" circumstances (i.e., low chance of being caught and relatively mild penalties), cheating might be the rational decision, especially if measured against the time saved by not studying (see Woessner, 2004 for a more detailed discussion of the issues).

A more detailed example (see Figure 7.2) might be helpful to illustrate the nature of a utility calculation of cheating. Imagine that Bob is prepar- ing for a final exam. He has to prepare for four others and has little confi- dence in his ability to score well on this one. As the exam approaches, he considers bringing a "cheat sheet" into the test, which he feels will make a big difference in his grade. Moreover, it will give him additional time to study for classes he believes he can do well in and cares more about. In the figure, this decision is represented by the diamond, and the shaded rect- angles represent his two options. The lines on the right represent the pos- sible outcomes of each option, with the percentages indicating the likelihood of each outcome. What does utility theory say about this decision? First Bob generates options. In this case, he is choosing between more study or preparing, or using his "cheat sheet." Next, he must determine the possible outcomes of each option. If he studies, Bob estimates the chances of getting a grade of "A" at 5 percent, of "B" at 10 percent, of "C" at 60 percent, and of "D" or lower at 25 percent. If he opts to cheat, these chances improve dramatically, to 30, 30, 30, and 9 percent, respectively, with a 1 percent chance of being caught. Next, Bob must subjectively value these outcomes. Many factors come into play, including his desire for high grades, self- esteem lost by cheating, neutralizing and moral attitudes, and the time

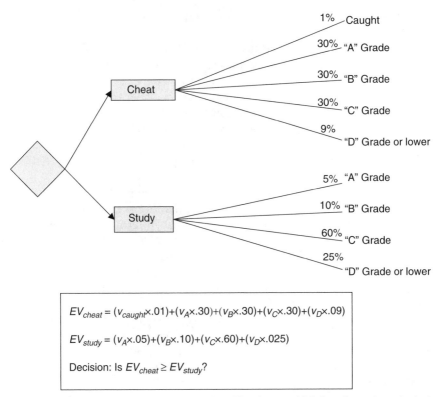

FIGURE 7.2 A decision tree generated by utility theory, which describes a hypothetical decision in which a student must choose between cheating and studying. The possible outcomes of these choices are represented at the right, and the formulas for evaluating the choices and selecting one are below.

saved by cheating. In a less hypothetical example, utility theory claims that Bob will act as a rational agent and (often without awareness) assign numbers to these values. Once Bob has computed the values for each outcome, they are multiplied by the probabilities of each outcome. The results are called "expected utilities." The choice with the higher expected utility is, under utility theory, deemed to be the best one. This calculation is described by the formula in the figure. The value of each possible outcome is multiplied by the probability of that outcome and then summed for each option. As the bottom equation indicates, the option with the highest summed value (called the expected value) is chosen.

These calculations are not entirely conscious. Just as most of us typically do not calculate the parabolic arc of a baseball in flight in order to catch it, we often perform complex utility calculations automatically. Only sometimes do we make these calculations consciously, just as a baseball player may consider where in the outfield to play for a particular hitter or may position himself without thinking.

What does this analysis tell us about integrity decisions? First, it points out the importance of consequences. As Woessner (2004) points out, allowing a student caught cheating to rewrite a paper for a lower grade (or retake an exam in our example) penalizes him or her only mildly on both the grade dimension and the time dimension. Giving a student a 0 on the exam is a more severe grade penalty but represents a savings of time for the student. The only effective penalties will be those that strongly affect the expected utility of cheating. The problem with this is fairness. Draconian measures are often rejected by both faculty and students as unreasonable, and might even increase cheating by fostering neutralizing attitudes. It is therefore essential to devise penalties that are seen as reasonable by all involved, but still penalize those who cheat in terms of both time and academic standing.

A more sensible but far more difficult alternative is to increase students' estimated (and actual) likelihood of being caught. This can be accomplished in two ways. First, increased vigilance on the part of proctors and instructors generally, including technological methods like Internet search engines in the case of plagiarism, will serve as a deterrent. A second and more interesting option is to generate a perception among students that being caught is more likely. This is effective due to a decision-making strategy called *availability* (Tversky & Kahneman, 1973). By this strategy, decision makers estimate probabilities of future events by sampling their memory. Ease of retrieval is used as an estimate of future likelihood. Cheating might be made to seem less attractive if the salience of being caught were greater. The means of achieving this include involving student leaders in a publicity campaign, publishing a "police blotter" of integrity offenses (anonymously), and encouraging admitted integrity violators to write public apologies for publication in campus newspapers. These methods would make cheating and plagiarism less attractive options for potential rules violators.

Utility theory provides an excellent basis for understanding decisions, but is limited by some crucial underlying assumptions. One assumption in particular, the assertion that decision makers will always act in a rational manner, including choosing the option that has most utility to them, is problematic. A more modern behavioral decision theory called prospect theory (Kahneman & Tversky, 1984) elaborates on both the value calculation and the likelihood estimation process in ways that address this concern and have profound implications for integrity decisions. Under prospect theory, values are calculated relative to a reference point, not the total value as in utility theory. In a student's context, the reference point might be the status quo, but it might also be the GPA necessary for admission to medical school, or the exam grade needed to pass a course in order to graduate. This means that a student contemplating an integrity violation will not compare the absolute values of outcomes and choose the highest one, but instead will prefer the one that has the highest value relative to the reference point. For an example, consider Bob, the hypothetical student whose plight was chronicled above and in Figure 7.2. Under utility theory, he would choose to cheat (or not) based on his evaluation of the expected value of each of those options. This might lead him to cheat in a case where cheating would earn him 57 percent on average, whereas not cheating would earn a 55 percent grade. Notice that cheating does provide a higher expected value, but neither one would lead to a passing grade. Bob's decision under utility theory is not informed by this crucial reference point. This isn't a very realistic assessment of student behavior, however. In reality, Bob might decide to cheat only if he believes that there is no chance of passing the course without cheating and that there is a greater chance of passing if he does, even if the possibility of being caught reduces the expected value of cheating. If a score of 65 percent is required for passing, Bob wouldn't accept a very likely score of 63 percent (expected value of 63 percent), but would choose a dishonest option that had a 60 percent chance of scoring 83 percent, even if the chance of being caught was great enough to reduce the expected value to 50 percent. This example highlights the major distinction between prospect and utility theories—the reference point. Reference points are particularly crucial because of Kahneman and Tversky's (1984) findings that losses (results less than the reference point) are treated differently than gains.

This differential treatment is the result of the value function that prospect theory proposes. The value function is the relationship between the psychological experience of an event and the value of the event on an objective gain/loss scale. This function is described in Figure 7.3 (based on Tversky & Kahneman, 1981). The x-axis of this graph is the objective worth of an outcome relative to the reference point, with gains on the right and losses on the left. The y-axis is the value the decision maker would experience if that outcome came to pass. The shape of the function has two distinctive features. First, it flattens out at the high and low ends. This means that the proportional impact of losses and gains is diminished as their magnitude increases. A $2 gain is psychologically much larger than a $1 gain, whereas a $1,000,001 gain is psychologically identical, more or less, to a $1 million gain.

The second feature of the function is that it is much steeper for losses than for gains. This means that the negative experience of a loss is greater than the positive experience of a gain of the same magnitude. These features combine to create a behavioral phenomenon relevant to academic integrity: gain/loss framing. Gain/loss framing refers to the well-documented (see Kühberger, 1998 for a review) finding that people tend to prefer taking risks to avoid losses, but they are risk averse when evaluat-

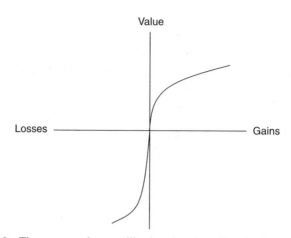

FIGURE 7.3 The prospect theory utility function, from Tversky & Kahneman (1981). It represents the relationship between the objective value of a decision outcome, measured as incremental gains or losses from a reference point (on the *x*-axis), and the personal value one places on that outcome (on the *y*-axis).

ing gains. The most famous example of this is the Asian disease problem (Tversky & Kahneman, 1981):

> Imagine that the United States is preparing for the outbreak of an unusual Asian disease, which is expected to kill 600 people. Two alternative programs to combat the disease have been proposed. Assume that the exact scientific estimate of the consequences of the programs are as follows:
> Program A_1: If this program is adopted, 200 people will be saved.
> Program B_1: If this program is adopted, there is one-third probability that 600 people will be saved and two-thirds probability that no people will be saved.
>
> or
>
> Program A_2: If this program is adopted, 400 people will die.
> Program B_2: If this program is adopted, there is one-third probability that nobody will die and two-thirds probability that 600 people will die.

Program A is the same in both cases: 200 people live and 400 die. The same is true for program B: one-third of the time everyone lives and two-thirds of the time everyone dies. However, in Tversky and Kahneman's original study, 72 percent of people chose A_1 and 78 percent of different people chose B_2. This is an example of gain/loss framing, in which preferences for identical outcomes are reversed as a function of the description, or frame, of the problem. It turns out that we tend to prefer risky choices when faced with losses and sure things when dealing with gains. This is an example of how a reference point can influence preferences, since losses and gains must be relative.

This concept of prospect theory can be applied to integrity decisions by considering the frame within which students are making decisions. It is often the case that a student's reference point is a high grade (relative to his or her current state), and so the status quo would be evaluated as a loss. This reference point can be created as a result of family pressures, a desire for admission to a competitive graduate program, or any of the other external motivations that have been described in the integrity literature (Anderman, Greisinger, & Westerfield, 1998; Dweck, 1986; Dweck & Leggett, 1988; Eison, 1981; Eison, Pollio, & Milton, 1986; Rettinger, Jordan, & Peschiera, 2004; Wryobeck & Whitley, 1999). When students are choosing between studying and cheating (or plagiarizing), they may consider studying to be a sure loss, based on their previous grades. Faced with that, evidence indicates that the typical decision is to take a risk, even a long shot, in order to avoid that loss. This interpretation is supported by findings that students who expect lower grades are more likely to cheat, as are students who are motivated by grades (Rettinger, Jordan, & Peschiera, 2004). The policy implication of this finding is in line with those stemming from the motivation literature—it is essential to help students to frame their current grades as gains and not set unrealistic expectations. When students see the value of the learning as a gain, and the grade as a byproduct, or even their current grade as a point of reference, they will be risk averse because getting caught cheating will jeopardize their status quo.

As discussed earlier, expected values are made up of both the subjective value of an outcome and the decision maker's expectation of the outcome's likelihood. Prospect theory's handling of probability judgment also can inform our understanding of integrity decisions. Judgments of future probabilities are not simply affected by availability, as we mentioned earlier, but rather are estimated according to prospect theory's *decision weight function*. This happens because it is clear that people do not treat all probabilities appropriately. For example, very low probability events are often eliminated from consideration, while the chances of slightly higher probability events occurring are overestimated. Most future possibilities are underweighted, meaning that we underestimate their chance of occurrence. One consequence of this function is the certainty effect. This manifests itself as a strong preference for certain (i.e., definite) events and a willingness to eliminate small probabilities of failure to achieve certainty. For a student considering graduation or medical school, even the most remote probability of receiving an "unacceptable grade" might be too much to bear. The certainty effect, therefore, can explain why high-achieving students are as willing as they are to cheat (Whitley, 1998). The small probability of failure looms greater for a student whose chances of success are near certain than for one whose chances are lower to begin with. This can certainly be expected to lead to cheating and plagiarism.

Another consequence of the decision weight function is the rounding of very small probabilities to zero. This happens as a means of cognitive economy (see Payne, Bettman & Johnson, 1993, for a discussion of this issue) because it is often impossible to consider all possible outcomes in a complex environment. However, some events have much higher probabilities than we estimate and more serious consequences. People engaging in risky behaviors, for example, often ignore the possibility of the most serious consequences (e.g., an overdose when using drugs), because those outcomes are relatively rare. For integrity decisions, the likelihood of being caught is often very slim, as in the case of furtive exam cheating. If a student underestimates this probability, or rounds it to zero, it can strongly influence the utility for cheating.

A final consequence of these approaches is an effect called temporal discounting. Economists have observed this phenomenon in market situations, and it has been replicated in the laboratory by psychologists as well (Frederick, Loewenstein, & O'Donoghue, 2002). Temporal discounting takes place whenever a decision maker chooses a smaller immediate payoff instead of a larger one in the future. They essentially value an outcome less the farther away that outcome is in time. This applies equally to positive and negative outcomes, such that the bad does not seem as bad or the good as good, the farther away the event occurs. This principle has been used to explain political behavior such as the tax cuts of the 1980s, as well as children's preferences for a single marshmallow now over two marshmallows later (Mischel, Ebbesen, & Zeiss, 1972). The standard method of evaluating temporal discounting is to present subjects with a series of choices between a current state of affairs and a future one. Participants usually (see Frederick, Loewenstein & O'Donoghue, 2002, for examples and exceptions) require a larger payoff for future events, although the size of the difference often varies depending on the context.

Temporal discounting is often related to notions of self-control and cited as a contributor to risky behavior (Critchfield & Kollins, 2001). People are willing to sacrifice their long-term well-being for short-term gains, thus explaining behavior such as drug abuse, unsafe sex, and risky driving behavior. It is a short step to extend this to academic integrity violations. Like these other behaviors, it is often the case that no immediate harm comes to the decision maker as the result of the choice, and those consequences that may arise will be felt in the future. Even if one is caught cheating (or speeding), the immediate experience of being caught, though aversive, is not the most serious repercussion, which usually comes at a later time. The delay between an infraction and the punishment is often sufficiently great as to make the consequences seem remote. When combined with the often low probability of being caught, those consequences can seem virtually nonexistent. Therefore, when evaluating the utility of cheating, both temporal and probability discounting can lead students away from the choice of academic integrity because the benefits of cheating seem close at hand and the potential consequences seem quite distant.

To review, deliberative theories of decision making as a whole try to describe the choice process as a means of maximizing value to the decision maker, but these theories differ on how they conceive of the valuation and likelihood estimation processes. Utility theory is a simple weighting of possible outcomes based on their likelihood. Prospect theory is more sophisticated because it postulates that the decision maker recalibrates both the values and the likelihoods of events. These recalibrations (based on the value function and decision weight function) allow prospect theory to explain a wide range of decision behaviors, including gain/loss framing and temporal discounting.

EMOTION-BASED THEORIES

In recent years, the judgment and decision-making field has turned away from an exclusive emphasis on more deliberative decision models and has begun to include emotion as a substantial component (Clore, Schwarz, & Conway, 1994; Forgas, 1995; Schwarz, 2000 are useful reviews). Schwarz (2000) describes three ways that emotion can influence decisions. First, one's existing mood can be included in an unrelated decision task. Second, the anticipation of emotions such as regret and disappointment can hold sway when choosing among options. Finally, recall of previous emotions can be used to guide future decisions.

Existing mood is thought to act on decisions in two ways. First, mood changes the selection of cognitive processes. Schwarz and Clore (1996) reviewed a large body of literature that shows that positive moods tend to lead to top-down decision making. Decision makers tend to rely on current knowledge in these cases, making more sweeping generalizations rather than focusing on details. When one is in a negative mood, the reverse holds. Prior knowledge is downplayed in favor of current details. As Schwarz (1990) puts it, "we usually feel bad when things go wrong and feel good when we face no particular problems." Negative moods therefore signal a need to focus attention on the status quo with an eye toward fixing it, while positive moods give the "all clear" to continue as normal. One bit of evidence for this claim comes from Hertel, Neuhof, Theuer, and Kerr (2000), who had participants play a game analogous to "chicken," in which two players must choose between two courses of action. If both players choose the same one, they lose the game. This is the game made famous by two drivers speeding headlong toward one another. The driver who swerves loses the game, but not nearly as completely as if neither driver swerves. Hertel et al. (2000) found that participants who were in a good mood tended to simply mimic other players in a laboratory version of the game, while those in a bad mood were more deliberative.

Second, emotion is a source of information. For many decisions, as part of the overall strategy, people consider their "gut" feelings (see the later discussion of Damasio's view of how this works). Because emotions are conceived as global states of the brain and mind, it is difficult to separate preexisting feelings from feelings about a contemplated course of action. This can result in the foiling of the normal emotional decision check; because people are aware of the multiple sources of emotions, they discount emotions as a source of information. It can also result in "seepage" between prior emotions and the judgment at hand. Positive moods lead to the overestimation of positive outcomes, and the reverse is true for those in sad moods (Johnson & Tversky, 1983).

A more complex view of emotions and their role in decision making comes from Lerner and Keltner (2000). They point out that while the decision literature typically simplifies emotions as either positive or negative (Elster, 1998; Forgas, 1995; and Higgins, 1997 as cited in Lerner & Keltner, 2000), less attention is paid to how particular emotions might affect decisions. They choose to focus on the distinctions between anger and fear, two negative emotions that would have the same effect on decisions if valence is the dominant factor. This contrast is of particular interest to integrity decisions because many of them are made in the context of anxiety over performance in a class. Lerner and Keltner (2000) claim that in addition to the valence of an emotion, the appraisals that people make will tend to be different in different emotional states. For example, fear will tend to lead to the perception of negative events as unpredictable and controlled by the situation. By contrast, angry people will appraise negative events as predictable and caused by others. Lerner and Keltner test these claims empirically by examining participants' anger and fear dispositions and evaluating their risk assessments, all using questionnaires. The results supported their hypothesis: The risk assessments of angry people were generally more positive, and those of fearful people were more negative. While it is a large leap to say that people who are in a transitory fearful state will also show these negative assessments of risk, it is a distinct possibility based on these findings. If that is the case, it is likely that students whose concern about being caught actually engenders fear are less likely to cheat than those who are not afraid. Lerner and Keltner (2000) propose that this is due to a change in their assessments of risky events generally rather than their utility for this particular risk, which is a new, testable claim.

Of particular relevance to our topic of academic integrity decisions is the role that anticipated emotions play in our decisions (according to Schwarz, 2000, the second way in which emotions affect decision making). The two most commonly discussed emotions in these contexts are regret and disappointment, both of which are considered to be the result of an outcome that is not as desirable as possible. Regret happens when the outcome is thought to be due to the choices of the decision maker, whereas disappointment is the result of the random nature of future events. Zeelenberg, van Dijk, Manstead, and van der Pligt (2000) propose a theory of anticipated emotions that explains how the possibility of these feelings drives decision making by adding the concepts of regret and disappointment to traditional utility theory (see Mellers, Schwartz, & Ritov, 1999 for another view). Possible regret is estimated by taking the difference between the chosen outcome and the actual outcome and weighting that difference by the probability of the event occurring. For example, when deciding whether to plagiarize a paper, a student estimates the chance of being caught at 5 percent, and knows that the resulting grade on the paper would be a zero. If he or she writes the paper him or herself, he or she expects to earn an 80 percent. Therefore, the anticipated regret for cheating and being caught would be $(0-80) \times .05$, which equals 4. Of course, these numbers are just an example, because the calculations would be made based on the utility of the points, not the number of points themselves, but they provide an example of how the theory works. By contrast, disappointment is calculated by subtracting the difference between the actual and the expected outcome. For example, a student who expects to score 90 points on an exam but actually earns 75 will feel substantial disappointment. This is different from regret, because it is not seen as the consequence of a particular choice, but as the result of a more global situation.

This theory makes interesting predictions for integrity decisions. Because avoiding disappointment is a strong motivator, and disappointment is determined based on the expected outcome, students who expect to do well are therefore most susceptible to disappointment. This gives them motivation to do well, possibly by cheating. Given the high probability of a good student performing well, under utility theories it is hard to explain why these good students would choose to cheat, given the marginal benefit and large potential losses. However, adding this emotional component makes it easy to explain cheating by otherwise high-performing students: They are avoiding disappointment.

Anticipated regret is also likely to drive students to cheat. Students imagine a situation in which they earn a low score based on their own work and compare it to the case in which they cheat to earn a good grade. They therefore anticipate regret and make the decision to cheat. However, the converse is also true. Students who are contemplating cheating may imagine the situation in which they are caught and suffer a large penalty. This would lead to substantial anticipated regret and might lead away from the cheating option. A testable prediction based on this work would be that students who cheated spent more time and cognitive effort considering the first situation and those who did not spent time considering the latter.

According to Schwarz (2000), the third contribution of emotion to decision making is the emotional consequences of previous decisions on later ones. Damasio and his colleagues (Bechara, Damasio, & Damasio, 2000) have proposed the *somatic marker hypothesis* as an explanation of how memories of previous emotional states impact decision making. This theory has its basis in the biological responses of the body to situations. In particular, the biological responses that are caused by emotional states become associated with complex situations. For example, the physical anxiety a student faces during testing can become associated with the testing situation. This happens as the result of brain activity in a system that includes the orbitofrontal cortex, amygdala, and other brain areas. Decisions are then made at least partially based on these learned emotional responses. This allows for the emotional state to play into or even control a person's decision process.

There is strong evidence for this viewpoint from the neuropsychology literature. By comparing patients with damage to the orbitofrontal cortex to patients with damage to other areas of the brain, it is possible to see the unique effects of damage to this area. In order to support the somatic marker hypothesis, the orbitofrontal patients must show an inability to use emotional cues to predict future consequences in making decisions while performing other tasks normally. The particular decision task used to assess this situation is the so-called Iowa gambling task (Bechara et al., 1994). In that task, decision makers must choose cards from one of four decks placed in front of them. Each card contains an outcome that can be either positive (winning money) or negative (losing money). They choose one card at a time and may choose as many as they want from any deck, alternating as they please. Of the four decks, A and B have larger payouts but even larger losses, so the net result is a loss over any 10 card interval (the decks are not randomly shuffled before play). C and D, on the other hand, have smaller payouts than A and B, but also smaller losses, so that 10 cards will lead to a net gain. Participants are expected to learn this by trial and error, for which a functioning somatic marker system is required.

Because the task is complex, it is very difficult to see these patterns consciously. Nonetheless, normal controls and patients with lesions in areas other than the frontal cortex gradually drift toward decks C and D over the course of the 100 trial experiments. Although they initially prefer the high payouts, over time they realize that they are losing money because of the larger losses. By contrast, patients with frontal cortex lesions (in the ventromedial area) make the shift for a while, but tend to return to the higher payout/bigger loss decks. Furthermore, when skin conductance data are collected using GSR methods, normal control subjects develop a stronger response to the "bad" decks over time, while frontal lobe patients do not. Since skin conductance is a well-established physical correlate of emotional activity, Bechara et al. (2000) concluded that their patients' somatic marker system is not functioning. Combined with the propensity of these patients to be insensitive to the consequences of real-life actions, this conclusion has been taken as strong support for their hypothesis. The frontal lobe helps create emotional responses to decision situations, which in turn allows decision makers to respond appropriately to the likely out-comes of their choices, thus giving emotions a role in the decision process.

This theory can be applied to all decisions, including integrity decisions. A student who chooses between doing his or her own work and copying the work of another will have somatic markers indicating the emotional states associated with these behaviors. For example, a student who has not cheated in the past and has learned strong negative associations with that behavior might feel a pang of anxiety or disgust when the opportunity to cheat is presented. However, if the student has received low marks in the past, then not cheating might instigate an even stronger response. On the positive side, a student who is rewarded for doing well may choose the action associated with those rewards. Unfortunately, that may be cheating. While these results may seem trivial, the somatic marker hypothesis now provides us with a strong theoretical explanation for the mechanism that drives them.

In summary, emotions can have three kinds of effects on decisions. First, our emotions influence our biases. Positive moods let us know that the situation is good, leading to a status quo bias. Negative moods warn us that things are not as we would like, which can bias us toward decisions that lead to change. Second, emotions are a source of information. The mood we are in affects the decisions we make, as Lerner and Keltner (2000) demonstrated. Fear and anger, for example, can lead to very different decisions. Finally, the emotional consequences of previous decisions can affect later ones. Decision makers are able to use "gut feelings" to help determine the appropriateness of a choice. As Damasio and his group have shown, people with brain damage to certain emotion centers of the brain are unable to do this, typically with negative consequences.

SPECIALIZED MODELS OF INTEGRITY DECISIONS

Within the management literature a set of theories specifically intended to describe dishonesty in the workplace (for example, Treviño, 1986) has developed. These theories often emphasize lying in various contexts (e.g., to bosses, customers, shareholders), but can be applied to classroom cheating in a straightforward way. Many have been applied already. For example, the cognitive portion of Treviño's (1986) interactionist model is based on Kohlbergian principles. While an entire book could be dedicated to these theories themselves, of most interest to us is Lewicki's behavioral model (1983, as cited in Treviño, 1986), because it has a cost/benefit flavor consonant with more general decision theory. In discussing workplace deception, Lewicki proposed, as in prospect theory, systematic biases in the judgments that make up a so-called deception decision. In particular, the costs of deception are underestimated. Lewicki argued that this is because of something like rationalization or self-deception about the real harms of deception to others. This once again seems quite similar to the neutralizing attitudes (Whitley, 1998) that are a major component of the academic integrity literature.

In recent work by Schweitzer, Ordóñez, and Douma (2004), Lewicki's ideas have been extended to include the notion that a decision maker's goals might lead to increased dishonesty. Just as a reference point in prospect theory may lead to increased risk taking in order to avoid a relative loss, the prospect of missing a goal might lead to dishonesty, particularly when the goal is externally motivated. Events at companies like Enron, AIG, and others bear this out in the marketplace. In the laboratory, Schweitzer et al. (2004) collected data using an anagram task and different goal conditions. Participants were asked to create as many anagrams as possible and in the reward condition were allowed to take money from an envelope (anonymously) for each trial in which they met the predetermined goal. The test was self-scored, and so lying in this study was quite akin to cheating. Participants were most likely to cheat on this task in the reward condition, even though they could take the money without cheating. This seems to indicate that participants were willing to cheat but not steal. It is probably the case that they saw this as merely "rounding off" and used this as a rationalization for taking the payment. This inference is based on the fact that most cheating took place when participants were close to achieving the goal.

EVOLUTIONARY APPROACH

One of the more interesting shifts in the last quarter century has been the consideration of human cognition in light of evolutionary constraints. A guiding principle of this work has been that cognition (including decision making) is not general, but rather is based on special-purpose cognitive abilities. For example, Cosmides (1989) proposed that social relationships have required special mental mechanisms because they have been so crucial to human survival, and these mechanisms have in turn been recruited for general-purpose use. In order to survive in groups, people must be able to rely on one another to fulfill social obligations without cheating. This sort of social structure could evolve only if the groups' members had a mechanism for detecting when someone cheats.

This hypothesis is supported in the laboratory. In a famous set of studies using the Wason selection task (Cosmides, 1989), it was convincingly demonstrated that rule violations are much more easily detected when those rules reflect social contracts as opposed to logical or legal strictures. These were part of the "cheater detection" mechanism from which our decision processes may have developed. In the Wason task (1983), participants are presented with four cards, two face up and two face down. They are told that each card is two-sided and that there are rules governing the relationship between the two sides. The information on the cards varies depending on the design of the particular study. The task for participants is to choose which cards to flip over in order to verify that the relationship rules are being followed. For example, in the original problem structure, the only rule was that a card with an even number on the front must have a vowel on the back (see Figure 7.4). Participants were then shown cards displaying 2, 5, R, and E on each side. The logically correct answer is to check the 2 and the R card. The 2 must be checked to make sure that it has a vowel on the back, and the R must be checked to make sure that it does not. Participants using similar versions had great difficulty with this problem, answering it correctly between 4 and 25 percent of the time (Cosmides, 1989).

What bearing does this have on cheating? Not much, and that, argues Cosmides, is why people have such difficulty with the task. She proposed that problems of the same structure describing social contracts will be intrinsically much easier to solve, even beyond the fact that they are more familiar, concrete, and otherwise practical (see the 1989 paper for evidence supporting this claim). For instance, Griggs and Cox (1983) changed the rule to be, "If a person is drinking beer, then he must be over 21 years old," and the card labels were "beer," "soda," "25," and "19." In this configuration, 75 percent of participants were able to recognize that one must test the "beer" and "19" cards. While simply a variation on the Wason selection task, this particular variation uses socially relevant rules rather than arbitrary ones, and in general people have been shown to be very effective at recognizing violations of social contracts.

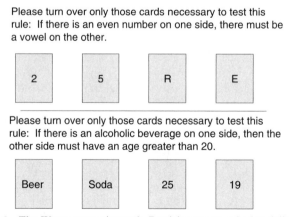

FIGURE 7.4   The Wason reasoning task. Participants are asked to follow the instructions in both panels of the figure. This demonstrates the difference between abstract reasoning and the identical problem placed in a socially relevant context.

This is not the same as moral reasoning. Participants in these studies are revealing an ability to detect violations of social contracts, not determine the moral validity of those contracts. Their performance in the social contract version of the Wason task demonstrates that they understand the socially agreed upon behavioral norms and are able to apply them to the behavior of others. It stands to reason that they are able to make the same judgments about their own prospective behavior. When applying this ability to their own behavior, as when contemplating an act that might be considered an integrity violation, a student is weighing the force of the social contract against the benefit accrued from breaking that contract.

In the case of classroom cheating, this means that students should be expected to indicate their belief that cheating is wrong, especially when others do it. They should be particularly sensitive to the unfairness of cheating when its social nature is emphasized. According to this view, students should recognize cheating as a problem most clearly and be disinclined to do it most strongly when they see it as a violation of the social contract among students and the community of scholars. They will be most inclined toward cheating when it does not violate a social contract, as when many of their peers are perceived to be (and may even be) cheating, or when given social permission in other ways. For example, teachers who leave the room while administering an exam may be seen as violating the social contract, as might parents whose help on assignments crosses the line from helpful to inappropriately collaborative. Once students see the social contract against cheating as violated, this neutralizes their internalized social prohibitions against cheating. They do not see the decision to copy another student's homework as one of academic integrity because the social prohibition against it is not in force.

RECONCILING THESE APPROACHES

Along with this evolutionary style of reasoning about cheating and fairness has come an increasing interest in the biological mechanisms that are responsible for it. Given that evolution acts on organisms, if one believes in an evolutionary account, one must expect to find neurological evidence for the processing of fairness. This is, in fact, the case, and it has interesting implications for academic integrity decisions. Two types of neuroscientific evidence will be brought to bear here. First is the report by Stone et al. (2002) of a brain injury patient who is able to reason normally, with the notable exception of the social exchange tasks described above. The second type of evidence comes from Sanfey et al. (2003), who used an fMRI (brain activity scanner) to study healthy people playing a social exchange game. They found that people who are perceived as taking advantage of a social contract are punished. While delivering the punishment, the punisher's brain shows an activity pattern that is often associated with negative emotions, including disgust. Taken together, this evidence lends support for the idea that the cognitive and emotional aspects of our integrity decision process are moderated by brain activities that are specialized for those purposes.

Stone and her colleagues (2002) studied a patient, R. M., who suffered damage to his brain's limbic system (orbitofrontal system, amygdala, and temporal pole). This is the area that handles, among other things, emotional reactions to our environment. They asked him and a set of control subjects to complete Wason task variants like those described above, in three conditions. They all completed descriptive problems, which had no social content and acted as control problems; precaution problems, which are more concrete and provide a pragmatic, not a social, explanation for the problem solver; and social contract problems, which activate the "cheater detection" mechanism. It is of great interest that R. M. performed normally on the descriptive and precaution problems when compared to control subjects (some healthy and others with brain injuries), but his performance on the social contract problems was much worse. His difference in percent correct between the precaution and social contract was 31 percent. This is staggering, compared to the 1.2 percent average difference for the controls. The authors use this evidence to stake a claim for the limbic system as responsible for managing cheating detection.

A related set of findings comes from the experimental literature. Sanfey et al. (2003) used fMRI imaging techniques to study healthy people in a situation called the Ultimatum Game. There are two players in the game. The first player is given an amount of money, often $10. He or she may choose to share this money with the second player in any ratio he or she chooses. The second player has two choices, to accept or reject this offer. If player 2 accepts, both players are paid according to the deal. If player 2 rejects the offer, neither player is paid. Economic theory suggests that player 1 offer as little money as possible, because it is in his/her own best interest to give up as little as possible and because player 2 should accept (preferring some small gain to no gain at all). This rarely occurs. In studies of the game, participants reject offers that are too low (20 percent of the total) roughly half the time (Roth, 1995, as cited in Sanfey et al., 2003). It is helpful to note that the players are made aware that this is a one-time game, because this behavior could be seen as a perfectly rational part of an ongoing negotiation in the context of a long run.

Why do people violate this seemingly obvious economic principle? The most likely hypothesis is that they believe that low offers are unfair (Pillutla & Murnighan, 1996). Thus they are punishing those players who treat them unfairly, even when it costs them money to do so. From a neuroscientific standpoint, this behavior is interesting because it points to a case where neuroimaging techniques can illuminate the separable cognitive and emotional factors that are presumably being weighed in order to reach a decision (Sanfey et al., 2003). As it turns out, the brain area most closely related to rejection of unfair offers is the anterior insula, which has been associated with negative emotions, including disgust. Moreover, the other area that is most active during the task is the prefrontal cortex (specifically the dorsolateral portion). This is an area often associated with executive function and decision processes, as described earlier. However, Sanfey, et al. (2003) found no relationship between the activity in this area and rejection of unfair offers, which led them to conclude that while this portion of the brain is crucial for decision making, it is probably being overshadowed by the emotional response in those cases. One possible inference from this finding is that participants had a viscerally unpleasant reaction to the offers (based on the perception of them as unfair) and responded accordingly by rejecting the offers and punishing the offending players who had made them.

Sanfey's (2003) finding that brain activity consistent with deliberation is happening concurrently with activity related to emotional memory provides a means of reconciling the emotional and deliberative approaches to academic integrity decision making. Decision makers are simultaneously calculating the utility of their options (cheating or not cheating) and receiving visceral cues. The utility calculation probably includes prospect theory concepts like decision weights and the value function as well as memory retrievals of similar circumstances. It might also include forecasts of both regret and disappointment should the final outcome not match their values. This would occur if they chose to reject any offer, since they would get no money. If these deliberations were the only processes, it is likely that the participants would have been much more likely to choose the option with the highest utility and accept the low payment. However, emotion does factor in. As Sanfey et al. found, when activity in the insula is greatest, their participants were most likely to reject unfair offers. In a similar vein, when a student is deciding whether or not to cheat, a somatic marker might steer the student away from such risky behavior. Thus both emotional and deliberative processes interact in complex ways to determine students' integrity decisions.

INTEGRITY THEORIES

Having discussed deliberative and emotional perspectives on decision making in general, we now turn to their application to integrity decisions. In particular, it will be useful to look at some of the major effects and theories in the literature and reconcile them with the generic decision theories above. This section examines the role of neutralizing attitudes, intrinsic and extrinsic motivation, penalties, and enforcement through the lens of decision theory.

Neutralizing attitudes, as discussed by Whitley (1998), refer to those attitudes that allow a student to engage in behavior that violates his or her ethical code. They can be conceptualized in the decision framework in three different ways. First, neutralizing attitudes can be seen as reducing the negative value of the cheating option. Violating one's ethical code would have some cost in terms of utility to all decision makers, with the amount varying based on individual differences. The strength of neutralizing attitudes represents one of the ways in which utility for cheating varies. At the simplest level, neutralization is a reduction in the disutility of violating one's moral code by cheating.

Second, neutralizing attitudes may affect the emotions involved in making integrity decisions. Self-disapproval (for lack of a better term) is an emotion that students may anticipate as a result of cheating. This would motivate some to avoid cheating. Neutralizing attitudes, however, can act to alleviate self-reproach, and if a student decision maker can assuage guilt with neutralizing attitudes, then holding those attitudes will serve to increase the incidence of cheating.

Neutralizing attitudes might impact the cheating decision in a third way, in those cases where moral codes are the first (or only) means of judgment. In that case, there is no chance that students will choose to cheat unless the code can be circumvented. Neutralizing attitudes allow that to happen and open the door for the other sorts of decision strategies that we have been discussing.

As discussed in other chapters in the present volume, student motivation is a crucial element in their cheating behavior. One of the most robust findings in the integrity literature is that students who are extrinsically motivated are more likely to cheat than those who are not, while the reverse holds true for intrinsically driven students (Anderman et al., 1998; Rettinger, Jordan, & Peschiera, 2004). These motivations can be seen as input to utility-based decision theories, and integrity decisions fall out logically. Intrinsic motivation means that utility is placed on knowledge, understanding, and the learning process. The disutility of a low grade is minimal, but not learning has a great deal of disutility. Cheating on an exam or plagiarizing an assignment would have very little utility in this case, because the gains on those tasks are measured in grades and the losses in learning. An intrinsically motivated student is searching for the opposite and thus would be unlikely to cheat. Naturally, the reverse is true for the extrinsically motivated student. Because grades are the main source of utility for these students, their behavior will seek to maximize them. Because learning and the educational process do not have as much value to a student in this state of mind, forgoing them will not be seen as a loss. From an emotional standpoint, these students will be highly motivated to reduce the disappointment that results from getting a lower grade than expected. The possible regret that would occur if they are caught cheating would be of less concern than that disappointment, which would also pre-dispose them to cheating and plagiarism behaviors.

FUTURE RESEARCH

This chapter calls into stark relief the opportunities to apply basic decision research techniques and theories to academic integrity decisions. It is possible to extend our theorizing about integrity decisions by applying domain-general decision-making principles and practices to them. Prospect theory, for example, provides a framework for evaluating students' assessments of the likelihood of being caught and their valuation of possible outcomes. Evolutionary approaches to fairness detection cause us to ask whether students feel that cheating, or not cheating (in the case of helping another student), is the more fair course of action. Our understanding of the roles of emotion in decision making can be applied to integrity decisions. Anticipating regret or disappointment could lead to either more or less cheating, depending on the context. A student's emotional state at study or exam time can also be expected to color his or her decision process. Emotion-based decision theories lend insight into such situations. Furthermore, a neuroscientific approach could be useful in highlighting and contrasting the roles of emotional and deliberative aspects of the decision process.

Turning first to prospect theory, it is easy to see that its general-purpose principles can be fruitfully applied to many aspects of the integrity literature. For example, it would be interesting to see how students value grade points generally. It is reasonable to hypothesize that the value function in prospect theory still applies, although this is an empirical question. Furthermore, the effect of grading systems on the grade value function might be an area for exploration. The decision weight function should also be studied in the integrity context. Do students, for example, round the probability of being caught cheating down to zero in cases when it seems remote? If so, this alone might have a significant impact on cheating behavior.

There are standard research methodologies (e.g., Camerer, 1989) for finding the appropriate parameters of the prospect theory (or utility theory) functions. The basic strategy is to present a series of similar choices to participants, varying either the values or the likelihoods of occurrence for at least one outcome. This provides researchers with data necessary to understand the value and decision weight functions that are being brought to bear on a decision. Another method is one popularized by Tversky and Kahneman (1973, 1981), who presented different groups of participants with functionally identical decisions, varying only in their description. The classic Asian disease problem is an example of this method. As you may recall, all participants were asked to choose between a risky and a sure response to the disease. Half read about these options described in terms of deaths, and the other half in terms of people saved, although the options were actually identical. By observing the difference between these frames, Tversky and Kahneman were able to find support for the reference point in prospect theory.

For our purposes, prospect theory can be used as a framework to conceptualize the factors that the integrity literature has shown to affect students' choices, like learning/grade orientation and neutralizing attitudes. While these constructs clearly have been shown to be associated with differences in cheating behavior, prospect theory allows us to see the mechanism leading to those differences within the minds of individual students. For instance, neutralizing attitudes can be conceived of as simply increasing the value of cheating directly. Alternatively, neutralization might be a means of changing the negative value that a student places on being caught by giving him or her a rationalization. By presenting students with a series of slightly varied hypothetical cheating decisions or by using an Asian disease method variant, it would be possible to determine which of these theoretically different accounts of neutralization is a stronger explanation. Learning and grade orientations can be theorized about and tested in the same way. It seems likely that these orientations manifest themselves in different values for certain outcomes, such as increased value for grades by a student with a grade orientation.

Prospect theory reference points may play a large role in integrity decisions. As Schweitzer et al. (2004) point out, a decision maker's goals can lead to cheating if the difference between meeting and not meeting them is great enough. A series of cheating decisions that vary only by the magnitude of this difference would allow us to see the effect of goals on cheating. A similar approach would be effective for determining whether cheating decisions are affected by temporal discounting. The farther removed a decision maker is from the consequences (both good and bad) of cheating, the less we would predict that those consequences would be factored into the decision. In short, prospect and utility theory methods and concepts can be applied to the theoretical integrity literature in endless ways, with the goal of demonstrating what effects each construct has on an individual's integrity decision process.

Further research on emotion-based decision approaches to integrity decisions would also be fruitful. For example, given the finding that one's emotional state acts as a decision cue, it would be interesting to see if hypothetical cheating decisions could be influenced by mood induction. One would predict that fearful students would apply their negative assessments of risk to cheating, and opt not to do so, while students who are more angry or frustrated would seek to alter the status quo by cheating.

Anticipated regret and disappointment are promising factors for exploration in future research. While it is straightforward to stay that students may anticipate regret at being caught cheating, it is also possible that a student might regret not cheating in certain circumstances. One who feels taken advantage of by classmates who cheat might anticipate regret for not cheating. Others might regret being denied admission to medical school, for instance, much more than compromising their integrity. As we discussed earlier, anticipated disappointment principles create an explanation for why good students who seemingly have much to lose and little to gain by cheating would still do so. Vignette research on this topic would help determine whether this interpretation does in fact hold. Although previous research (Rettinger, Jordan, & Peschiera, 2004) indicated that students expecting to do worse were considered higher cheating risks, a vignette study that specifically manipulates anticipated disappointment might lead to different results.

A particularly intriguing possibility in the area of emotions and integrity decisions comes from Damasio's use of GSR (galvanic skin response) methods. Previous use of this method demonstrated that orbitofrontal patients have problems making decisions because their emotional responses to bad decisions are absent. While it would be impractical to present integrity decisions to this specialized population, it would be fascinating to find students who admit to previous integrity infractions and compare their GSR responses to noncheaters when exposed to integrity decisions in the lab. We might predict a smaller emotional response by cheaters than by noncheaters. Given the later development of the frontal lobe of the brain, a comparison of high school and college students in a study such as this one would be of particular interest.

Finally, taking a cue from the evolutionary approach to decision making, we see that an examination of the role of fairness in cheating decisions is needed. Cosmides (1989) pointed out the fundamental role that fairness and its detection play in all sorts of tasks. Integrity decisions are a likely domain in which this issue would emerge. The fundamental question to be answered empirically is, "Do students who think of cheating in terms of fairness act differently?" The answer cuts more than one way. As research on neutralization tells us, students consider *not* helping a fellow student to be unfair in some cases. Appropriate research could therefore point to those cases in which cheating is determined to be fair rather than unfair, and shed light on whether that determination influences cheating behavior.

As this section makes abundantly clear, the application of decision theory to integrity decisions is an intellectually diverse endeavor with wide-ranging theoretical implications. In the future, we may expect to understand the mechanisms that allow motivation, neutralizing attitudes, emotions and fairness to determine why a student cheats or plagiarizes in a particular instance. This understanding can help direct teachers, parents, and policymakers in the fight for academic integrity.

PRACTICAL ISSUES

Enforcement and penalties are the major tools used to combat violations of integrity code. How they are perceived by students contemplating integrity decisions has a substantial influence on how those decisions are approached and ultimately made. As mentioned earlier, very low probability events are often ignored, whereas the likelihood of slightly higher probability events is overestimated. This means that from a practical standpoint, increasing students' perception of the probability of being caught is a powerful tool in preventing cheating. In that context, a modest increase in the perceived chance of being caught will have the greatest effect if students initially believed they had no chance of being caught. Given the power of availability and salience to influence probability judgments, we should strive to make students aware of cases of integrity violations as much as possible while maintaining privacy.

From the perspective of prospect theory, we must strive to help students get away from thinking about an honestly earned grade as a "loss," no matter what that grade is. With that negative frame, they are likely to be risk seeking in avoiding that sure loss. Classes that are scored by accumulating points for assignments and success on exams, rather than as percentages of total possible points, can contribute to setting a gain frame for students, which will help them see cheating as a risk to be avoided. Students who view the current grade as the status quo, or even as points earned (gained) throughout each course, will on average be risk averse. In the same vein, the penalties for cheating or plagiarizing must represent a loss in all senses. Students must lose time and points and must be made to feel regret as a result of an integrity violation. These consequences should be made plain early and often, with the preventative goal of reducing the expected utility of cheating.

At an educational institution, some might argue that revolutionizing the integrity culture, not intervening with individual decision makers, is the key to decreasing academic dishonesty on campus. This is a false dichotomy. A culture is built of individuals, and in order to create a persuasive, communitywide message, it is crucial to understand how each member of that community is most effectively persuaded. Clearly, an individual's values (i.e., utilities) and history play a fundamental role in his or her choices. For individual integrity decisions (and thus the campus integrity culture) to shift fundamentally, the values that students bring to those decisions must be affected. Decision theory provides us with a framework for understanding how students make integrity decisions. From that basic knowledge we can predict and implement the most effective ways to make the case for integrity and the educational process, which is the first step in a more widespread sea change in cheating in academic settings.

REFERENCES

Anderman, E. M., Greisinger, T., & Westerfield, G. (1998). Motivation and cheating during early adolescence. *Journal of Educational Psychology, 90*(1), 84–93.

Bechara, A., Damasio, A. R., Damasio, H., & Anderson, S. W. (1994). Insensitivity to future consequences following damage to human prefrontal cortex. *Cognition, 50,* 7–15.

Bechara, A., Damasio, H., & Damasio, A. R. (2000). Emotion, decision making and the orbitofrontal cortex. *Cerebral Cortex, 10*(3), 295–307.

Camerer, C. (1989). An experimental test of several generalized utility theories. *Journal of Risk and Uncertainty, 2,* 61–104.

Clore, G. L., Schwarz, N., & Conway, M. (1994). Cognitive causes and consequences of emotion. In G. L. Clore, N. Schwarz, M. Conway, R. S. Wyer, & T. K. Srull (Eds.), *Handbook of social cognition: Vol. 1: Basic processes.* Hillsdale, NJ: Lawrence Erlbaum.

Cosmides, L. (1989). The logic of social exchange: Has natural selection shaped how humans reason? Studies with the Wason selection task. *Cognition, 31,* 187–276.

Critchfield, T. S., & Kollins, S. H. (2001). Temporal discounting: Basic research and the analysis of socially important behavior. *Journal of Applied Behavior Analysis, 34,* 101–122.

Damasio, A. R. (1994). *Descartes' Error: Emotion, Reason, and the Human Brain.* New York: Putnam.

Dweck, C. S. (1986). Motivational processes affecting learning. *American Psychologist, 41*(10), 1040–1048.

Dweck, C. S., & Leggett, E. L. (1988). A social-cognitive approach to motivation and personality. *Psychological Review, 95*(2), 256–273.

Eison, J. A. (1981). A new instrument for assessing students' orientations towards grades and learning. *Psychological Reports, 48,* 919–924.

Eison, J. A., Pollio, H. R., & Milton, O. (1986). Educational and personal characteristics of four different types of learning- and grade-oriented students. *Contemporary Educational Psychology, 11,* 54–67.

Elster, J. (1998). Emotions and economic theory. *Journal of Economic Literature, 36,* 47–74.

Forgas, J. P. (1995). Emotion in social judgments: Review and a new affect infusion model (AIM). *Psychological Bulletin, 117,* 39–66.

Frederick, S., Loewenstein, G., & O'Donoghue, T. (2002). Time discounting and time preference. *Journal of Economic Literature, 40*(2), 351–401.

Griggs, R. A., & Cox, J. R. (1983). The effects of problem content and negation on Wason's selection task. *Quarterly Journal of Experimental Psychology, 35A,* 533.

Hertel, G., Neuhof, J., Theuer, T., & Kerr, N. L. (2000). Mood effects on cooperation in small groups: Does positive mood simply lead to more cooperation? *Cognition and Emotion, 14*(4), 441–472.

Higgins, E. T. (1997). Beyond pleasure and pain. *American Psychologist, 52*(12), 1280–1300.

Johnson, E. J., & Tversky, A. (1983). Affect, generalization and the perception of risk. *Journal of Personality and Social Psychology, 45,* 20–32.

Kahneman, D., & Tversky, A. (1984). Choices, values, and frames. *American Psychologist, 39*(4), 341–350.

Kühberger, A. (1998). The influence of framing on risky decisions: A meta-analysis. *Organizational Behavior and Human Decision Processes, 75*(1), 23–55.

Lerner, J. S., & Keltner, D. (2000). Beyond valence: Toward a model of emotion-specific influences on judgment and choice. *Cognition and Emotion, 14*(4), 473–493.

Mellers, B., Schwartz, A., & Ritov, I. (1999). Emotion-based choice. *Journal of Experimental Psychology: General, 128*(3), 332–345.

Mischel, W., Ebbesen, E. B., & Zeiss, A. R. (1972). Cognitive and attentional mechanisms in delay of gratification. *Journal of Personality and Social Psychology, 21,* 204–218.

Payne, J. W., Bettman, J. R., & Johnson, E. J. (1993). *The adaptive decision maker.* Cambridge: Cambridge University Press.

Pillutla, M. M., & Murnighan, J. K. (1996). Unfairness, anger, and spite: Emotional rejections of ultimatum offers. *Organizational Behavior and Human Decision Processes, 68,* 208–224.

Rettinger, D. A., Jordan, A. E., & Peschiera, F. (2004). Evaluating the motivation of other students to cheat: A vignette experiment. *Research in Higher Education, 45*(8), 873–890.

Sanfey, A. G., Rilling, J. K., Aronson, J. A., Nystrom, L. E., & Cohen, J. D. (2003, June 13). The neural basis of economic decision-making in the ultimatum game. *Science, 300,* 1755–1758.

Savage, L. J. (1954). *The foundations of statistics.* Wiley: New York.

Schwarz, N. (1990). Feelings as information: Informational and motivational functions of affective status. In E. T. Higgins & R. Sorrentino (Eds.), *Handbook of motivation and cognition: Foundations of social behavior* (pp. 527–561). New York: Guilford.

Schwarz, N. (2000). Emotion, cognition and decision making. *Cognition and Emotion, 14*(4), 433–440.

Schwarz, N., & Clore, G. L. (1996). Feelings and phenomenal experiences. In E. T. Higgins & A. Kruglanski (Eds.), *Social psychology: Handbook of basic principles* (pp. 433–465). New York: Guilford.

Schweitzer, M. E., Ordóñez, L., & Douma, B. (2004), Goal setting as a motivator of unethical behavior. *Academy of Management Journal, 47*(3), 422–432.

Stone, V. E., Cosmides, L., Tooby, J., Kroll, N., & Knight, R. T. (2002). Selective impairment of reasoning about social exchange in a patient with bilateral limbic system damage. *Proceedings of the National Academy of Sciences, 99*(17), 11531–11536.

Treviño, L. K. (1986). Ethical decision making in organizations: A person situation interactionist model. *Academy of Management Review, 11*(3), 601–617.

Tversky, A., & Kahneman, D. (1973). Availability: A heuristic for judging frequency and probability. *Cognitive Psychology, 5,* 207–232.

Tversky, A., & Kahneman, D. (1981). The framing of decisions and the psychology of choice. *Science, 211,* 453–458.

Von Neumann, J., & Morgenstern, O. (1947). *Theory of game and economic behavior,* 2nd ed. Princeton, NJ: Princeton University Press.

Wason, P. C. (1983). Realism and rationality in the selection task. In J. StB. T. Evans (Ed.), *Thinking and reasoning: Psychological approaches.* London: Routledge & Kegan Paul.

Whitley, B. E. (1998). Factors associated with cheating among college students: A review. *Research in Higher Education, 39*(3), 235–274.

Woessner, M. C. (2004, April). Beating the house: How inadequate penalties for cheating make plagiarism an excellent gamble. *PS-Political Science and Politics,* 313–321.

Wryobeck, J. M., & Whitley, B. E. (1999). Educational value orientation and peer perceptions of cheaters. *Ethics & Behavior, 9*(3), 231–242.

Zeelenberg, M., van Dijk, W. W., Manstead, A. S. R., and van der Pligt, J. (2000). On bad decisions and disconfirmed expectancies: The psychology of regret and disappointment. *Cognition and Emotion, 14*(4), 521–541.

MORAL AND SOCIAL MOTIVATIONS FOR DISHONESTY

8

REAPING WHAT WE SOW: CHEATING AS A MECHANISM OF MORAL ENGAGEMENT[1]

THERESA A. THORKILDSEN,
COURTNEY J.GOLANT, AND L. DALE RICHESIN

Many successful people attribute their accomplishments, in part, to valuable educational experiences. They tell stories of putting forth high effort under the watchful eye of teachers, mentors, or tutors and of receiving rewards for the fruits of their labor. Like individuals who have not achieved high levels of personal and professional success, highly accomplished people can also offer stories of when they were tempted or enticed into cheating. If even successful individuals sometimes view academic dishonesty as beneficial, controversial questions emerge about when cheating is and is not likely to occur and whether individuals benefit from taking dishonest shortcuts in their academic work. Some investigators, for example, have asked whether students assume that cheating is always wrong (Guttman, 1984; Johnson, Hogan, & Zonderman, 1981; Leming, 1978; Turiel, 1983), defined conditions under which it would be adaptive to cheat (Johnson, 1981), and questioned definitions of such transgressions across settings (Turiel, 1983). Others have focused on when and why dishonest behavior might cause harm to communities and how members might respond when cheating is detected (Dienstbier, 1995; Dienstbier, Kahle, Willis, & Tunnell, 1980; Johnson et al., 1981).

[1] We would like to thank the students and faculty of Togiak School and Elmwood Park High School for their contributions to this project. Correspondence concerning this article should be addressed to Theresa A. Thorkildsen, Education and Psychology, M/C 147, 1040 West Harrison, University of Illinois at Chicago, Chicago, IL 60607–7133, e-mail: thork@uic.edu.

Copyright © 2007 by Academic Press, Inc.
All rights of reproduction in any form reserved.

These questions remain controversial in large part because investigators have not agreed on a universal definition of cheating and have found that most individuals are likely to engage in some forms of dishonesty when unavoidable situational pressures clash with personal agendas (Hartshorne & May, 1928). When external requirements threaten students' feelings of competence, for example, especially when the consequences of failure seem far reaching, students may decide that cheating is their only recourse. Many individuals will also be dishonest and justify their behavior as situationally appropriate, even in experimental settings where there are no long-term consequences and cheating is simply an expedient means of completing a task. Different definitions and the resulting empirical findings highlight tensions between personal interests and respect for communal values.

Cheating is not the only way in which moral functioning is detected in school, but it highlights an especially complex means by which adolescents define and act on their experience. To better understand this phenomenon, we tested a model of moral engagement that accounts for why some individuals cheat even though such behavior can be harmful. Our model also clarifies somewhat vague references to *personal expectations* or *aspirations* found in many theories of motivation. Using survey results, we illustrate how the personal standards of adolescents' engagement can compel them to work hard or to take shortcuts and in turn affect the quality of their performance (Figure 8.1). Our model is predicated on the idea that motivation is an intentional system that includes choices and actions (Lewin, 1951; Nicholls, 1989; Weiner, 1979), and we extend earlier work on agentic cycles of success and failure responses to include more enduring personal standards.

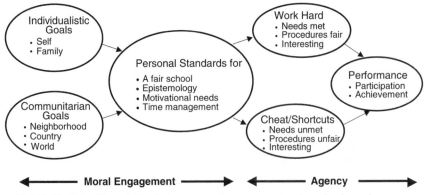

FIGURE 8.1   The relation between moral engagement and academic performance.

In this model, expectations that are central to engagement differ from those that drive agency, but the two are often coordinated in the motivational force. In this view, success and failure do not depend on outcomes alone, but on the relation between outcomes and the agent's aspirations and expectations (Lewin, 1936). Terms like *aspirations* and *expectations* reflect the whole of someone's functioning and represent the integration of engagement and agency mechanisms. Such general motivational forces are formed out of deeply rooted values, in response to the parameters of particular situations, or by the coordination of these general and specific reactions. In school, the nature and form of adolescents' effort are affected when they detect the likelihood that their expectations will remain unmet and cheating may seem to be one of several options for fulfilling task demands.

Whereas most research on motivation emphasizes proximal decisions such as how to do well on a particular task, our evidence suggests that adolescents' life goals and personal standards for how an ideal school ought to be organized are part of a more holistic intentional force that guides their academic behavior. Individuals and groups whose life goals emphasize personal advancement may be more inclined to cheat in situations where they could maintain personal esteem and avoid detection. Those who are preoccupied with forming and sustaining communal values may be more likely to cheat when dishonesty preserves group boundaries, trust among ingroup members, the exclusion of outgroup members, and group norms. Students' commitments to individualistic and communitarian values may also foster negative views about cheating as well as different visions of an ideal school.

In our model, adolescents are responsible agents who decide if and when to cheat. Their general values are part of *engagement*, a knowledge-driven force comprised of thoughts, emotions, and reactions situated within each agent (Bandura, 1999; Thorkildsen, 2004). Indicators of engagement are apparent when adolescents report internal standards such as those associated with fairness and learning, but such beliefs may or may not foster action. Only after adolescents' sense of agency is activated will we see that internal standards are translated into intentions and choices. *Engagement* involves active thoughts about how the world does and ought to function, and *agency* involves specific intentions. Both engagement and agency beliefs guide adolescents' behavior and affect their success and failure experiences.

Although our research emphasizes the structural characteristics of moral engagement, practical inferences can be drawn by comparing our findings with existing research on academic dishonesty. Therefore, in addition to justifying our current moral engagement model, we will offer prevention-focused suggestions on how to foster constructive beliefs about cheating. We also offer ideas about how educators might respond once cheating is detected.

MORAL AND ACADEMIC ENGAGEMENT

Engagement is complex in that individuals organize their experiences to form personal standards that are part of a force that also includes physical and emotional reactions. Because these aspects of engagement offer support for someone's sense of agency and guide personal decisions about when and how to act, adolescents' vision of how they *could* act becomes agency only if they formulate intentions and actually carry out their vision. Other problems are likely to involve different intentional models, but, to study cheating, we began by identifying the structure and form of standards that seem to affect students' academic participation.

WHY TWO TYPES?

In educational settings, we distinguish between two types of engagement, moral and academic, to illustrate the difference between students' knowledge about interpersonal and intrapersonal factors. *Moral engagement* involves questions of justice, ethical conduct, and reactions to interpersonal circumstances. When adolescents decide to cheat because they assume that schools are inherently unfair and teachers will discriminate against them, their sense of moral engagement is likely to be activated. Such individuals rely on an ethical vision of how the world ought to function and use their personal standards to regulate interpersonal behavior. *Academic engagement* involves questions of personal achievement and regulates the acquisition and use of new knowledge. Adolescents who cheat simply to earn higher test scores may have little regard for the effect of their behavior on a community and focus instead on their internal struggle to seem competent. Academic engagement regulates individualistic approaches to intellectual tasks because adolescents use personal achievement standards to formulate a plan for attaining particular performance outcomes and are not concerned about how their behavior affects others.

Cheating because institutional practices are unfair or because one desires to get high test scores suggests that adolescents are engaged rather than disengaged; each person has imagined attainable goals and evaluates the implications of those goals in light of either ethical or academic agendas. This point may be easier to see if you imagine how students might protest against particular activities. Protesters who are engaged with an issue will outline why their arguments are better than those of others, evaluate both agreeable and disagreeable ideas, and sanction themselves when they are tempted to behave in ways that are incongruent with their internal standards.

Protesters can also be disengaged. Disengaged protesters would knowingly perpetuate moral harm or willfully avoid intellectual opportunities while consciously absolving themselves of responsibility for these decisions

(Bandura, 2004); they are likely to cause harm or otherwise exhibit behavior that deviates from their personal standards without experiencing ordinary feelings of self-condemnation. They form goals but do not connect their actions and values. Disengaged cheaters, such as those who believe they want to do well in school but consciously avoid opportunities to learn, would not acknowledge the incongruity between their internal standards and their avoidance behavior. Cheaters or protesters who damage property as a means of objecting to a lack of opportunity seem unable to notice that they have contributed to the problem and do not align their general ethical beliefs with their goal-directed actions.

Students can also be apathetic, in which case they will not formulate clear personal standards even though they can generate and describe more fleeting mental representations when asked to do so. These students may detect moral and intellectual challenges but lack interest and avoid exerting any kind of effort. Their behavior is not guided by internal standards because the formation and activation of such standards require at least a basic commitment to thinking about available opportunities. Apathy is apparent when adolescents avoid completing tasks, forming consistent goals, or sustaining personal commitments. Students with such motivational states may or may not complete those tasks assigned to them and, even when they do, are not likely to use such activities to organize a life for themselves. Their behavior is disorganized in large part because their mental models of the world remain so.

Our studies of cheating did not include adolescents who expressed apathy or disengagement because students with these forms of motivation were unwilling to participate or offer meaningful responses. Despite this limitation, we can learn much about cheating by taking seriously the views of adolescents who voluntarily reveal their ideas. Adolescents indicated that moral engagement definitely plays a role in their academic performance, and despite individual differences in the content of students' beliefs, detectable structural features were articulated. Adolescents' moral engagement in school was also associated with their agency to work hard and cheat, but at least two organizational pathways were detected. Finally, it was possible to identify a few mechanisms by which adolescents excuse their dishonest behavior.

EMPHASIZING MORAL STANDARDS

We emphasize moral engagement when studying cheating because dishonesty can harm individuals as well as communities, but in educational contexts, learners often synthesize standards of moral and academic engagement. When thinking about how to teach controversial and noncontroversial science topics, for example, many adolescents reported different views on how the two topics ought to be taught (Thorkildsen, Sodonis, & White-McNulty, 2004). They were asked to think about whether various practices promoted learning (academic engagement) and whether these practices would foster a fair classroom (moral engagement). Everyone's personal standards contained references to justice, epistemology, motivation, and learning strategies, but individuals coordinated these differently depending on the nature of the situation. This was also apparent when children and adolescents explored fair contest practices and determined how learning, test, and contest situations ought to be prioritized (Thorkildsen, 2005; Thorkildsen & White-McNulty, 2002). Together, these studies suggest that adolescents easily move between moral and academic priorities when organizing their beliefs about school, but our emphasis on moral engagement acknowledges that dishonesty typically calls forth moral standards.

A WORKING DEFINITION OF
MORAL ENGAGEMENT

We have theorized about moral engagement in educational settings by offering descriptions of students' personal standards and drawing inferences about how these standards are used to make meaning of common classroom events. Children and adolescents have been reliable informants on how an ideal classroom should be organized, why everyone should work toward particular curricular ends, how best to meet the needs of different learners, and which strategies ultimately foster personal and academic success. Nevertheless, this work only touches the surface of students' beliefs about justice as well as the ethics associated with forming standards for epistemology, identity, and learning.

One possible cause of this knowledge shortage is a resistance to exploring students' understanding of the moral implications of education, a bias that is also evident in how teachers converse with adolescents. Educators and students commonly discuss the knowledge and emotions associated with academic engagement, but students' moral engagement is commonly left to evolve through a process of habituation. Students who can easily habituate to educators' moral expectations and imagine a meaningful role for school in society are likely to find their sense of engagement rewarded because they can use these values, be they pro- or anti-school in nature, to activate their agency beliefs. Those who are less introspective may feel isolated and apathetic or actively nurture their own disengagement (Bandura, Barbaranelli, Caprara, & Pastorelli, 1996; Dishion & Owen, 2002; Graham, 2004; Pellegrini, Bartini, & Brooks, 1999; Ryan, 2001). In responding to the knowledge shortage, we found two emergent dimensions to be central features of moral engagement in school: adolescents' life goals and beliefs about how an ideal school ought to be organized.

The Structure and Importance of Life Goals

Life goals such as those associated with improving one's personal and familial relations or contributing to the betterment of humankind are important features of engagement because they offer a guiding sense of purpose. The distinction between becoming a virtuous person who looks out for personal interests and becoming a virtuous citizen who participates in collective activities has been with us at least since Aristotle (1985). Individualistic and communitarian goals can be expressed by focusing on proximal agendas such as how to organize daily activities or distal agendas in which future ends are represented as something to aspire to. Proximal goals such as how to contribute to a classroom community may feel like personal resolutions, whereas distal goals such as how to become an active citizen may feel like hopeful daydreams, and students may differ in their willingness to imagine and articulate either type of goals.

Despite differences in students' willingness to articulate life goals, everyone's behavior is affected by their vision of how immediate activities align with future outcomes (Weiner, 1979). Some students seem preoccupied with their personal success, whereas others are also interested in contributing to collective agendas. These personal values flourish independently of individuals' ability to articulate them, but are most likely to foster personal well-being when they are used to select life tasks that facilitate participation in social groups (Cantor, Kemmelmeier, Basten, & Prentice, 2002). Life goals differ from personal standards as to how an ideal school ought to be organized, but are part of the force that compels or undermines students' participation in school activities and helps individuals sustain educational commitments even when faced with unpleasant situations or disappointments.

Standards for an Ideal School

In the early 1980s, when concerns about equal educational opportunity were dominant in conversations about education, at least one researcher began asking questions about the effects of these conversations on students' motivation (Thorkildsen, 1994; Thorkildsen & Nicholls, 1991). Before describing the role of fairness beliefs in motivation, it seemed logically necessary first to verify the structure of students' beliefs about an ideal school, and then to determine if these beliefs differ across age groups or reflect individual differences in attitudes. To do this evaluation, two lines of research were evaluated: one which relied on methods commonly used in social psychology and the other which relied on methods commonly used in developmental psychology. The social psychological approach focused on individual differences and involved identifying central constructs found in the research on equal educational opportunity; conducting focus group discussions of these themes; and designing surveys for children, adolescents, and young adults (Thorkildsen, 1989c; Thorkildsen &

Nicholls, 1998). The developmental approach involved mastering the Piagetian method of critical exploration, designing several fairness puzzles to administer to students, and documenting the structure of students' reasoning across different age groups (Thorkildsen, 1989a, 1989b, 1991, 1993; Thorkildsen & Schmahl, 1997).

Results obtained using methods from social psychology were inconclusive (Thorkildsen, 1989c). Fairness was seen as a controversial topic, but methods from social psychology did not foster new knowledge about the conditions under which students were affected by activities promoting equal educational opportunity. Only 1 percent of students in grades 9, 11, and college (n = 314, 276, and 258, respectively) asserted that they always found schooling to be fair and that the practices used in school were fair to all students. Similarly, only 2 percent of these students said that school was never a fair place either for themselves or other students. Most of these students said they thought about fairness often (72 percent) and got very upset when noticing unfair events (77 percent). Furthermore, a large number of these students reported talking regularly about fairness with their friends (63 percent) and noticing when students did something unfair (65 percent). Similarly, research with fifth graders suggested that preteens are sensitive to the diverse needs of students with high and low ability, but they assumed that teachers should not consider ability differences when deciding how to best help students learn (Thorkildsen & Nicholls, 1998). Together, these findings suggest that students are likely to see schooling as fair for some of the students some of the time, but this research offers little information on how students' fairness beliefs correspond with personal standards about an ideal school.

In contrast, meaningful and replicable findings were apparent in structured interviews, and from that work it is now possible to identify systematic ideas found across age groups (6 to adult). For our study of motivation and cheating, we wrote survey items to represent visions of an ideal school evident in earlier conversations with adolescents (ages 10 to 18). A brief review of Thorkildsen's earlier developmental work illustrates the origin of these ideas, although it is helpful to remember that references to age serve only to illustrate the point at which particular ideas are common and should not be seen as evaluation benchmarks.

First, it is important to remember that adolescents' beliefs about an ideal school reflect their understanding of why they participate in such institutions. These beliefs differ depending on whether individuals consider learning, test, or contest situations (Thorkildsen, 1989b) and involve the coordination of justice, epistemology, identity, and aesthetic concerns (Thorkildsen, 2005). Despite these commonalities, there are age-related differences in how students define fairness, which types of knowledge they find most valuable, and their understanding of motivational and identity themes.

Studies of reasoning about learning situations, for example, suggest that ages 10, 12, and 16 are typical points at which reasoning becomes more complex (Thorkildsen, 1989a, 1993; Thorkildsen & Schmahl, 1997). Prior to age 10, most children justify their decisions by focusing either on the ratio of work to playtime or on how many assignments each student completes. They assert that, regardless of students' abilities, fair practices should ensure that all distributions of such concrete entities are equal. Between ages 10 and 16, students assume that everyone can and should learn the same things and that fair practices should ensure this kind of equality. Around age 16, adolescents start to acknowledge that individual differences in performance may reflect more enduring talents and tastes and begin to think about equity in the distribution of learning opportunities. High school students also incorporate complex representations of epistemology, motivation, and other identity dimensions into their justifications for how educational goals ought to be attained (Thorkildsen, Sodonis, & White-McNulty, 2004).

Age trends for testing situations show a different pattern (Thorkildsen, 1991), and students report different purposes for testing (Thorkildsen, 2000). Before about age 8, many children define copying as legitimate and see this as one means by which all students can attain the same test scores. They assume that teachers can determine which students understand schoolwork, even if they help one another, because teachers know what students are capable of doing. Between about ages 8 and 10, most students assume that fair tests will ensure that everyone scores equally well even though individuals do their own work. These children describe individual differences in attainment as reflecting differences in effort and support rather than ability. By about age 12, most students fully understand that tests are intended to detect individual differences in performance, and they judge practices that undermine the accuracy of such evaluations to be unfair.

Regardless of these age-related patterns, children differ in their beliefs about how much testing is fair (Thorkildsen, 2000). Some students equate testing and learning, assuming that one process will not occur without the other; other students see tests as a form of feedback that can facilitate learning; and a third group seems to assume that tests interfere with learning because they offer a paper trail of mistakes. As is the case for learning situations, adolescents have also incorporated themes of epistemology, identity, and aesthetics in such decisions about fair testing (Thorkildsen, Sodonis, & Weaver, 2005).

Age differences in reasoning about contest situations follow yet another trajectory, although most children and adults assume that contests should not be common in a fair school (Thorkildsen & White-McNulty, 2002; Thorkildsen & Weaver, 2003). Prior to age 12, students assume that luck-based decision making such as pulling winners' names from a jar would be a fair way to judge a science contest. The youngest students assume that skill and luck-based decisions are interchangeable, but after about age 8 students distinguish both types of contests and plausible reasons for each. On the other hand, 12- to 14-year-olds are likely to say that luck-based academic contests are unfair because the manifest purpose of such contests is to evaluate achievement. Like adults, older students show no difficulty defining lotteries as fair forms of entertainment, but they assume that academic contests should reward talent.

Age patterns were also apparent when students considered which types of knowledge should be emphasized in school and how teachers ought to teach each type (Nelson, Nicholls, & Gleaves, 1996; Nicholls & Nelson, 1992; Nicholls, Nelson, & Gleaves, 1995; Nicholls & Thorkildsen, 1988, 1989). Younger students are learning to distinguish the parameters of various tasks and see school as a place that should emphasize correct answers. Older students are more interested in controversial topics and offered richer elaborations on how opinions should be respected in school. To most students, controversial topics and matters of logic should be taught using discovery-focused methods, whereas constitutive and regulative conventions should be taught in a more didactic style.

The complexity of these beliefs suggests that students are unlikely to offer the same vision of an ideal school. It also suggests that students who appear to cheat may not be intentionally avoiding their intellectual responsibilities. Behavior that educators define as cheating could occur because students do not fully understand the norms and demands of activities in which participation is obligatory.

ENGAGEMENT IS NOT AGENCY

Although forms of engagement and agency are often highly integrated, individuals do not always coordinate these two aspects of motivation. Individuals may formulate personal standards without feeling the necessary levels of empowerment to act on their beliefs. One second grader, for example, assumed that school was a place that did not respect his intellectual interests; he developed a rich intellectual life at home but would not share his interests in school (Thorkildsen & Nicholls, 2002). Other individuals may behave in ways that suggest a sense of agency, but lack the necessary self-understanding to formulate more enduring personal standards. This was evident for another second grader who did not reveal clear agendas and whose social and intellectual behavior in school lacked consistency (Thorkildsen & Nicholls, 2002). Both of these second graders fractured their intellectual lives but differed in whether they experienced low levels of agency or engagement. Such gaps in students' functioning suggest the value of distinguishing engagement and agency, even if, ideally, both are integrated in a more general motivational force.

On average, moral engagement, agency, and academic performance seem to be integrated in the minds of adolescents, but their sense of agency is most directly associated with their proximal goals as well as success and failure on particular tasks. This led us to define *engagement* as students' general understanding and willingness to participate in activities and *agency* as students' specific aspirations. Claims from expectancy-value frameworks and our ethnographic accounts of students' lives suggest that academic progress is controlled as much by students' specific aspirations as by their more general educational standards (Eccles et al., 1993; Lewin, 1936; Thorkildsen & Nicholls, 2002; Wigfield & Eccles, 2000), and our exploration of motivation to cheat accounts for this level of organization.

CHARACTERISTICS OF AGENCY

Individuals' sense of agency has most often been represented as an inference drawn from observations of behavior; when individuals act in certain ways, agency is assumed to be present, and when they fail to act, agency is assumed to be absent (Bandura, 1977). Researchers who try to synthesize research findings find that agency is multifaceted and dependent on the parameters of particular situations. We deviated from this observational norm in our cheating study because adolescents are reliable informants on the nature of their experience. Agency, in this work, is represented using adolescents' beliefs about when they feel smart and inclined to work hard or cheat and take shortcuts; each form of agency is comprised of three goal-specific dimensions.

Agency to Work Hard

Agency is often equated with effort when researchers rely on observational measures. The nature and meaning of effort can be quite controversial, but three dimensions are especially consistent with the quest to understand moral functioning. We have found that students will work hard to fulfill their basic needs for competence, self-determination (or power), and affiliation. Students also work hard when they assume that fair procedures will allow them to attain their goals as well as when they find the task interesting and personally meaningful. Therefore, our survey contained items that reflect all three agency dimensions.

More specifically, the assumption that adolescents will work hard to meet basic needs emerges from evidence that individuals formulate positive agency beliefs when their effort leads to success (Atkinson, 1958). Three primary needs for competence, power, and affiliation, represented in work with adults, are also evident in modified forms among children (Thorkildsen & Nicholls, 2002). Some researchers assume that the competence needs form a superordinate category within which achievement striving, relatedness, and self-determination are subordinate dimensions

(Deci & Moller, 2005; Skinner, Zimmer-Gembeck, & Connell, 1998), but we see three distinct needs that evolve from individuals' emotional, cognitive, and behavioral functioning. In our work, students' *need for competence* has been comprised of a profile including task orientation in which individuals feel successful when they master particular tasks, and ego orientation in which individuals feel successful if they show superiority over others or avoid appearing inferior (Nicholls, 1989). The need for power in adults corresponds to a *need for self-determination* in youth who are dependent on expert guidance. This need has sometimes been delineated as freedom from the oppression that undermines goal attainment and freedom to set personal goals (Franklin, 1993; Fromm, 1942). In school, *freedom from* is associated with individuals' efficacy to complete assigned tasks, and *freedom to* reflects autonomy in setting personal goals and agendas; both forms of freedom reflect individuals' sense of control in decision making. In addition, young peoples' sense of belonging reflects their *need for affiliation*; this need reflects successful involvement in exchange relationships wherein students collaborate with others and intimate relationships in which they form close ties with adults and peers (Hill, 1987). Whereas advocates of cognitive evaluation theory seem to agree with White (1959) that competence is the supreme motive in life (Ryan & Deci, 2000), we see it as only one essential impetus for behavior.

Like researchers who coordinate utilitarian ideas about motivation and moral functioning (Graham, 2004), we also assessed whether adolescents' sense of agency incorporated the belief that *fair procedures* will foster the attainment of personal and professional goals. This decision was consistent with Lewin and Lewin's (1941) assertion that even young children should participate in democratic decision making, but that such participation will inevitably be constrained by the child's psychological and sociological world. It is also consistent with versions of expectancy-value theory that allow for the formulation of utilitarian reasons for working hard (Wigfield & Eccles, 2000) and with evidence that effort can often be misapplied in ways that render success unlikely (Nicholls, 1976, 1984).

The third dimension of agency to work hard incorporated values commonly associated with intrinsic motivation (Deci & Ryan, 1987). Whereas individuals with a strong task orientation are involved in task mastery even if the process is not pleasurable, intrinsic motivation places strong emphasis on ensuring that individuals find their activities interesting and emotionally pleasing. In an ideal world, the two approaches to effort are integrated, but in practice they often are not. Therefore, in addition to measuring a task-oriented approach to meeting competence needs, agency to work hard included an *aesthetic interest* scale to evaluate adolescents' commitment to interesting and aesthetically pleasurable tasks.

Agency to Cheat or Take Shortcuts

Three parallel dimensions that evaluated when individuals feel inclined to be dishonest were used in our study to evaluate adolescents' agency to cheat. Consistent with previous findings on who is likely to cheat (Johnson, 1981), we assumed that cheating was most likely to occur when adolescents were more preoccupied with goal attainment than with the means by which they attain such ends. Our investigation extends earlier work in that items reflect general self-determination and affiliation needs as well as competence needs. Consistent with research suggesting that students tend to blame others while cheating (McCabe, 1999), we assumed that cheating would be more likely if individuals thought educational practices would be incongruent with their personal needs or if such routines were seen as ineffective and unfair. These concerns with personal needs and utilitarian values could elicit effort directed toward cheating as easily as hard work because adolescents would not reflect on the long-term or communal consequences of their actions.

The inclusion of a scale for agency to cheat out of personal enjoyment sustains our commitment to understanding intrinsic motivation. Adolescents could cheat because it is interesting and enjoyable and may not care about the quality of their performance. Some individuals, for example, cheat because they enjoy the risk of getting caught (Nagin & Pogarsky, 2003; Tibbetts, 1997). The proliferation of computer viruses, spam, spyware, and other forms of electronic cheating offers daily reminders of such approaches to life, and new technologies offer alternative ways to gratify such dishonest impulses (Szabo & Underwood, 2004; Underwood & Szabo, 2003). This led us to include general statements indicating that cheating might be fun, interesting, or pleasurable on our agency to cheat scale.

VERIFYING THE IMPORTANCE OF MORAL ENGAGEMENT

Accounts of expectancy-value theory that emphasize academic engagement imply that adolescents' vision of how schools ought to function may not play an important role in their performance (Wigfield & Eccles, 2000). Although it is easy to see that adolescents who become overly preoccupied with political and ethical issues may have difficulty attending to more immediate classroom pressures, it also seems likely that moral engagement is at least as important as academic engagement to the facilitation of agency. We sought to evaluate the role of moral engagement while describing individual differences in who is likely to cheat, but this agenda required careful consideration of what to measure and how to best capture key constructs.

MEASURING ADOLESCENTS' LIFE GOALS

Rather than compare and contrast the particular details of adolescents' evolving life goals, as has been done in studies of life tasks (Zirkel & Cantor, 1990), we used indicators of common present and future aspirations to evaluate whether students were committed to individualistic or communitarian agendas (Table 8.1). Our intent was to see if the content of life goals was orthogonal to the temporal features of these goals; individualistic and communitarian values may be expressed as present or future aspirations.

Existing claims about adolescents' tendency to adopt individualistic and communitarian agendas remain controversial. Research on moral development, for example, sometimes suggests that adolescents are so preoccupied with their own immediate experience that they cannot fully consider communitarian values (Power, Higgins, & Kohlberg, 1989). Findings from service learning programs contradict this assumption, offering many cases in which adolescents become highly committed to group activities facilitating broader societal goals (Youniss & Yates, 1997). Rather than claim that adolescents are focused either on themselves or on community agendas, our plan was to determine if existing contradictory conclusions reflect unexamined cultural or temporal differences in adolescents' representations of their identity. We assumed that when motivation is optimal, individuals balance both personal and collective concerns.

To compare temporal and cultural possibilities, we obtained interesting samples of adolescents from two very different subcultures within American society. The entire high school student body ($n = 36$) in a remote part of Alaska, many of whom indicated that Yupik was their native language, responded to our battery of measures. A corresponding sample of Caucasian students living in a Chicago area suburb was selected using gender, age, and typical grades as matching variables.[2] These adolescents were selected because they came from vastly different communities with strikingly different social norms and daily practices; high levels of collective action and community involvement are essential for survival in remote parts of Alaska, whereas this is not the case for individuals living in the Chicago area. Both groups of students reported whether they held communitarian and individualistic life goals with proximal and distal parameters.

Our questions about cultural differences in life goals reflect a convergence of work in sociology, psychology, and education. When designing

---

[2] We selected these variables to account for the fact that the Alaskan sample was restricted to residents in a single community who did not reflect a normal distribution on demographic variables of age and gender. Our measure of performance did not favor one sample over another because both groups had the same grade distribution. The range of typical grades was from A to F, with the median and mode of B.

TABLE 8.1 Scale Names and Sample Items

Scale name	Sample items
Individualistic goals (*14 items*, α = .87)	I will feel most successful if I earn enough money to live comfortably; build a strong family; discover more about my abilities; make my parents proud.
Communitarian goals (*23 items*, α = .93)	I will feel most successful if I help solve global problems; make my neighborhood a better place to live; keep my country safe; learn what is important in the world today; do what my community expects of me.
Fair school (*35 items*, α = .81)	Students should receive opportunities that match their abilities; students should have an equal chance to pass tests; the most talented person should win contests.
Epistemology (*18 items*, α = .90)	Teachers should make sure students learn useful facts; spelling rules; different points of view; scientific methods.
Sustain high motivation (*39 items*, α = .93)	Students will stay involved if they solve problems by working hard; study with smart people; can monitor their progress; have teachers who help them see which ideas are right; put interesting ideas together.
Improve low motivation (*39 items*, α = .95)	Students will become involved if they learn things that make sense; feel like part of a community; set their own schedule; have teachers who help with confusing work; do work that is important to them.
Time management (*4 items*, α = .88)	I set hours each day for homework; study at the same time everyday; make a list of things to do; think about how much time I have to do each task.
Agency to work hard (*33 items*, α = .95)	I work hard when . . . *My needs are met* (*15 items*, α = .88) I see how I might use the ideas; my teachers care about me; I can figure out problems on my own. *Procedures are fair* (*10 items*, α = .89) rules for behavior are fair. *Tasks are aesthetically interesting* (*8 items*, α = .85) the material fascinates me.
Agency to cheat or take shortcuts (*34 items*, α = .97)	I take shortcuts or cheat when . . . *My needs are unmet* (*16 items*, α = .92) I am not sure how to do the work; I don't like my teacher; I can't see why the work is worth doing. *Procedures are unfair* (*7 items*, α = .85) the teacher never gave us directions. *Misbehaving is aesthetically interesting* (*11 items*, α = .92) cheating is exciting.
Performance (*8 items*, α = .85)	*Participation* (*4 items*, α = .90) At school I usually look forward to discovering new things; enjoy learning; work as hard as I can; get involved in working. *Achievement* (*4 items*, α = .76) I think my ability at schoolwork is (excellent to very poor). . . . Compared to most other students my ability in schoolwork is (top to bottom). . . . I am usually satisfied if my grade is (A to F). . . . Normally I do (all, some, a little, none) of my homework.

Note: Names in italics reflect subscales that were compared in the analyses. There were 72 participants; 36 Alaskan Natives (21 males, 15 females) from a remote part of Alaska matched by age, sex, and self-reported grades with 36 Caucasians living in a Chicago suburb (21 males, 15 females). All the high school students from the Alaskan school participated. Because mean scale scores did not differ for these two groups, they are combined in these analyses.

measures of individuals' proximal and distal life goals, we evaluated whether adolescents endeavor to become virtuous persons, virtuous citizens, or people with multidimensional values. Incorporating adolescents' responses to the proximal and distal levels of communitarian and individualistic scales into a cluster analysis, we found two reliable clusters that did not involve a temporal distinction. Some adolescents endorsed only strong individualistic life goals regardless of whether they reflected present or future aspirations ($n = 19$ Alaskan and 14 Illinois students); this is consistent with aiming to become a virtuous person. Other adolescents endorsed both individualistic and communitarian goals, again synthesizing temporal dimensions ($n = 17$ Alaskan and 22 Illinois students); this option is not apparent in Aristotle's philosophical distinction, but it seems useful for maintaining citizenship in today's world.

By including adolescents from communities with two very different assumptions about citizenship, we could also evaluate foundational ideas in sociology. Sociological distinctions between *gemeinschaft* (local relations that include kinship, friendship, and neighborhood) and *gesellschaft* (instrumental relations that involve broader notions of one's neighborhood, country, and world) offer a clear means of defining variation in communal norms while accepting that everyone is dependent on others (Tönnies, 1957/1887, 1971/1925). In communities formed around local relations, the preservation of close relational ties often fulfills a stronger role in decision making than the more expedient methods of preserving instrumental ties. This implies that, out of practical necessity, communities like those functioning in remote parts of Alaska may be preserved as organic ends in themselves. If so, students should be preoccupied with obligations to themselves, family members, and peers because their survival depends on maintaining close ties. Communities like those in the greater Chicago area can be formed around instrumental goals that need not include close ties. In such communities, students would be preoccupied with obligations to their neighborhoods, country, or global events because they are not as dependent on specific persons, and environmental pressures allow for more impersonal interactions.

We did not find community differences, but our sampling procedures allowed for a generous test of these sociological possibilities. Differences in relational ties were maximized in that responses from all 14- to 17-year-olds in the Alaskan school were compared to a matched group of adolescents from Illinois who simply live in the same neighborhood and may or may not know one another. Counter to the assumptions of sociologists, the distribution of life goals looked similar across the two samples. There were no significant differences between the life goals of Alaskan and Illinois residents, even though the observed power of this test was adequate (.69).

Failure to find significant sociological differences was consistent with our psychological prediction that adolescents are likely to formulate personal impressions of the cultural norms they experience. By suggesting that students within each community differ in whether they embrace individualistic or communitarian life goals, our findings support theories that bridge sociological and psychological accounts of human functioning (Triandis, 1995).

A further caveat is raised by the fact that all our respondents were citizens of the United States. The Alaskan and Illinois samples could offer a vertical comparison of two similar subcultures rather than a horizontal comparison of groups that can be ranked along a more global scale (Triandis & Gelfand, 1998). Daily life, personal interests, and future life prospects seem to be entirely different for these two groups. For example, Alaskan students may miss school to participate in walrus hunting and herring or halibut fishing, whereas the Illinois students are unlikely to even imagine such community obligations. Illinois students might take for granted their trips to the mall, the anonymity of riding the METRA, and a plethora of opportunities to meet individuals from different subcultures, but the Alaskan students would probably find such complex choices surprising. Despite such differences, the distribution of educational standards across the two groups is remarkably similar. These findings contradict the views of educators and psychologists who have argued that academic achievement is primarily affected by communal norms (Aronson, Steele, Salinas, & Lustina, 1999; Fordham & Ogbu, 1986) and allow for individual variation in the formation of life goals. We could not rule out the possibility that because of their age and position in the life cycle, adolescents from these two states share more cultural commonalities than differences.

MEASURING ADOLESCENTS' BELIEFS ABOUT AN IDEAL SCHOOL

Sociological claims concerning adolescents' beliefs about an ideal school were also unsupported. Adolescents evaluated a range of ideas, extracted from interviews with individuals ranging in age from 6 to 74, about how an ideal school ought to be organized (Table 8.1). They evaluated how often common practices should be used in a fair school, the types of knowledge teachers should emphasize, how to sustain motivation among students who are already highly motivated, how to improve the motivation of students who are not engaged, and the effectiveness of particular time management strategies. As was the case for life goals, there were no differences between Alaskan and Illinois residents on these scales, even though the observed power was adequate (.85).

MEASURING ADOLESCENTS' SENSE OF AGENCY

Recall that adolescents' sense of agency reflects their decisions to act and that three themes are likely to influence moral action. Students' sense of agency is guided by their basic needs for competence, self-determination, and affiliation; their recognition that fair procedures will allow them to attain their goals; and their perception that tasks are interesting and personally meaningful (Table 8.1). When measuring agency to work hard, we found that the three dimensions were associated ($\alpha = .95$ for all 33 items in the sample reported here). The three dimensions of agency to cheat also formed an internally consistent scale ($\alpha = .97$ for all 34 items in the sample reported here), and that scale was significantly different from agency to work hard ($r = -.21$).

The moral engagement model could be tested by using aggregate agency scales or by isolating each dimension. Aggregate measures of agency to cheat and work hard were correlated with performance to verify their role in the moral engagement model, but the six agency dimensions were treated separately to verify that cheating is a mechanism by which moral engagement affects performance. Once again the two samples could be combined because there were no differences in the agency beliefs of Alaskan and Illinois residents, even though the test had adequate power (.57).

PREDICTING ACADEMIC PERFORMANCE

The first step in verifying that adolescents' personal standards and agency beliefs played a role in their academic behavior involved a measure of performance that included both self-reported participation and achievement behaviors. Consistent with Lewin's (1936) view that motivation is not dependent on outcomes alone, all adolescents rated themselves on four items that reflect the depth of their school participation and four indices of their typical achievement (Table 8.1). This performance measure was moderately associated with all the dimensions of engagement and agency, suggesting that each dimension plays some role in adolescents' academic functioning (Table 8.2). Sample restrictions prevented a meaningful interpretation of regression analyses and the use of more complex structural equation modeling, but the fact that measures of individualistic and communitarian life goals were only moderately correlated suggests that there may be at least two pathways in relations between moral engagement, agency, and performance.

ENGAGEMENT PATHWAYS

Relations between adolescents' life goals and their beliefs about an ideal school were explored in greater depth by assigning each participant to an individualistic or communitarian cluster. Using life goal orientation as a between-subjects variable, we used adolescents' responses to the ideal school scales as within-subjects variables in a repeated measures analysis of variance. The predicted interaction between life goals and beliefs about

TABLE 8.2 Correlations Between Life Goals, Beliefs About an Ideal School, Agency, and Performance

Scales by blocks	1	2	3	4	5	6	7	8	9
Life goals									
1. Individualistic	—								
2. Communitarian	.64	—							
Ideal school									
3. Fair school	.28	.42	—						
4. Epistemology	.60	.64	.50	—					
5. Sustain high motivation	.50	.52	.70	.62	—				
6. Improve low motivation	.38	.37	.42	.50	.63	—			
7. Time management	.41	.53	.28	.60	.36	.23	—		
Agency									
8. Agency to work hard	.70	.58	.47	.76	.71	.68	.44	—	
9. Agency to cheat	-.30	-.12	.03	-.11	-.09	.08	-.15	-.22	—
Performance	.59	.43	.43	.57	.55	.37	.43	.62	-.41

an ideal school was significant (Figure 8.2). Adolescents reporting communitarian goals showed more complex beliefs about school than those with individualistic goals, indicating a richer understanding of possible roles for school in society.

The likelihood that communitarian life goals may offer a more beneficial pathway through engagement is further strengthened when differences in performance are tested. Adolescents adhering to communitarian values ($M = 3.94$, $sd = .59$) showed significantly stronger levels of performance than those adhering to individualistic values ($M = 3.30$, $sd = .77$, $F_{(1, 70)} = 16.05$, $p < .001$, $\eta^2 = .19$).

These findings raise possibilities for how educators might maximize the fit between adolescents' understanding of the world and the environments in which they learn (Eccles et al., 1993). Given that adolescents may be committed to individualistic or communitarian life goals, perhaps schools could implement policies and practices that teach students the importance of collective action. Adolescents holding individualistic life goals can benefit from more detailed information about when and how they are dependent on others, while adolescents holding communitarian life goals could strengthen their knowledge of how communities preserve trust, personal responsibility, and integrity.

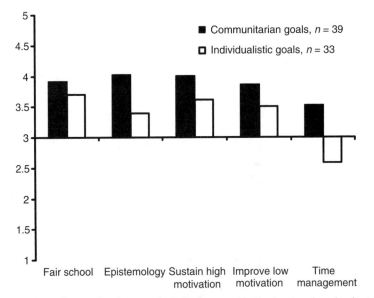

FIGURE 8.2   Interaction between beliefs about an ideal school and goal orientation, $F_{(4,280)} = 6.59$, $p < .001$, $\eta^2 = .09$. In all cases, means for communitarian and individualistic groups were significantly different from one another, $p < .05$.

These possibilities are exemplified in the use of honor codes in colleges and universities (McCabe, Trevino, & Butterfield, 2001), although such practices could be improved upon. Schools with honor codes find a lower incidence of cheating (McCabe & Trevino, 1993) and encourage students to think about the consequences of following individualistic and communitarian guidelines (McCabe, Trevino, & Butterfield, 1999). Available honor codes and corresponding practices imply that students should reflect on values appropriate for *gemeinschaft* notions of community (McCabe et al., 1999). They seem to illustrate how individualistic values of personal freedom and integrity are essential for the preservation of trust. Common codes also acknowledge that students should be responsible, vigilant moral agents who preserve communal values. Schools use different practices to maintain honesty codes, but many codes require individuals to sign a written pledge, accept responsibilities for reporting transgressions, take unproctored exams, and participate in the judiciary process once cheating is detected. Students who participate in such activities are expected to articulate personal standards for holding themselves and others accountable, but they are not asked to describe why such behavior is important for the larger community. Although communitarian values are not made explicit, students are expected to avoid dishonest behavior and to report those peers who do cheat, with the understanding that any violation of community values harms everyone.

This individualistic bias is consistent with studies of how classroom contexts affect academic dishonesty; researchers commonly explore local relations between students and their teachers. Some investigations have focused on how teachers could organize classrooms and build stronger levels of academic engagement (Anderman, Griesinger, & Westerfield, 1998; Murdock, Hale, & Weber, 2001). Others have offered details of who students blame for their actions or the conditions under which students are inclined to cheat (Jensen, Arnett, Feldman, & Cauffman, 2002; Murdock, Miller, & Kohlhardt, 2004). Together these findings suggest that students who cheat may benefit from discussions about communal norms because they are more likely to use surface-level learning strategies and avoid recognizing their responsibilities to others.

Put another way, these discussions could include more details relevant to *gesellschaft* agendas. Students could learn more about why schools are legitimate institutions and how to preserve the value of such institutions in the larger society. Our findings suggest that many adolescents are aware of their collective responsibilities; more than half our sample (54 percent) endorsed life goals consistent with participation in a global world and had little trouble offering reliable visions of an ideal school. Honor codes and school policies could be expanded to openly acknowledge such responsibilities.

If honor codes were to reveal more about the role of school in society, the resulting conversations could help students strengthen personal standards that are central to moral engagement. Rather than advocating private pledges to report cheating, we can suggest that educators openly involve students in conversations about the role of school in society and have successfully done so in a wide range of settings. These discussions would include comparisons between specific classroom practices and the degree to which they simulate professional responsibilities in the larger community. For example, student involvement in the judiciary process when responding to cheating could be expanded to include discussions of why legal systems are important, comparisons of different legal systems, and information about why different roles are needed within a particular system. Conversations about the details of particular academic subjects such as mathematics or science could be expanded to include reflection on when and where key ideas are used in the world and how individuals contribute to such professions. Details of particular strategies for fostering learning and motivation could be compared to those used by individuals working in a variety of careers, while promoting discussions about how and why such career paths evolved in our world. When educators offer vivid portraits of how intellectual ideas are used in society and allow students to enact such possibilities, adolescents can formulate stronger beliefs about how the world does and ought to function. Helping students see when and where they can participate in such activities can help them see why honesty, reliability, and integrity matter to those around them, even if dishonest individuals sometimes achieve their goals.

LIFE GOALS AND AGENCY

A communitarian pathway to connecting school and personal life goals can also foster more optimal forms of agency than does the individualistic pathway. When the three indices of agency to work hard and the three indices of agency to cheat were included as within-subjects variables in a repeated measures analysis of variance, the interaction between these measures and adolescents' life goal classifications was significant (Figure 8.3). Adolescents revealing communitarian life goals were more likely than adolescents reporting individualistic life goals to report all three types of agency to work hard. The two groups did not differ in reported agency to cheat, although adolescents with communitarian life goals were more likely to disagree with each dimension.

These findings are consistent with previous research suggesting that students who are inclined to cheat may generally adopt expedient strategies for getting through their assignments (Anderman et al., 1998; Murdock et al., 2001; Smith, Ryan, & Diggins, 1972; Vitro & Schoer, 1972). When individuals feel pressured to show high levels of ability or experience fear

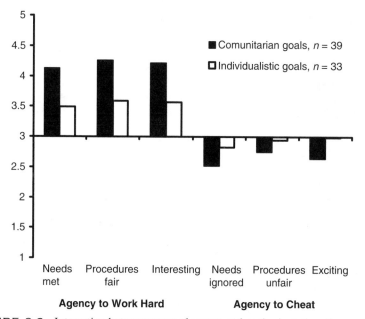

FIGURE 8.3 Interaction between types of agency and goal orientation, $F_{(5,350)} = 11.31$, $p < .001$, $\eta^2 = .14$. Mean differences between life goal orientations were significant for hard work subscales, but not for cheating subscales.

of failure, they are likely to depend on surface-level learning strategies that include taking shortcuts in their work. Similarly, individuals who cheat are likely to show higher levels of impulsivity than individuals who do not cheat (Nagin & Pogarsky, 2003). Preoccupation with individualistic goals increases the self-consciousness that strengthens such worries and encourages adolescents to develop standards for avoiding meaningful academic work. The activation of these low standards in the engagement force can undermine students' sense of agency or fracture their ability to draw connections between schoolwork and life's work.

Our findings also suggest that adolescents may benefit from richer conversations about how schooling can facilitate personal and communal growth. Clearly, students can benefit from classroom practices that minimize distracting pressures, but there is minimal evidence in support of practical alternatives. When children and adolescents are asked for advice on what educators might do to help them learn, they consistently seek acknowledgment of their personal contributions to communal goals (Thorkildsen, 2004; Thorkildsen & Jordan, 1995; Thorkildsen et al., 2004). Most students intuitively understand that meaning-making involves the coordination of their own interests and those of their parents or teachers, but only some see the relations between local and societal demands. As students with individualistic life goals are asked to consider communal agendas, they may learn to situate themselves in a broader *gesellschaft*. Students with communitarian life goals could also learn from thoughtful comparisons of their personal standards and the complex communities in which they function. In other words, discussions about agency can encourage adolescents to draw connections between schoolwork and life's work and could be part of the manifest rather than the hidden curriculum of school.

DETERRING CHEATING

Research on deterrence suggests that individuals are less likely to cheat when the risk of being caught is high, but not when the penalty of being caught is severe. [See Nagin & Pogarsky (2003) for a review.] Intrapersonal factors such as a preoccupation with immediate goals and a tendency for individuals to shade judgments in a manner favorable to themselves have also been associated with cheating. As Festinger (1957) noted, when individuals' beliefs and behavior diverge, they are likely to assuage the resulting pressure by changing one or the other.

A subset of our findings supports possible approaches to changing adolescents' agency to take shortcuts. Participants were asked to evaluate a set of excuses students often invent for their academic behavior, only some of which focused on cheating. Using the aggregate agency to cheat measure, we designated those individuals whose mean scale score reflected agreement with most items ($M > 3.0$) as indicating high agency to cheat ($n = 11$ Alaskan and 19 Illinois students) and those with lower scores as indicating low agency to cheat ($n = 25$ Alaskan and 17 Illinois students). Figure 8.4 compares adolescents' agency to cheat and the three excuses we measured. Adolescents with a strong agency to cheat were most likely to support a palliative comparison and minimize the consequences of cheating; those disagreeing with these positions indicated that they would harm themselves if they cheated. Despite practical evidence in support students' reliance on groupthink (Jensen et al., 2002), relatively few students endorsed this excuse.

Taken together, findings on students' justifications for cheating suggest that educators who want to minimize the likelihood of academic dishonesty could help students formulate strong justifications for the importance of honorable work. One option would be to undermine the validity of specific palliative comparisons invented by students to justify cheating. If students reported copying because they ran out of time, educators could foster class discussions on time management. Complaints about the difficulty of particular assignments could be interpreted as opportunities for exploring successful learning strategies. When such corrective discussions are conducted with humor, actual behavior is likely to change (Dienstbier,

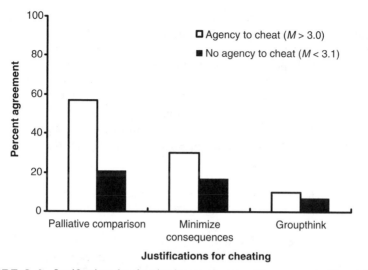

FIGURE 8.4   Justifications for cheating by agency status. There were significant differences between agency groups for acceptance of a palliative comparison ($\chi^2_{(2)} = 13.33$, $p < .001$), but not for minimizing consequences or groupthink.

1995; Dienstbier et al., 1980). As students' confidence in their dishonest strategies is undermined with information on honest alternatives, educators and peers learn more about the impracticalities of cheating or about how shortcuts can delay learning. Adolescents can even benefit from straightforward illustrations of how most learning follows a predictable curve in which change is initially fast but eventually slows before hitting a plateau. Effort can be strengthened by helping students reflect on where in the learning process their current activities fit.

A second option might be to elaborate on some of the long-term and short-term consequences of cheating using language that reinforces the potential harm of dishonesty. Rather than repeat trite phrases, educators can help adolescents see how tasks are interrelated and show them that shortcuts on current tasks can undermine their readiness for the next step. Labeling who is harmed by each cheating strategy can call attention to consequences adolescents may be too inexperienced to anticipate. Such discussions could also encourage individuals to reflect on their impulsive behavior and move beyond a preoccupation with immediate outcomes.

A third consideration was not as detectable in our survey data as we had anticipated, but it has been evident in qualitative studies of students' beliefs. We expected students to support groupthink as an excuse for their own cheating and to agree that they saw cheating as an acceptable norm, but this did not occur. Students' unwillingness to describe group pressure is consistent with findings from focus group discussions of cheating in which adolescents comfortably introduce cheating as a group norm when describing the behavior of their peers, but not in relation to themselves (McCabe, 1999). Similarly, students will indicate that their peers might reduce effort when faced with possible failure, but *they* certainly would not do so (Jagacinski & Nicholls, 1990). Perhaps educators could use groupthink to highlight the honest effort put forth by most students along with the personal and societal benefits of such action. In doing so, educators should be careful *not* to support the myth that many or most students cheat in school; groupthink can also foster moral disengagement (Bandura, 2004).

CHEATING AS A MECHANISM OF MORAL ENGAGEMENT

Adolescents are busy formulating personal standards that play a key role in their sense of agency, and discussions about daily experiences can help them formulate commitments to virtue as persons and as citizens. Rather than ignore cheating with the assumption that everyone does it or become overly preoccupied with identifying and punishing cheaters, our findings suggest that frequent discussions of communal norms can reinforce positive features of moral engagement. Details of what constitutes cheating can be situated alongside thoughtful alternatives. In this respect, individuals can reflect on when they are tempted to take shortcuts in their work without becoming disengaged in the academic enterprise.

Our moral engagement model suggests two pathways for fostering strong academic performance. Adolescents reporting communitarian life goals showed the most complex views about an ideal school and the strongest sense of agency to work hard, but positive signs of moral engagement were also evident among students with individualistic life goals. It is helpful to remember that we evaluated moral engagement and not moral disengagement or apathy, although the latter reflect conceptually viable approaches to school (Bandura et al., 1996). With this caveat in mind, it is still possible to draw generalizations from our findings. Educators could remember that schools promoting communitarian values also facilitate moral engagement by emphasizing the benefits of education for individuals, neighborhoods, countries, and the global community. Discussions about communal values offer helpful language and practices for encouraging students to resist the temptation to cheat. Such discussions can teach students to take collective pride in their achievements while promoting conduct-focused values of honesty and mutual respect.

Educators may also promote moral engagement in several ways. Establishing fair routines for achieving classroom goals and drawing connections between those routines and situational demands found in the larger society can help adolescents imagine responsibilities to local and global communities. With help in drawing connections between the details of particular lessons and more general questions of epistemology, students can clarify when and where to direct their efforts. Asking adolescents to reflect on their own motivational orientations as well as the orientations of their peers fosters an awareness of the dilemmas associated with collective activities. Furthermore, when academic pressures become overwhelming, adolescents benefit from realizing that others also face such pressures and from sharing the details of strategies they find effective. Taking time to describe connections between daily routines and activities found in various professions allows adolescents to acquire global knowledge and imagine comfortable standards for discovering their place in the world.

Some schools have started the process of fostering moral engagement by instituting honor codes. Whereas such practices teach students to accept responsibility for maintaining trust in their school community, most codes could be expanded to include *gesellschaft* as well as *gemeinschaft* perspectives on the world. Our findings support three suggestions. First, it seems important to remind students about the importance of schooling in the larger community by including vivid portraits of how individuals use intellectual skills outside of school. Second, it seems important to highlight, in detail, many of the ways in which cheating introduces personal and collective harm by undermining trust and collective values. Third, it seems important to show students how deep, conceptual learning and high-quality outcomes are likely to carry them further in the world than would be possible if they focused impulsively on the present and sought expedient, superficial indices of achievement.

The fact that nearly everyone in our sample could agree with at least one item on the cheating scale suggests that most adolescents are at least tempted to take such shortcuts. Our brief exploration of students' excuses supports some suggestions for responding to cheating. Accepting that cheating incidents offer teachable moments to individuals as well as classes, educators can tactfully talk about alternative methods of achieving personal goals. Using information from students' justifications, we can see whether individuals fully understand the nature of their transgression. Should they offer palliative comparisons, minimize the consequences of their actions, or emphasize groupthink, educators can act in ways to help students rethink their motives. First, creating obstructions to ensure that cheating is not an expedient way to function can undermine the validity of palliative comparisons between cheating behavior and academic outcomes. Second, highlighting the negative consequences of cheating by using humor to describe long-term and short-term effects can ensure that effects on the community are fully understood. Finally, it seems important to avoid supporting the myth that most people cheat and that individuals are justified in their actions because community values are already corrupt.

Seasoned educators will notice that we are reminding them of practices they can easily implement but may sometimes forget. This is intentional. We do not see how a new "anti-cheating" campaign could effectively deter the rising rates of cheating. Instead, we see moral engagement as an ever-evolving component of everyone's motivation that may be best compared to a muscle that constantly needs to be exercised. If adults remember that adolescents are formulating life goals and personal standards, sometimes for the first time in their lives, it may be easier to value conversations that evaluate personal and collective alternatives. It will also be easy to remember that opportunities to act on personal convictions allow adolescents to exercise moral agency and improve performance.

REFERENCES

Anderman, E. M., Griesinger, T., & Westerfield, G. (1998). Motivation and cheating during early adolescence. *Journal of Educational Psychology, 90*, 84–93.

Aristotle. (1985). *The Nichomachean ethics* (T. Irwin, Trans.). Indianapolis, IN: Hackett.

Aronson, J., Steele, C. M., Salinas, M. F., & Lustina, M. J. (1999). The effect of stereotype threat on the standardized test performance of college students. In E. Aronson (Ed.), *Readings about the social animal* (8th ed.). New York: Freeman.

Atkinson, J. W. (1958). Toward an experimental analysis of human motives in terms of motives, expectancies, and incentives. In J. W. Atkinson (Ed.), *Motives in fantasy, action, and society* (pp. 288–305). New York: Van Nostrand.

Bandura, A. (1977). Self-efficacy: Toward a unifying theory of behavioral change. *Psychological Review, 84*, 191–215.

Bandura, A. (1999). Moral disengagement in the perpetration of inhumanities. *Personality and Social Psychology Review, 3*, 193–209.

Bandura, A. (2004). Selective exercise of moral agency. In T. A. Thorkildsen & H. J. Walberg (Eds.), *Nurturing morality* (pp. 37–57). New York: Kluwer Academic.

Bandura, A., Barbaranelli, C., Caprara, G. V., & Pastorelli, C. (1996). Multifaceted impact of self-efficacy beliefs on academic functioning. *Child Development, 67*, 1206–1222.

Cantor, N., Kemmelmeier, M., Basten, J., & Prentice, D. A. (2002). Life task pursuit in social groups: Balancing self-exploration and social integration. *Self and Identity, 1*, 177–184.

Deci, E. L., & Moller, A. C. (2005). The concept of competence: A starting place for understanding intrinsic motivation and self-determined extrinsic motivation. In A. J. Elliot & C. S. Dweck (Eds.), *Handbook of competence and motivation* (pp. 579–597). New York: Guilford.

Deci, E. L., & Ryan, R. M. (1987). The support of autonomy and the control of behavior. *Journal of Personality and Social Psychology, 53*, 1024–1037.

Dienstbier, R. A. (1995). The impact of humor on energy, tension, task choices, and attributions: Exploring hypotheses from toughness theory. *Motivation and Emotion, 19*, 255–267.

Dienstbier, R. A., Kahle, L. R., Willis, K. A., & Tunnell, G. B. (1980). The impact of moral theories on cheating: Studies of emotion attribution and schema activation. *Motivation and Emotion, 4*, 193–216.

Dishion, T. J., & Owen, L. D. (2002). A longitudinal analysis of friendships and substance use: Bidirectional influence from adolescence to adulthood. *Developmental Psychology, 38*, 480–491.

Eccles, J. S., Midgley, C., Wigfield, A., Buchanan, C. M., Reuman, D., Flanagan, C., & McIver, D. (1993). Development during adolescence: The impact of stage-environment fit on young adolescents' experiences in schools and in families. *American Psychologist, 48,* 90–101.

Festinger, L. (1957). A theory of cognitive dissonance. Stanford, CA: Stanford University Press.

Fordham, S., & Ogbu, J. (1986). Black students' school success: Coping with the burden of acting white. *Urban Review, 18,* 176–206.

Franklin, V. P. (1993). *Black self-determination: A cultural history of African-American resistance.* Brooklyn, NY: Lawrence Hill Books.

Fromm, E. (1942). *Fear of freedom.* London: Routledge.

Graham, S. (2004). The role of perceived responsibility in nurturing morality. In T. A. Thorkildsen & H. J. Walberg (Eds.), *Nurturing morality* (pp. 19–36). New York: Kluwer Academic.

Guttman, J. (1984). Cognitive morality and cheating behavior in religious and secular school children. *Journal of Educational Research, 77,* 248–254.

Hartshorne, H., & May, M. A. (1928). *Studies in the nature of character. Vol. 1. Studies in deceit.* New York: Macmillan.

Hill, C. A. (1987). Affiliation motivation: People who need people . . . but in different ways. *Journal of Personality and Social Psychology, 52,* 1008–1018.

Jagacinski, C. M., & Nicholls, J. G. (1990). Reducing effort to protect perceived ability: "They'd do it but I wouldn't." *Journal of Educational Psychology, 82,* 15–21.

Jensen, L. A., Arnett, J. J., Feldman, S. S., & Cauffman, E. (2002). It's wrong, but everybody does it: Academic dishonesty among high school and college students. *Contemporary Educational Psychology, 27,* 209–228.

Johnson, J. A., Hogan, R., & Zonderman, A. B. (1981). Moral judgment, personality, and attitudes toward authority. *Journal of Personality and Social Psychology, 40,* 370–373.

Johnson, P. B. (1981). Achievement motivation and success: Does the end justify the means? *Journal of Personality and Social Psychology, 40,* 374–375.

Leming, J. S. (1978). Cheating behavior, situational influence, and moral development. *Journal of Educational Research, 71,* 214–217.

Lewin, G., & Lewin, K. (1941). Democracy and the schools. *Understanding the child, 10,* 1–7.

Lewin, K. (1936). Psychology of success and failure. *Occupations, 14,* 926–930.

Lewin, K. (1951). Intention, will and need. In D. Rapaport (Ed.), *Organization and pathology of thought: Selected sources* (pp. 95–153). New York: Columbia University Press.

McCabe, D. L. (1999). Academic dishonesty among high school students. *Adolescence, 34,* 681–687.

McCabe, D. L., & Trevino, L. K. (1993). Academic dishonesty: Honor codes and other contextual influences. *Journal of Higher Education, 64,* 522–538.

McCabe, D. L., Trevino, L. K., & Butterfield, K. D. (1999). Academic integrity in honor code and non-honor code environments. *Journal of Higher Education, 70,* 211–234.

McCabe, D. L., Trevino, L. K., & Butterfield, K. D. (2001). Dishonesty in academic environments: The influence of peer reporting requirements. *Journal of Higher Education, 72,* 29–45.

Murdock, T. B., Hale, N. M., & Weber, M. J. (2001). Predictors of cheating among early adolescents: Academic and social motivations. *Contemporary Educational Psychology, 26,* 96–115.

Murdock, T. B., Miller, A., & Kohlhardt, J. (2004). Effects of classroom context variables on high school students' judgments of the acceptability and likelihood of cheating. *Journal of Educational Psychology, 96*(4), 765–777.

Nagin, D. S., & Pogarsky, G. (2003). An experimental investigation of deterrence: Cheating, self-serving bias, and impulsivity. *Criminology, 41,* 167–193.

Nelson, J. R., Nicholls, J. G., & Gleaves, K. (1996). The effect of personal philosophy on orientation towards school: African American students' views of integrationist versus nationalist philosophies. *Journal of Black Psychology, 22,* 340–357.

Nicholls, J. G. (1976). Effort is virtuous, but it's better to have ability: Evaluative responses to perceptions of effort and ability. *Journal of Research in Personality, 10,* 306–315.

Nicholls, J. G. (1984). Achievement motivation: Conceptions of ability, subjective experience, task choice, and performance. *Psychological Review, 91,* 328–346.

Nicholls, J. G. (1989). *The competitive ethos and democratic education.* Cambridge, MA: Harvard University Press.

Nicholls, J. G., & Nelson, J. R. (1992). Students' conceptions of controversial knowledge. *Journal of Educational Psychology, 84,* 224–230.

Nicholls, J. G., Nelson, J. R., & Gleaves, K. (1995). Learning "facts" versus learning that most questions have many answers: Student evaluations of contrasting curricula. *Journal of Educational Psychology, 87,* 253–260.

Nicholls, J. G., & Thorkildsen, T. A. (1988). Children's distinctions among matters of intellectual conventions, logic, fact, and personal preferences. *Child Development, 59,* 939–949.

Nicholls, J. G., & Thorkildsen, T. A. (1989). Intellectual conventions versus matters of substance: Elementary school students as curriculum theorists. *American Educational Research Journal, 26,* 533–544.

Pellegrini, A. D., Bartini, M., & Brooks, F. (1999). School bullies, victims, and aggressive victims: Factors relating to group affiliation and victimization in early adolescence. *Journal of Educational Psychology, 91,* 216–224.

Power, F. C., Higgins, A., & Kohlberg, L. (1989). *Lawrence Kohlberg's approach to moral education.* New York: Columbia University Press.

Ryan, A. M. (2001). The peer group as a context for the development of young adolescent motivation and achievement. *Child Development, 72,* 1135–1150.

Ryan, R. M., & Deci, E. L. (2000). Intrinsic and extrinsic motivations: Classic definitions and new directions. *Contemporary Educational Psychology, 25,* 54–67.

Skinner, E. A., Zimmer-Gembeck, M. J., & Connell, J. P. (1998). Individual differences and the development of perceived control. *Monographs of the Society for Research in Child Development, 63,* 1–220.

Smith, C. P., Ryan, E. R., & Diggins, D. R. (1972). Moral decision making: Cheating on examinations. *Journal of Personality, 40,* 640–660.

Szabo, A., & Underwood, J. (2004). Cybercheats: Is information and communication technology fuelling academic dishonesty? *Active Learning in Higher Education, 5,* 180–199.

Thorkildsen, T. A. (1989a). Justice in the classroom: The student's view. *Child Development, 60,* 323–334.

Thorkildsen, T. A. (1989b). Pluralism in children's reasoning about social justice. *Child Development, 60,* 965–972.

Thorkildsen, T. A. (1989c). Fairness and motivation among high school and college students. Unpublished raw data.

Thorkildsen, T. A. (1991). Defining social goods and distributing them fairly: The development of conceptions of fair testing practices. *Child Development, 62,* 852–862.

Thorkildsen, T. A. (1993). Those who can, tutor: High ability students' conceptions of fair ways to organize learning. *Journal of Educational Psychology, 85,* 182–190.

Thorkildsen, T. A. (1994). Toward a fair community of scholars: Moral education as the negotiation of classroom practices. *Journal of Moral Education, 23,* 371–385.

Thorkildsen, T. A. (2000). The way tests teach: Children's theories of how much testing is fair in school. In M. Leicester, C. Modgil, & S. Modgil (Eds.), *Education, culture, and values, Vol. III: Classroom issues: practice, pedagogy, and curriculum* (pp. 61–79). London: Falmer Press.

Thorkildsen, T. A. (2004). Moral functioning in school. In T. A. Thorkildsen & H. J. Walberg (Eds.), *Nurturing morality* (pp. 137–156). New York: Kluwer Academic.

Thorkildsen, T. A. (2005). Personal standards for school among children and adolescents. Unpublished manuscript, University of Illinois at Chicago, Chicago IL.

Thorkildsen, T. A., & Jordan, C. (1995). Is there a right way to collaborate? When the experts speak can the customers be right? In J. G. Nicholls & T. A. Thorkildsen (Eds.), *Reasons for learning: Expanding the conversation on student-teacher collaboration* (pp. 137–161). New York: Teachers College Press.

Thorkildsen, T. A., & Nicholls, J. G. (1991). Students' critiques as motivation. *Educational Psychologist, 26*, 347–368.

Thorkildsen, T. A., & Nicholls, J. G. (1998). Fifth graders' achievement orientations and beliefs: Individual and classroom differences. *Journal of Educational Psychology, 90*, 179–201.

Thorkildsen, T. A., & Nicholls, J. G. (with Bates, A., Brankis, N., & DeBolt, T.). (2002). *Motivation and the struggle to learn: Responding to fractured experience*. Boston: Allyn & Bacon.

Thorkildsen, T. A., & Schmahl, C. (1997). Conceptions of fair learning practices among low-income African American and Latin American Children: Acknowledging diversity. *Journal of Educational Psychology, 89*, 719–727.

Thorkildsen, T. A., Sodonis, A., & Weaver, A. (2005). Adolescents' conceptions of fair ways to assess controversial and noncontroversial knowledge. Unpublished manuscript, University of Illinois at Chicago, Chicago, IL.

Thorkildsen, T. A., Sodonis, A., & White-McNulty, L. (2004). Epistemology and adolescents' conceptions of procedural justice in school. *Journal of Educational Psychology, 96*, 347–359.

Thorkildsen, T. A., & Weaver, A. (2003, April). Developing conceptions of the role of school in society: Prioritizing learning, test, and contest situations. Presented at the biennial meeting of the Society for Research in Child Development, Tampa, FL.

Thorkildsen, T. A., & White-McNulty, L. (2002). Developing conceptions of fair contest procedures and the differentiation of skill and luck. *Journal of Educational Psychology, 94*, 316–326.

Tibbetts, S. G. (1997). Gender differences in students' rational decisions to cheat. *Deviant Behavior, 18*, 393–414.

Tönnies, F. (1957). *Community and society* (C. P. Loomis, Trans.) New York: Harper & Row. (Original work published 1887.)

Tönnies, F. (1971). The concept of Gemeinschaft. In W. J. Cahnman & R. Heberle (Eds.), *Ferdinand Tönnies on sociology: Pure, applied, and empirical* (pp. 62–72). Chicago: University of Chicago Press. (Original work published in 1925.)

Triandis, H. C. (1995). The self and social behavior in differing cultural contexts. In N. R. Goldberger & J. B. Veroff (Eds.), *The culture and psychology reader* (pp. 326–365). New York: New York University Press.

Triandis, H. C., & Gelfand, M. J. (1998). Converging measurement of horizontal and vertical individualism and collectivism. *Journal of Personality and Social Psychology, 74*, 118–128.

Turiel, E. (1983). *The development of social knowledge: Morality and convention*. New York: Cambridge University Press.

Underwood, J., & Szabo, A. (2003). Academic offences and e-learning: Individual propensities in cheating. *British Journal of Educational Technology, 34*, 467–477.

Vitro, F. T., & Schoer, L. A. (1972). The effects of probability of test success, test importance, and risk of detection on the incidence of cheating. *Journal of School Psychology, 10*, 269–277.

Weiner, B. (1979). A theory of motivation for some classroom experiences. *Journal of Educational Psychology, 71*, 3–25.

White, R. W. (1959). Motivation reconsidered. *Psychological Review, 66*, 297–333.

Wigfield, A., & Eccles, J. S. (2000). Expectancy-value theory of achievement motivation. *Contemporary Educational Psychology, 25*, 68–81.

Youniss, J., & Yates, M. (1997). *Community service and social responsibility in youth.* Chicago: University of Chicago Press.

Zirkel, S., & Cantor, N. (1990). Personal construal of life tasks: Those who struggle for independence. *Journal of Personality and Social Psychology, 58*, 172–185.

9

THE "SOCIAL" SIDE OF SOCIAL CONTEXT: INTERPERSONAL AND AFFILIATIVE DIMENSIONS OF STUDENTS' EXPERIENCES AND ACADEMIC DISHONESTY

LYNLEY H. ANDERMAN, TIERRA M. FREEMAN, AND CHRISTIAN E. MUELLER

A number of studies on students' cheating and academic dishonesty refer to aspects of the social context present in the particular educational institution or setting under study (e.g., Eisenberg, 2004; Jordan, 2001; McCabe & Trevino, 1993). As such, social contexts can be considered at a range of levels of analysis—for example, at the level of the entire school or university or at the level of individual classrooms. The term *social context* however, is extremely broad, and the range of dimensions and characteristics captured by that umbrella term is vast (see Anderman & Anderman, 2000). Thus, in this chapter we address a particular subset of social context variables: those that relate to the interpersonal and affiliative dimensions of students' perceptions and experiences in educational settings. Furthermore, we suggest that there are at least three somewhat distinct approaches to what can be considered social characteristics of educational settings. These can be distinguished in terms of the meaning assigned to the word "social."

Copyright © 2007 by Academic Press, Inc.
All rights of reproduction in any form reserved.

One approach refers to the understanding that schools, universities, and colleges are institutions that exist within and reflect the larger society, and that have established organizational structures, procedures, and norms. This definition of *social as related to society* would include the study of objective behavioral rules and expectations in an institution. In relation to academic dishonesty, such studies would consider cheating as deviance, as violation of a societal norm.

A second approach refers to the interpersonal relationships that persons within an educational setting may develop with one another, and the perceptions, values, and obligations that come with those relationships. This definition of *social as interpersonal* would include the study of teacher–student relationships and peer relationships among students. In relation to academic dishonesty, such studies would examine associations between qualitatively different interpersonal relationships and perceived relationships and students' cheating-related attitudes and behaviors.

The third approach refers to individuals' perceptions of their relationship to the institution or group as a whole. This definition of *social as perceived affiliation* would include the study of students' sense of subjective belonging to or alienation from a setting, their perceived degree of attachment, acceptance, or "psychological fit." In relation to academic dishonesty, such studies would examine the extent to which students' sense of affiliation to an institution might protect against or promote academic dishonesty.

To date, much research on academic dishonesty and cheating, conducted within sociological or ethical frameworks, has focused on the first of these three approaches (e.g., Alschuler & Blimling, 1995; Hendershott, Drinan, & Cross, 2000; Vowell & Chen, 2004; Whitley & Keith-Spiegel, 2001). In this chapter, however, we focus on the latter two approaches, about which much less is known. We begin by reviewing empirical studies that have examined academic dishonesty in relation to social variables, organized within the categories of *social as interpersonal* and *social as perceived affiliation*. In addition to reviewing extant findings, we also discuss some of the psychological theoretical underpinnings for assuming that each group of variables might be associated with attitudes toward cheating and cheating behaviors. The chapter concludes by outlining a number of potentially fruitful directions for future research and discussing the implications of the research findings for educational practice.

SOCIAL AS INTERPERSONAL: PERCEIVED RELATIONSHIPS AMONG INDIVIDUALS AND GROUPS

In considering students' social experiences in educational settings, perhaps the most obvious focus for study includes the interpersonal relationships they form while in those settings. Indeed, the study of both teacher–student relationships and interactions, and peer relationships and influence, has a long tradition within literature on educational processes (e.g., see Anderman, Patrick, Hruda, & Linnenbrink, 2002; Brown, 1990; Davis, 2003; Pianta, 1999). Furthermore, the immediacy and proximity of both teachers and peers to students' experiences might be expected to increase the importance of such influences on students' academic beliefs and behaviors. Thus, in this section, we review studies that have examined various aspects of students' relationships with their teachers and fellow students, respectively. In terms of teacher–student relationships, these aspects include students' subjective perceptions of their teachers' personal characteristics and pedagogical practices. In particular, students' perceptions of their teachers' own attitudes toward cheating and likely behavioral response to incidents of dishonesty are included; these beliefs may be especially important in shaping the degree of mutual respect perceived between students and instructor. In terms of peer relationships, studies have focused variously on potential peer influence and modeling of cheating-related behaviors, students' perceptions of their peers' beliefs and behaviors, and potential social benefits that students might perceive in some forms of academic cheating.

TEACHER-STUDENT RELATIONSHIPS

Although many factors have been linked to student cheating behavior at all grade levels, relatively little research has explored the importance of the teacher–student relationship and its potential associations with cheating. This is surprising given the link that has been made between this important contextual factor and other motivational and achievement outcomes (e.g., Anderman et al., 2002; Harter, 1996; Murdock, Anderman, & Hodge, 2000; Turner & Meyer, 2004; Wentzel, 1997). As Stearns (2001) noted,

> Relatively little work has examined the influence of instructor behavior on academic integrity, a topic that is particularly intriguing because instructors are in a unique position: They not only control the classroom environment in which most cases of academic dishonesty occur, but they are also largely the creators of that environment. (pp. 275–276)

In general, the literature that does examine teacher characteristics and academic dishonesty falls into three general categories: students' perceptions of instructors' personality and/or pedagogy, students' perceptions of their instructors' attitudes toward cheating and other forms of academic dishonesty, and an examination of specific teaching methods that may help to prevent or decrease the occurrence of student cheating.

Perceptions of Teacher Characteristics and Pedagogy

Studies that have examined students' perceptions of teacher characteristics and pedagogical practices have focused primarily on a fairly limited number of variables. In particular, students' perceptions of whether teachers are "fair," particularly in terms of their assessment practices and grading, have been related to the likelihood of cheating (e.g., Genereux & McLeod, 1995; Graham, Monday, O'Brien, & Steffen, 1994; Robinson, Amburgey, Swank, & Faulkner, 2004). Similarly, students' general perceptions that teachers are friendly, approachable, and kind have been examined, along with teachers' communicated enthusiasm for and valuing of their course content and use of appropriate self-disclosure (Genereux & McLeod, 1995; Robinson et al., 2004; Stearns, 2001). Findings related to such perceptions are mixed and difficult to evaluate; this difficulty is exacerbated by limitations in the research methodologies employed in a number of the studies in question. Students' perceptions often appear to have been selected for study with little guiding theoretical framework and, in some instances, measured with single-item indicators (e.g., Genereux & McLeod, 1995; Robinson et al., 2004).

A more systematic approach was taken by Pulvers and Diekhoff (1999), who utilized an established multidimensional measure of the psychosocial climate of college classrooms, The College and University Classroom Environment Instrument (CUCEI; Fraser, Treagust, & Dennis, 1986). This measure, based on Fraser's extensive research on classroom climate in schools (e.g., Fraser, 1986), includes seven scales that measure different dimensions of students' perceptions of their class climate: personalization, involvement, student cohesiveness, satisfaction, task orientation, innovation, and individualization. In a discriminant analysis, Pulvers and Diekhoff found that three scales discriminated between self-reported cheaters and noncheaters. Cheaters reported perceiving their classes as less personalized (i.e., less characterized by individualized attention), less satisfying (enjoyable), and less task oriented (in terms of clear, well-organized class activities) than did noncheaters.

An alternative approach was taken by Murdock and her colleagues. Drawing on research on teachers' communication of "pedagogical caring" (Wentzel, 1997) and communicated respect toward students (e.g., Murdock, Anderman, & Hodge, 2000), Murdock, Hale, and Weber (2001) developed two separate scales related to teacher–student relationships with middle school students. These were students' perceptions of teachers' competence and commitment, defined as their "willingness, preparedness, and competence to teach" (p. 102), and perceptions of respect from teachers within the school. Logistic regression analyses, predicting self-reported cheating versus noncheating, found that both perceived competence and commitment, as well as teacher respect, were negatively associated with cheating.

Interestingly, a similar pattern of student reports emerges in McCabe's (1999) focus group study of high school students. In that research, participants cited teachers' lack of caring about their teaching and poor commitment to student learning as reasons for cheating. As one participant explained:

> A lot of teachers that I've dealt with are always talking about how they can't wait to go home . . . acting like they don't want to be there. Their job is to teach me, and if they can't do that for me, then I'm going to do what I can to move up in the world. If cheating is what I have to do, then that's what I'm going to do. (McCabe, 1999, p. 685)

Interestingly, in a more recent study utilizing hypothetical scenarios with high school students, Murdock, Miller, and Kohlhardt (2004) distinguished between perceived teacher caring and pedagogical competence. They found that both variables were associated with students' judgments of the justifiability of cheating, but that the two dimensions of teacher practice interacted in association with the likelihood of cheating actually occurring. That is, students reported that cheating was less likely to occur in classes where teachers displayed both pedagogical competence and pedagogical caring than in any other scenario. Across two studies, however, pedagogical competence emerged as the most consistently important teacher characteristic. When pedagogy was described as poor, the likelihood of cheating was high. Good pedagogy alone, however, did not guarantee lower rates of cheating. Rather, a combination of high pedagogical competence and either a mastery goal orientation or teacher caring was necessary before a lower probability of cheating was reported. In other words, instructional competence in terms of being able to facilitate students' learning was found to be a necessary but not sufficient condition for lowering the likelihood of cheating. Competence coupled with an adaptive psychosocial context, in either motivational or interpersonal terms, provided the optimal classroom conditions for academic honesty.

A related but distinct construct from teachers' pedagogical caring is the perception of a democratic class climate. Johnston (1996) distinguished between classes in which the teacher holds all of the power for decision making and is seen as the authority figure, and those in which the teacher emphasizes more egalitarian relationships and participatory decision making by the students. Based on qualitative interviewing of her own former students, Johnston suggests that students in democratic classes where their opinions are respected and welcomed report that they are less likely to cheat and find cheating more difficult. Such a democratic climate was viewed as reflecting a greater degree of teacher respect for students and a closer teacher–student relationship that, in course, made cheating less likely. Conversely, students who viewed the teacher as authoritarian were considered more likely to cheat because they viewed it as the teacher's responsibility to stop cheating from occurring. One important point made by Johnston was that even in her classroom, which she attempted to make a democratic one, some students viewed it as her responsibility to stop other students from cheating. So, although a positive relationship and a democratic and respectful class environment may be important inhibitors for some students, it cannot be assumed that a positive teacher–student relationship will end students' cheating behavior altogether.

Interestingly, Murdock et al. (2001), in the study described earlier, found that a democratic participation structure was positively associated with self-reported cheating, when other aspects of the social context, including classroom goal orientation and perceived teacher competence and commitment, were taken into account. In fact, an overreliance on positive relationships with students may lead some instructors into a false sense of immunity from students' dishonesty. Graham et al. (1994) report a finding that may be particularly important to note. In their study of faculty and student attitudes about cheating at smaller colleges, they found that faculty may tend to overestimate the importance of their interpersonal relationships with students. In that study, faculty (46.3 percent) were considerably more likely than students (29.3 percent) to report that students would not cheat because they did not want faculty to think less of them. A further complication is illustrated in a study by Cummings and Romano (2002) who found that the introduction of a class-level honor code was associated with less positive student perceptions of their instructor. In the absence of an institutional-level honor code, a class-level code left students perceiving their instructors as less affinity-seeking—that is, as making less of an effort to get students to like them. At the same time, the class-level honor code did not appear to be related to students' feelings of being trusted by their instructor and was associated with the belief that their instructor viewed cheating seriously. Thus, evidence from multiple studies presents a complex set of findings with relation to students' perceptions of their instructors. On one hand, the perceptions of pedagogical caring, fairness, and enthusiasm and provision of a democratic classroom environment all have been associated with students' avoidance of dishonesty. On the other hand, positive instructor–student relationships and class environments alone are not sufficient to prevent student cheating. Similarly, students report believing that instructors have a responsibility to ensure that others do not cheat, and yet they view instructors who enact honor codes as less supportive and less interested in forming positive relationships with students. Clearly, factors other than instructors' personal and pedagogical characteristics need to be considered in understanding students' beliefs and behaviors relating to academic integrity.

Teacher Attitudes toward Cheating

Beyond students' perceptions of their teachers' personal characteristics and general pedagogical approaches, several studies have pointed specifically at the importance of students' perceptions of their teachers' attitudes toward cheating and academic dishonesty. For example, in McCabe's (1999) interview study, high school students reported the belief that their teachers were aware of and accepted high levels of cheating in their schools. They reported that "the teachers don't care; they let it happen. . . . The students keep on doing it because they don't get into trouble" (p. 683). Similar beliefs have been reported in college populations, with students reporting that opportunities to cheat are common and that instructors make little effort to hold students accountable for dishonest behaviors (e.g., Ng, Davies, Bates, & Avellone, 2003).

Such student perceptions may, in fact, represent accurate observations of many instructors' practices. Despite McCabe and Trevino's (1997) well-cited finding that students are less likely to cheat in those classes where faculty include statements about university honor codes in their syllabi and discuss these policies openly, it may be that relatively few instructors follow these suggestions. Graham et al. (1994) interviewed faculty members in small colleges to determine their attitudes toward student dishonesty and their practices related to ensuring academic integrity in their classrooms. Surprisingly, these authors found that, in general, faculty did not seem to consider academic integrity a high priority, with approximately 20 percent reporting that they did not proctor exams in their classes and 64 percent reporting they did not mention cheating policies in their syllabi. These findings are particularly striking because of the relative availability of literature that focuses on practical teaching methods that teachers use to "promote academic integrity and to minimize cheating" (Stearns, 2001, p. 276). This literature describes specific instructional and classroom management practices, such as recommendations related to the physical layout of the classroom during examinations (Tibbetts, 1997) and use of multiple versions while administering exams (Whitley, 1998). Taken together, such practices may communicate to students that instructors view academic dishonesty seriously. In contrast, a lax approach to preventing cheating may lead some students to feel that teachers do not care about them or their learning. Interestingly, this may actually lead some students who would not have normally cheated to engage in cheating, in the absence of preventative methods. In the qualitative examination of college students' beliefs about cheating described above, Johnston (1996) found that classroom practices (i.e., not proctoring exams) can send a strong message to students about how individual instructors feel about cheating. In addition, her finding that many students believed it was the individual teacher's responsibility to prevent cheating from occurring in their classrooms is consistent with other studies that have reported students expressing support for tougher stances by professors and cheating policies that were enforced in the classroom (e.g., McCabe, Trevino, & Butterfield, 1999, 2001).

In examining potential explanations for why professors choose to ignore instances of cheating, Keith-Spiegel, Tabachnick, Whitley, and Washburn (1998) reported a number of deterrents to instructor action. These included being concerned about the strength of the evidence that cheating has occurred; the difficulty of broaching the topic with students; the amount of time and effort involved in reporting instances of cheating; the fear of reprisals; and lastly, denial that one's students would even cheat in the first place. Thus, it appears that factors at multiple levels of analysis contribute to students' perceptions of teachers' attitudes toward academic dishonesty. Students' observations that instructors do not act on instances of dishonesty may lead to increased cheating and acceptability of dishonest practices, but instructors may not act because they perceive a lack of support or too heavy practical and psychological demands placed on them by the larger institutional context. This multilevel pattern of effects has implications for both educational practice and research. First, if schools and universities want to reduce their students' academic dishonesty, they may need to consider the psychosocial context experienced by their instructors, as well as that of their students. From a research perspective, this pattern calls for multilevel examinations of students' dishonest behaviors, incorporating both classroom- or individual teacher-level variables and characteristics of the larger school context. To date, such complex examinations of the context of student dishonesty have not been common.

RELATIONSHIPS WITH PEERS

Beyond students' perceptions of their relationships and interactions with their teachers, a second important interpersonal dimension of educational settings involves relationships, interactions, and perceptions of the norms of peers. A review of studies of peer influence on academic dishonesty and cheating reveals a number of similarities within the research currently available. First, the majority of studies focus on undergraduate college samples, with a smaller number including students in high school. Almost no studies of peer relationships and influence have been conducted with younger populations (cf., Murdock et al., 2001), despite an extensive literature on the developmental importance of peers for early adolescents (e.g., Brown, 1990; Furman & Buhrmester, 1992; Hartup, 1993). In addition, almost all of the studies reviewed rely on survey methodology, including self-reports of students' own attitudes and behaviors as well as their perceptions of their peers' behaviors and beliefs. In contrast, very few studies have utilized focus group discussions (e.g., Chapman, Davis, Toy, & Wright, 2004; McCabe, 1999) and interview methods (Ng et al., 2003) to provide a more exploratory approach to understanding associations among students' peer relationships and their academic dishonesty. Finally, all of the studies reviewed were fundamentally correlational in nature and relied on data collected at a single point in time.

Beyond the similarities noted, the studies reviewed differed in terms of the specific peer-related variables examined. These differences reflect, in part, differences in the underlying assumptions or theoretical frameworks invoked for understanding why there might be associations between students' academic cheating or beliefs and peer characteristics. The studies also differ in terms of the specific peer relationships examined; that is, the developmental literature on peer relationships in childhood and adolescence makes very clear distinctions between, for example, friendship and acceptance (e.g., Asher, Parker, & Walker, 1996) and friendship and peer group affiliation (e.g., Brown & Lohr, 1987). Although these distinctions tend not to be made explicitly in the literature on academic dishonesty, they do emerge in some of the studies reviewed. The following discussion is organized around the ways in which peer influence or interaction has been examined.

Group Membership

Several studies have examined potential peer influence on college students' academic dishonesty by including items that tap into students' involvement in a range of extracurricular activities (e.g., Baird, 1980; Diekhoff et al., 1996; Kerkvliet, 1994; Michaels & Miethe, 1989). Although the specific activities and group memberships included vary somewhat across studies, a wide range of activities have been examined, including political, musical, and cultural organizations and religious activities, as well as the more common focus on membership in fraternities and sororities and athletic teams (McCabe & Trevino, 1997). In addition, these variables typically are treated as demographic variables, at the same conceptual level as students' gender, age, or grade point average (e.g., Whitley, 1998). Interestingly, McCabe and Trevino argued that, while participation in other types of extracurricular activities should be considered demographic characteristics of the individual, fraternity/sorority membership should be conceptualized as a contextual variable. This suggestion is based on the idea that fraternity/sorority life provides a context within which the "norms, values, and skills associated with cheating" (p. 383) can be easily acquired and in which access to resources and support for facilitating cheating behavior is available. In a meta-analysis of studies of college students' academic dishonesty, Whitley (1998) found that involvement in both fraternities/sororities and other forms of extracurricular activity did have small but significant effects on students' cheating. Importantly, however, he found no evidence to suggest that these effects were any stronger for fraternity and sorority membership than for other forms of activity.

In a recent study, Pino and Smith (2003) also found a positive association between students' membership in a fraternity or sorority and their self-reported academic dishonesty, including cheating and plagiarism. Interestingly, however, these authors conducted a hierarchical regression analysis and found that such membership, though significant in an early step of the analysis, became nonsignificant in later steps when other variables were taken into account. When students' locus of control, self-reported class attendance, and tendency to focus on learning for its own sake, rather than simply on grades, were included in the analysis, fraternity/sorority membership no longer made an independent contribution. This pattern of results suggests that the associations between extracurricular activity and cheating may be mediated through several cognitive and behavioral variables that provide greater explanatory power than do simple demographic descriptors. That is, membership in a fraternity/sorority undoubtedly influences students' academic cheating behavior through several mechanisms: membership may provide material assistance and resources for cheating (such as files of exam questions or term papers to be copied) and also may lead to lower levels of class attendance and less focus on learning. Alternatively, students' choices about joining fraternities and sororities, and about class participation and studying, may all reflect some more general, underlying orientation toward attending college. In other words, Pino and Smith's findings suggest differing potential models of causal effects that could be examined empirically through causal modeling techniques; to date, however, such models are not common in the literature on academic dishonesty. On a more general level, these findings highlight the need for studies that look beyond simple associations between descriptive and demographic variables and cheating, to really understand the processes that underlie such associations.

Perceptions of the Peer Culture

An alternative approach to understanding peer influence has focused on students' beliefs about their peers' attitudes toward cheating and frequency of cheating behavior. Several authors have reported that students perceive a peer "culture of cheating" that influences their own attitudes toward the acceptability of academic dishonesty, and the likelihood of their engaging in dishonest behaviors (e.g., Ameen, Guffey, & McMillan, 1996; McCabe, 1999; Schab, 1991).

One commonly cited phenomenon is the finding that students tend to overestimate the degree of cheating in which their peers and classmates engage (e.g., Jordan, 2001; McCabe & Trevino, 1993; Ng et al., 2003). Furthermore, those students who report more frequent cheating themselves are also more likely to overestimate cheating by others (Jordan). Two theoretical explanations for this overestimation have been proposed. The first is that overestimation is an example of the "false consensus effect" (Ross, Green, & House, 1977)—that is, the tendency of individuals to see their own behavior and attitudes as fairly typical in the larger population.

This belief in a consensus then provides a form of self-justification for the individual's own dishonest behavior (see Chapman et al., 2004). Thus, in terms of college students, this hypothesis assumes that students who are likely to engage in academically dishonest behaviors tend to believe that "everyone else does it," in order to avoid feeling guilty about their own transgressing of formal institutional rules.

A second possible explanation stems from differential association theory (see Vowell & Chen, 2004), a sociological theory that suggests that individuals learn unconventional behavior, just as more conventional behaviors are learned, through their associations with others. That is, individuals learn attitudes, strategies, motivations, and rationalizations for their behavior depending on the values, beliefs, and behaviors of those with whom they form social relationships. Thus, students who cheat may have experienced differential exposure to cheating behavior, compared to their classmates, based on the specific peer group with which they are affiliated. This suggestion provides an explanation for the frequently reported association between academic dishonesty and membership in fraternities and sororities, discussed earlier. In addition, Jordan (2001) provided direct evidence for this phenomenon, in that undergraduates in his study who reported higher levels of actual cheating behavior themselves also reported having seen someone else cheat more often. Similarly, at the institutional level, McCabe and Trevino (1993) stated that students in colleges and universities with lower overall levels of cheating tended to report an institutional-level culture of disapproval of cheating and believed they would be embarrassed to be caught cheating by a classmate. In comparison, students in institutions with high levels of cheating were more likely to report that cheating was acceptable.

Finally, beliefs about friends' cheating behavior may have a more pronounced effect on some students' own intentions to cheat than others. Tibbetts (1997) found that perceptions of friends' cheating had a more pronounced effect on male college students' own intentions than on females'. Few studies have reported exploring such interactions of peer-related variables with the student characteristics, but this seems another important direction for future research. In addition to examining interactions with students' gender, for example, it may be fruitful to consider whether peer characteristics are similarly important for students of different ages or class standing, and achievement (e.g., Finn & Frone, 2004).

Cheating in the Service of Peer Relationships

In contrast to the suggestion that students may come to view cheating as acceptable because of peers' modeling and acceptance of dishonesty, an alternative explanation is that students may engage in cheating behavior, even if they perceive it to be unacceptable, out of concern for maintaining interpersonal relationships with peers. Several researchers have described various types of social goals that students can hold for relationships with peers in educational settings (e.g., Anderman, 1999a, 1999b; Ryan, Keifer, & Hopkins, 2004; Urdan & Maehr, 1996; Wentzel, 1994). Furthermore, students' reported social goals have been linked to a range of academic beliefs and behaviors (e.g., Anderman & Anderman, 1999; Ryan, Hicks, & Midgley, 1997; Wentzel, 1994, 1997). Students' engagement in dishonest academic behaviors also may be associated with their social goals. Although social goals have not been examined explicitly in relation to cheating, the desire and intention to form and maintain positive peer relationships are implicit in several studies. This effect may occur because students come under direct pressure from peers to cheat in order to retain acceptance within the larger peer group (Robinson et al., 2004), or because "passive" cheating, such as working with a peer on an assignment or allowing a classmate to copy one's work, may promote friendship. There also is some evidence that students may decide that "helping out" a peer may contribute to future reciprocity of favors, when that student is in need (Ng et al., 2003). These findings reflect the students' attitude that decisions made about cheating are basically pragmatic; that cheating is seen not as a problem, but more as a reasonable strategy for getting better grades (McCabe, 1999). Certainly, students are well aware that reporting a peer's cheating behavior carries social consequences, and so they are unlikely to do so (McCabe), at least in part out of fear of reprisal (Evans & Craig, 1990).

If institutional and peer norms clash, students may prefer to protect their relationship with their peers at the expense of violating school rules (Eisenberg, 2004). For example, Chapman et al. (2004) reported that undergraduate students in focus groups differentiated between "self-interest cheating" in which the person cheating is the beneficiary (a concept similar to "active cheating," as in Eisenberg) and "social-interest cheating" in which the person is trying to help a friend. Many students in this study did not see social-interest cheating as particularly unethical, and they even reported feeling good about being able to help a friend. The distinction between self-interest and socially motivated types of cheating is reinforced by Chapman et al.'s finding that students reported that they were much more likely to cheat with a friend (75 percent) than with an acquaintance (45 percent). Chapman et al.'s study is unusual in the literature on cheating in distinguishing between qualitatively different types of peer relationships (such as distinguishing friends from acquaintances or classmates), and their findings reinforce the importance of doing so. As noted earlier, future studies also might benefit from explicitly examining the role of students' various social goals in association with their cheating behavior. For example, students who have particularly strong desires to form friendships or gain social acceptance (e.g., Anderman, 1999a; Anderman & Anderman, 1999) may be more vulnerable to engaging in social-interest cheating behaviors, whereas students who endorse social status or performance approach goals (e.g., Anderman, 1999a; Ryan et al., 2004) may be more likely to engage in self-interest cheating. These questions and hypotheses remain to be addressed by future studies.

Cheating Because of Social Comparison

A final category of peer influence on cheating occurs in situations where students may cheat to keep up with their peers academically, so as not to appear incompetent (Ng et al., 2003; Schab, 1991). A number of studies have linked students' individual-level competitiveness or desire to achieve with cheating (e.g., Robinson et al., 2004; Whitley, 1998). Similarly, perceived context-level performance goal orientation and pressure for grades have been linked to cheating (Ameen et al., 1996; Anderman, Griesinger, & Westerfield, 1998; Evans & Craig, 1990; Jordan, 2001; McCabe, 1999). Thus, a focus on social comparison and competition with peers, at both the individual and context level, may increase the likelihood of students' engaging in academically dishonest behaviors. In terms of individual differences, Pulvers and Diekhoff (1999) have suggested that a tendency toward cheating may be linked to the individual's level of need for approval. Similarly, in a meta-analysis, Whitley (1998) found that variables he labeled "interpersonal processes," including fear of negative evaluation by others, high self-monitoring, and sensitivity to others' opinions, had a small but nontrivial effect on college students' academic dishonesty.

In summary, then, a number of studies have documented different peer-related influences on students' cheating-related attitudes and behaviors. The literature in this area, however, reflects multiple different ways in which such influences may occur, and any synthesis of findings is difficult at this point. Future research would benefit from a clearer explication of the specific peer relationships under study and the assumptions underlying hypotheses about social relationships in association with students' dishonesty.

SOCIAL AS PERCEIVED AFFILIATION: RELATIONSHIPS BETWEEN THE INDIVIDUAL AND THE INSTITUTION

A second approach to considering students' affiliative experiences in education examines students' sense of social connectedness and "fit" at the larger setting and institutional level. That is, rather than focusing on specific dyadic or small-group relationships, research has established the importance of constructs such as students' subjective sense of belongingness, relatedness, connectedness, community, or bonding with school, in relation to a wide range of academic and well-being outcomes (e.g., Anderman, 2002; Finn & Frone, 2004; Goodenow, 1993a, b; Goodenow &

Grady, 1993; Resnick et al., 1997; Solomon, Watson, Battistich, & Schaps, 1996; Voelkl & Frone, 2000; also see Anderman & Freeman, 2004, for a review). Despite differences in terminology and definitions of constructs, a common assumption underlying this literature is the importance of students' affiliation with or attachment to the other members of their educational environments. The literature on attachment theory supports numerous positive outcomes associated with an individual's psychological feelings of attachment, beginning with parental relationships and continuing throughout the lifespan, including social relationships in school settings (see Davis, 2003). Similarly, educational researchers have described potential models for the ways in which students' attachment to social groups within their school promote the subjective sense of school belonging or connectedness, which, in turn, facilitates the internalization of the values promoted in those settings (Finn, 1989; Goodenow, 1993a, b). Thus, students' sense of belonging in school stems from their attachment to the people in those settings. Conversely, students' perceptions of alienation, defined by some as "nonattachment," especially in school settings, have been associated with such maladaptive outcomes as truancy, delinquency, sexual promiscuity, drug abuse, and cheating (see Calabrese & Cochran, 1990; Finn, 1989, for reviews).

Much of the cheating-related research in this area has been framed within social bond theory (Hirschi, 1969). Briefly, this theory proposes that a person's attachment to an institution serves as a limiting agent in terms of his or her participation in misconduct, as prescribed by the norms of the institution. More specifically, Hirschi contends that an individual's social bond has four components: attachment, involvement, commitment, and belief in the values adhered to within one's social networks. Once a student feels a sense of attachment to school, it is his or her time investment and commitment to the established relationships within this environment that either reinforce participation in valued activities or inhibit participation in activities contrary to the established belief system. "Thus the restraint on deviant behavior is individuals' concern for the feelings of others, and the desire to be viewed in a positive manner by those to whom they are emotionally close" (Vowell & Chen, 2004, p. 230). In relation to academic dishonesty, therefore, this theory would hypothesize that students' sense of attachment and belonging should serve as a restraint against misconduct, whereas a lack of attachment or sense of alienation would disinhibit behaviors that contradict the norms of academic settings.

Initial investigations into the psychosocial nature of students' academic dishonesty tended to focus on students' alienation in their school environments, based on the assumption that feelings of alienation are the inverse of attachment or bonding to the environment. Such studies have revealed consistent positive associations between students' alienation and their engagement in cheating (e.g., Calabrese & Cochran, 1990; Newhouse,

1982; also see Whitley, 1998, for a review). In other words, the more alienated students feel or are perceived to be, the more likely they are to cheat in their academic endeavors. Alienation, however, has not been defined consistently across studies. For example, Newhouse operationalized alienation in freshmen college students as negative responses to survey items assessing their expectations toward the various groups on campus and views about themselves, their acquaintances, and school in general. In contrast, Calabrese and Cochran defined alienation in high school students as their perceptions of the unfairness of teachers, their school, and a general disliking of school. Finally, Eve and Bromley (1981) measured college students' "culture conflict orientation," a construct similar to alienation, using survey items that tapped students' views of campus rules as unimportant and attraction to "taking chances." Nevertheless, although a uniform definition of alienation has not been utilized across studies, the association between student alienation and cheating appears to be moderately robust (Whitley, 1998).

In contrast to the early focus on student alienation, a number of more recent studies of students' achievement motivation have focused instead on what may be viewed as the opposite perception, the sense of belongingness. The concept of belongingness, relatedness, or sense of community is found in many theories of human motivation. Indeed, Baumeister and Leary (1995) suggested that the need to belong should be conceptualized as a fundamental and universal human need. At least three separate bodies of educational research have included concepts of perceived belonging or relatedness within their theoretical frameworks. For example, self-determination theory (e.g., Deci, Vallerand, Pelletier, & Ryan, 1991) posits that achievement motivation is determined by a student's perceptions of relatedness, autonomy, and competence within educational settings. Similarly, Solomon and his colleagues suggest that belonging is a subcomponent of a student's sense of community, which also includes student participation in the decision making and norm establishment of the school community (e.g., Solomon et al., 1996). Finally, Finn (1989) and his colleagues propose that the two components of belonging and valuing make up a student's academic identity which, combined with participation in the activities of school, has been related to lower dropout rates (Osborne, 1997; Voelkl, 1996, 1997).

Regardless of differences in the theoretical underpinnings, students' subjective sense of class or school belonging has been linked to such positive outcomes as general achievement motivation, academic self-efficacy, positive affect, intrinsic motivation, task value, expectancy for success, and achievement (Anderman, 2002; Anderman, 2003; Freeman, Anderman, & Jensen, 2005; Goodenow, 1993a, b). In addition, belongingness (or connectedness) has been shown to be a protective factor against such negative outcomes as emotional distress, depression, suicidal involvement, violence, cigarette usage, alcohol usage, marijuana usage, early sexual debut, and alcohol and marijuana usage at school (Finn & Frone, 2000; Hoyle & Crawford, 1994; Resnick et al., 1997).

In contrast to the studies of student alienation and academic dishonesty, however, relatively few studies have examined the role of students' sense of belonging in relation to their cheating beliefs and behaviors. One exception is the study by Murdock et al. (2001), which used logistic regression to discriminate between middle school students who reported having cheated and those who did not, in relation to perceived classroom goal structure, personal academic motivation, and social perceptions, including the sense of school belonging. As has been reported elsewhere (i.e., Anderman, 2003; Anderman & Anderman, 1999; Freeman et al., 2005; Roeser, Eccles, & Sameroff, 2000; Roeser, Midgley, & Urdan, 1996), school belonging was significantly correlated with students' academic self-efficacy, personal mastery orientation, perceived classroom mastery goal structure, perceived classroom extrinsic goal structure, perceptions of teacher commitment, classroom participation structure, and perceptions of teacher respect. In the presence of the other variables examined, students' sense of school belonging was not significantly associated with self-reported cheating behavior, although the association was in the expected direction.

In contrast, Finn and Frone (2004) found significant associations between cheating and academic identity in their study of later adolescents. They examined the relationships between students' academic performance, school identification, and cheating in a sample of 16- to 19-year-old high school and college students. School identification was defined as "the extent to which students have a sense of belonging in school and value school and school-related outcomes" (p. 117). As had been shown previously, gender, age, and achievement were significantly related to cheating. Specifically, males, younger students, and lower achievers were more likely to report cheating behaviors. In addition, both academic self-efficacy and school identification were negatively related to cheating. In other words, students with lower levels of efficacy and identification with school were more likely to engage in cheating behaviors. Finn and Frone also examined potential interactions between students' achievement levels and their self-efficacy and school identification. In terms of students' self-efficacy, students with low levels of efficacy did not differ in terms of self-reported cheating behaviors as a function of their academic performance. Among students with high levels of self-efficacy, however, poorer levels of performance were associated with the highest level of cheating, whereas high performance was associated with the lowest level of cheating.

The interaction of students' level of academic performance and school identification was not found to be statistically significant ($p = .08$), but the pattern of results is suggestive. Students in this study who reported high levels of school identification were less likely to cheat, regardless of their level of academic achievement. Among students with low levels of school identification, however, level of academic performance was related to cheating. That is, students who were low in school identification and academic performance reported cheating more often than did other students. In contrast, students with low identification but higher performance were less likely to cheat. Thus, this study supports the importance of school identification in relation to students' cheating behavior and provides at least limited support for the suggestion that this association may be particularly important for students with low levels of academic performance. Further investigations of the moderating effects of prior achievement and expectations of success are warranted.

Recently, Vowell and Chen (2004) expanded the current literature by examining the explanatory power of four prominent sociological theories used to address academic cheating. They compared the results of six different regression models, designed to reflect the hypotheses of the various theories, in the prediction of college students' self-reported cheating behaviors. Each theory-based model was compared to a baseline model, which included only the control variables of students' gender, year in college, age, and marital status. Based on the additional amount of variance explained in the outcome variable, the authors found the strongest support for differential association theory—that is, that students learn attitudes favorable to cheating through their peer relationships and interactions. In addition, some limited support was found for social bond theory, primarily in relation to students' reported attachment to school.

Taken together, therefore, the literature that has examined aspects of students' academic dishonesty in relation to their perceived affiliation with their larger educational contexts, presents a somewhat mixed picture. On one hand, earlier studies that examined various operationalizations of student alienation reported quite robust associations. On the other hand, more recent studies that have examined students' sense of belonging or attachment seem less consistent. One possible reason for this pattern of findings might be some confusion about the definitions of students' alienation, sense of belonging, and academic identity. We propose that these constructs, though related, are conceptually and empirically distinct and may not be equally important in understanding students' cheating beliefs and behaviors.

In addressing students' withdrawal from school, Finn (1989) reviewed prior research that conceptualized alienation as "nonattachment" and "noninvolvement" in school-related activities. According to Seeman (1975, as cited by Finn), however, the experience of alienation encompasses a much more complex set of perceptions than just the absence of belonging. That is, "The essential components of alienation . . . are powerlessness, meaninglessness, normlessness, self-estrangement, social isolation, and cultural estrangement" (Finn, p. 124). Therefore, a conceptualization of belonging and alienation as the inverse of one another is somewhat limited, since social isolation is but one dimension of a more multifaceted construct.

Further complexity arises from Finn's (1989) suggestion that students' sense of belonging is one of two aspects of a student's academic identity. He proposes that academic identity is composed of two constructs: belonging and valuing. In his participation-identification model, Finn states that students' identification with school is facilitated through their active participation in school-related activities, which is reinforced by teachers. Since their academic behaviors are being reinforced, students' identification with and commitment to school are further strengthened, leading to students' continued and increased active participation in academically valued behaviors. Thus, this conceptualization does not focus on students' psychological sense of belonging in school uniquely, since valuing is considered simultaneously. Other researchers, working in an expectancy-value framework, also have identified students' task values as important predictors of their continued academic motivation (e.g., Meece, Wigfield, & Eccles, 1990; Wigfield & Eccles, 1992). In those models, however, students' values are not viewed as a subcomponent of their academic identity. Despite differences in conceptualization, however, both academic identity and school belonging have been linked empirically to such maladaptive academic outcomes as alcohol and marijuana usage at school and academic misconduct (Finn, 1989; Resnick et al., 1997; Roeser et al., 2000; Voelkl & Frone, 2000).

Goodenow (1993b) offers a somewhat different conceptualization of perceived school belonging. She defines school belonging as students' perception of their "psychological membership in the school or classroom, that is, the extent to which students feel personally accepted, respected, included, and supported by others in the school environment" (p. 80). This definition of belonging appears to capture a somewhat more complex set of experiences than simply the sense of social isolation, and yet may not be a direct inverse of alienation, as defined earlier. Thus, there appear to be several similar but potentially distinct constructs in the literature, including perceived alienation, belonging, values, and identity, all of which may have some association to students' academic dishonesty. To date, the extent to which these terms are redundant with one another is not clear, nor do we know which may be the most important in terms of explanatory power.

As a preliminary examination of this question, we recently conducted an explicit test of the factor structure of students' psychological sense of belonging, alienation, valuing, and academic identity (Anderman, Freeman, & Mueller, 2005). Survey items were adapted to the university level from Goodenow's (1993) Psychological Sense of School Membership questionnaire, Voelkl's (1996) Identification with School questionnaire, and Osborne's (1997) Academic Identity scale. Exploratory factor analysis identified five distinct factors, which accounted for over 48 percent of the variance in students' responses. This analysis suggested that items measuring students' belonging, alienation, academic valuing, and positive and negative academic identity were tapping different constructs. Confirmatory factor analysis, conducted on a separate subsample, was used to compare the fit of competing models. These analyses supported the suggestion that belongingness, alienation, valuing, and identity are moderately correlated but distinct constructs. Of particular interest is the separation of the measures of alienation and belonging, suggesting that these constructs have some unique meaning for respondents and are not simply the reverse of one another. In considering the literature on students' academic dishonesty, the mixed findings for students' alienation and belonging, described earlier, may reflect a tendency to use these terms interchangeably and some confounding of the specific constructs under examination.

DIRECTIONS FOR FUTURE RESEARCH

In our review of the literature, we found a reasonably large number of sources that draw connections between students' interpersonal relationships and perceptions and their cheating-related attitudes and behavior. Despite this emerging work, however, it is clear that there are a number of limitations in the empirical research available and that many questions remain to be addressed. First, there is a clear need for more studies that build on established knowledge about teacher–student and peer relationships in educational settings. A large body of research has established important dimensions in the interpersonal and pedagogical relationship between teachers and students, in relation to outcomes such as academic motivation and affective reactions to school (e.g., Patrick et al., 2001; Turner & Meyer, 2004; Wentzel, 1994, 1997). This work can provide a strong foundation for studies of students' academic integrity or dishonesty as well. Similarly, studies of students' relationships with peers and classmates would benefit from differentiating between qualitatively different types of peer relationships and peer-related goals (e.g., Asher et al., 1996; Brown, 1990; Urdan & Maehr, 1996). Integration of these developmental and psychological perspectives into the study of student cheating would give our understanding of social influences more specificity and clarity.

Grounding studies of students' academic dishonesty in psychological theory would also address a second limitation in much of the literature currently available. The examination of theoretical rationales for hypothesized associations between variables leads, almost inevitably, to a consideration of the psychological mechanisms that underlie and explain those associations. This type of rationale is notably lacking in many of the studies we reviewed. That is, much of the currently available research has documented simple relationships between pairs of variables but has done little to explicate *how* and *why* those variables might be related. Thus, in some studies, students' interpersonal relationships come to be viewed more as a demographic variable—that is, as a somewhat stable characteristic of the individual rather than as something dynamic that takes place between two or more people. In addition, treating relationships as demographic markers (e.g., this student does or does not belong to a certain social group) potentially omits important information about the individual's goals for being in that relationship or their perceptions of how they are treated within it (see Anderman, 1999b; Urdan & Maehr, 1996).

A third and related limitation in the extant literature relates to the need to explore individual and group differences in students' perceptions, interpretations, and reactions to the social context of their educational settings. As noted in the discussion of students' perceptions of their peers' beliefs, a small number of studies have reported that students' own characteristics such as their gender (e.g., Tibbetts, 1997) or level of achievement (e.g., Finn & Frone, 2004) can moderate the effects of their perceptions on their own beliefs and behaviors. Again, consideration of potential moderation effects would move the literature beyond reporting simple bivariate associations to provide a more sophisticated understanding of students' academic dishonesty. In particular, a psychological approach might examine the contribution of alternative moderators of the effects of contextual variables, such as students' motivational characteristics, beliefs about the purpose of schooling, and learning strategies.

One important implication of our call for a more complex conceptualization of students' interpersonal relationships in the study of cheating is that the methodological approaches utilized would have to become similarly complex. As noted throughout this chapter, many of the studies currently available rely on quite limited approaches in terms of overall design and data collection. A fuller understanding of the interpersonal context of educational settings requires moving beyond reliance on survey methodology to include data from multiple sources. For example, understanding teacher–student relationships and the class-level social climate may require combining self-report data with direct observation of classroom interactions. Similarly, a much fuller understanding of the reciprocal effects of peer group influences and students' cheating could be gained from tracking students' experiences over the course of an academic semester, as compared to data collected at a single point in time. In addition, the field would benefit considerably from researchers who do include survey data in their studies utilizing established measures or developing scales with testable psychometric properties, rather than relying on single items to represent quite complex constructs.

Finally, studies of social contextual influences on students' academic integrity and dishonesty would benefit from including more multilevel examinations of contexts. As has been demonstrated in other areas of educational research, simultaneous examination of individual and context-level variables through multilevel modeling techniques has tremendous potential to enrich our understanding of complex phenomena (e.g., Lee, 2000). To date, such approaches are relatively rare in studies of students' cheating. In summary, therefore, there seems to be a need for more sophisticated and nuanced studies of the interplay between students' interpersonal and affiliative experiences in educational settings, including their relationships with teachers and peers, and their sense of belonging to the institution, as well as their cheating-related attitudes and behaviors. Despite the burgeoning concern over rates of student dishonesty and the number of empirical articles available, much remains to be understood about the underlying causes of this phenomenon.

IMPLICATIONS FOR EDUCATORS

Despite the limitations in the available literature base on students' interpersonal and affiliative experiences in relation to academic dishonesty, some implications for educational practice do appear to be emerging. In terms of individual instructors, operating at the classroom level, the evidence seems fairly clear that certain pedagogical characteristics are quite robustly associated with a reduced likelihood of student cheating. Note that this does not mean that any one of these characteristics, or even a combination of them, will prevent all instances of student dishonesty. Contextual variables, even in optimal conditions, are not the only factors in play. This caveat, however, does not mean that instructor practices are unimportant.

Synthesizing across the available evidence, it seems that a combination of instructor competence, enthusiasm, and reduced emphasis on competition and public performance is most likely to reduce student dishonesty. In terms of competence, instructors who provide clear, well-organized learning activities and individualized attention, and who are fair in their assessment and grading practices, are less likely to promote student dishonesty. In addition, demonstrating a commitment to student learning and projecting one's own enthusiasm for and valuing of the content of one's courses may also act as a preventative measure (e.g., Murdock et al., 2001; Pulvers & Diekhoff, 1999). Beyond these general characteristics of competent and enthusiastic instruction, there also is evidence that students are more likely to cheat in situations they perceive as highly competitive or focused on grades and performance (Anderman et al., 1998). Thus, instructors may help to prevent academic dishonesty by deemphasizing competition and protecting students' privacy in terms of their achievement levels. Also, in view of the fact that students may be willing to violate institutional rules in the interests of helping friends and maintaining peer relationships (Eisenberg, 2004), instructors might consider incorporating at least some collaborative activities into their classes. Such activities could serve two purposes: first, they provide a legitimate opportunity to work with classmates and address peer social issues within the context of the class. Second, they provide an opportunity to discuss the distinctions between collaboration and dishonesty, and to clarify for students where the limits of acceptable behavior lie. Finally, several studies have shown that students view their instructors as responsible for preventing cheating and expect them to treat instances of dishonesty seriously (e.g., Johnston, 1996; McCabe et al., 2001). Lax monitoring of students' behavior or failing to act on cases of cheating will create an environment that encourages further dishonesty.

Beyond strictly pedagogical practices, instructors' relationships with students and the social climate created in their classes also have some relation to the likelihood of student dishonesty. As discussed earlier, students report being less likely to cheat when they perceive instructors to be friendly, approachable, and respectful in their interactions. It is also clear, however, that instructors cannot rely too heavily on having good rapport with students to prevent dishonest actions (Graham et al., 1994). Rather, instructors need to balance the creation of a positive class climate with their responsibility to provide high-quality instruction and to set expectations for academic integrity. As we have discussed elsewhere (Anderman et al., 2002), it is in this balancing of warm, appropriate interpersonal interactions with high-quality instruction and academic expectations that teachers create the optimal environment for students to thrive. In view of the currently available research, it appears that such an environment also is most likely to ameliorate the current increasing trend toward academic dishonesty in our students.

REFERENCES

Alschuler, A. S., & Blimling, G. S. (1995). Curbing epidemic cheating through systemic change. *College Teaching, 43*, 123–126.

Ameen, E. C., Guffey, D. M., & McMillan, J. J. (1996). Accounting students' perceptions of questionable academic practices and factors affecting their propensity to cheat. *Accounting Education, 5*, 191–205.

Anderman, E. M. (2002). School effects on psychological outcomes during adolescence. *Journal of Educational Psychology, 94*, 795–809.

Anderman, E. M., Griesinger, T., & Westerfield, G. (1998). Motivation and cheating during early adolescence. *Journal of Educational Psychology, 90*, 84–93.

Anderman, L. H. (1999a). Classroom goal orientation, school belonging and social goals as predictors of students' positive and negative affect following the transition to middle school. *Journal of Research and Development in Education, 32*, 89–103.

Anderman, L. H. (1999b). Expanding the discussion of social perceptions and academic outcomes: Mechanisms and contextual influences. In T. Urdan (Ed.), *Advances in motivation and achievement, Vol. 11*, pp. 303–336. Greenwich, CT: JAI.

Anderman, L. H. (2003). Academic and social perceptions as predictors of change in middle school students' sense of school belonging. *Journal of Experimental Education, 72*, 5–23.

Anderman, L. H., & Anderman, E. M. (1999). Social predictors of changes in students' achievement goal orientations. *Contemporary Educational Psychology, 25*, 21–37.

Anderman, L. H., & Anderman, E. M. (Eds.). (2000). The role of social context in Educational Psychology: Substantive and Methodological Issues [Special issue]. *Educational Psychologist, 35*(2).

Anderman, L. H., & Freeman, T. M. (2004). Students' sense of belonging in school. In M. Maehr & P. Pintrich (Eds.), *Advances in motivation and achievement: Vol. 13. Motivating students, improving schools: The legacy of Carol Midgley* (pp. 27–63). Oxford: Elsevier, JAI.

Anderman, L. H., Freeman, T. M., & Mueller, C. (2005). Factor structure of students' sense of university belonging, alienation and academic identity. Manuscript in preparation.

Anderman, L. H., Patrick, H., Hruda, L. Z., & Linnenbrink, L. (2002). Observing classroom goal structures to clarify and expand goal theory. In C. Midgley (Ed.), *Goals, goal structures, and patterns of adaptive learning* (pp. 243–278). Hillsdale, NJ: Lawrence Erlbaum.

Asher, S. R., Parker, J. G., & Walker, D. L. (1996). Distinguishing friendship from acceptance: Implications for intervention and assessment. In W. M. Bukowski, A. F. Newcomb, & W. W. Hartup (Eds.), *The company they keep: Friendship in childhood and adolescence*. New York: Cambridge University Press.

Baird, J. S., Jr. (1980). Current trends in college cheating. *Psychology in the Schools, 17*, 515–522.

Baumeister, R. F., & Leary, M. R. (1995). The need to belong: desire for interpersonal attachments as a fundamental human motivation. *Psychological Bulletin, 117*, 497–529.

Brown, B. B. (1990). Peer groups and peer cultures. In S. S. Feldman & G. R. Elliott (Eds.), *At the threshold: The developing adolescent* (pp. 171–196). Cambridge, MA: Harvard University Press.

Brown, B. B., & Lohr, M. (1987). Peer group affiliation and adolescent self-esteem: An integration of ego-identity and symbolic interaction theories. *Journal of Personality and Social Psychology, 52*, 47–55.

Calabrese, R. L., & Cochran, J. T. (1990). The relationship of alienation to cheating among a sample of American adolescents. *Journal of Research and Development in Education, 23*, 65–72.

Chapman, K. J., Davis, R., Toy, D., & Wright, L. (2004). Academic integrity in the business school environment: I'll get by with a little help from my friends. *Journal of Marketing Education, 26*, 236–249.

Cummings, K., & Romano, J. (2002). Effect of an honor code on perceptions of university instructor affinity-seeking behavior. *Journal of College Student Development, 43*, 862–875.

Davis, H. A. (2003). Conceptualizing the role and influence of student-teacher relationships on children's social and cognitive development. *Educational Psychologist, 38*, 207–234.

Deci, E. L., Vallerand, R. J., Pelletier, L. G., & Ryan, R. M. (1991). Motivation and education: The self-determination perspective. *Educational Psychologist, 26*, 325–346.

Diekhoff, G. M., LaBeff, E. E., Clark, R. E., Williams, L. E., Francis, B., & Haines, V. J. (1996). College cheating: Ten years later. *Research in Higher Education, 37*, 487–502.

Eisenberg, J. (2004). To cheat or not to cheat: Effects of moral perspective and situational variables on students' attitudes. *Journal of Moral Education, 33*, 163–178.

Evans, E. D., & Craig, D. (1990). Teacher and student perceptions of academic cheating in middle and senior high schools. *Journal of Educational Research, 84,* 44–52.

Eve, R. A., & Bromley, D. G. (1981). Scholastic dishonesty among college undergraduates: Parallel tests of two sociological explanations. *Youth and Society, 13,* 3–23.

Finn, J. D. (1989). Withdrawing from school. *Review of Educational Research, 59,* 117–142.

Finn, K. V., & Frone, M. R. (2004). Academic performance and cheating: Moderating role of school identification and self-efficacy. *The Journal of Educational Research, 97,* 115–122.

Fraser, B. J. (1986). *Classroom environment.* London: Croom Helm.

Fraser, B. J., Treagust, D. F., & Dennis, N. C. (1986). Development of an instrument for assessing classroom psychosocial environment at universities and colleges. *Studies in Higher Education, 11,* 43–54.

Freeman, T. M., Anderman, L. H., & Jensen, J. M. (2005). The psychological sense of belonging in college freshmen: Relations to motivation and instructor practices. Unpublished manuscript.

Fuhrman, W., & Buhrmester, D. (1992). Age and sex differences in perceptions of networks of personal relationships. *Child Development, 63,* 103–115.

Genereux, R. L., & McLeod, B. A. (1995). Circumstances surrounding cheating: A questionnaire study of college students. *Research in Higher Education, 36,* 687–699.

Goodenow, C. (1993a). Classroom belonging among early adolescent students: Relationships to motivation and achievement. *Journal of Early Adolescence, 13,* 21–43.

Goodenow, C. (1993b). The psychological sense of school membership among adolescents: Scale development and educational correlates. *Psychology in the Schools, 30,* 79–90.

Goodenow, C., & Grady, K. E. (1993). The relationship of school belonging and friends' values to academic motivation among urban adolescent students. *Journal of Experimental Education, 62,* 60–71.

Graham, M. A., Monday, J., O'Brien, K., & Steffen, S. (1994). Cheating at small colleges: An examination of student and faculty attitudes and behaviors. *Journal of College Student Development, 35,* 255–260.

Harter, S. (1996). Teacher and classmate influences on scholastic motivation, self-esteem, and level of voice in adolescents. In J. Juvonen & K. R. Wentzel (Eds.), *Social motivation: Understanding children's school adjustment* (pp. 11–42). New York: Cambridge University Press.

Hartup, W. W. (1993). Adolescents and their friends. In B. Laursen (Ed.), *Close friendships in adolescence* (pp. 3–22). San Francisco: Jossey-Bass.

Hendershott, A., Drinan, P., & Cross, M. (2000). Toward enhancing a culture of academic integrity. *NASPA Journal, 37,* 587–597.

Hirschi, T. (1969). *Causes of delinquency.* Berkeley: University of California Press.

Hoyle, R. H., & Crawford, A. M. (1994). Use of individual-level data to investigate group phenomena. *Small Group Research, 25,* 464–484.

Johnston, D. K. (1996). Cheating: Limits of individual integrity. *Journal of Moral Education, 25,* 159–172.

Jordan, A. E. (2001). College student cheating: The role of motivation, perceived norms, attitudes, and knowledge of institutional policy. *Ethics & Behavior, 11,* 233–247.

Keith-Spiegel, P., Tabachnick, B. G., Whitley, B. E., Jr., & Washburn, J. (1998). Why professors ignore cheating: Opinions of a national sample of psychology instructors. *Ethics & Behavior, 8,* 215–227.

Kerkvliet, J. (1994). Cheating by economics students: A comparison of survey results. *Journal of Economic Education, 25,* 121–133.

Lee, V. E. (2000). Using hierarchical linear modeling to study social contexts: The case of school effects. *Educational Psychologist, 35,* 125–141.

McCabe, D. L. (1999). Academic dishonesty among high school students. *Adolescence, 34,* 681–687.

McCabe, D. L., & Trevino, L. K. (1993). Academic dishonesty: Honor codes and other contextual influences. *Journal of Higher Education, 64,* 520–538.

McCabe, D. L., & Trevino, L. K. (1997). Individual and contextual influences on academic dishonesty: A multicampus investigation. *Research in Higher Education, 38,* 379–396.

McCabe, D. L., Trevino, L. K., & Butterfield, K. D. (1999). Academic integrity in honor code and non-honor code environments: A qualitative investigation. *Journal of Higher Education, 70,* 211–234.

McCabe, D. L., Trevino, L. K., & Butterfield, K. D. (2001). Cheating in academic institutions: A decade of research. *Ethics & Behavior, 11,* 219–232.

Meece, J., Wigfield, A., & Eccles, J. S. (1990). Predictors of math anxiety and its influence on young adolescents' course enrollment intentions and performance in mathematics. *Journal of Educational Psychology, 82,* 60–70.

Michaels, J. W., & Miethe, T. D. (1989). Applying theories of deviance to academic cheating. *Social Science Quarterly, 70,* 870–885.

Murdock, T. B., Anderman, L. H., & Hodge, S. (2000). Motivational context, student beliefs, and alienation: Stability and change from middle school to high school. *Journal of Adolescent Research, 15,* 327–351.

Murdock, T. B., Hale, N. M., & Weber, M. J. (2001). Predictors of cheating among early adolescents: Academic and social motivations. *Contemporary Educational Psychology, 26,* 96–115.

Murdock, T. B., Miller, A., & Kohlhardt, J. (2004). Effects of classroom context variables on high school students' judgments of the acceptability and likelihood of cheating. *Journal of Educational Psychology, 96,* 765–777.

Newhouse, R. C. (1982). Alienation and cheating behavior in the school environment. *Psychology in the Schools, 19,* 234–237.

Ng, H. W. W., Davies, G., Bates, I., & Avellone, M. (2003). Academic dishonesty among pharmacy students: Investigating academic dishonesty behaviours in undergraduates. *Pharmacy Education, 3,* 261–269.

Osborne, J. W. (1997). Identification with academics and academic success among community college students. *Community College Review, 25,* 59–69.

Patrick, H., Anderman, L. H., Ryan, A. M., Edelin, K., & Midgley, C. (2001). Teachers' communication of goal orientations in four fifth-grade classrooms. *Elementary School Journal, 102,* 35–58.

Pianta, R. C. (1999). *Enhancing relationships between children and teachers.* Washington, DC: American Psychological Association.

Pino, N. W., & Smith, W. L. (2003). College students and academic dishonesty. *College Student Journal, 37,* 490–501.

Pulvers, K., & Diekhoff, G. M. (1999). The relationship between academic dishonesty and college classroom environment. *Research in Higher Education, 40,* 487–498.

Resnick, M. D., Bearman, P. S., Blum, R. W., Bauman, K. E., Harris, K. M., Jones, J., Beuhring, T., Sieving, R. E., Shew, M., Ireland, M., Bearinger, L. H., & Udry, R. (1997). Protecting adolescents from harm: Findings from the National Longitudinal Study on Adolescent Health. *Journal of the American Medical Association, 278,* 823–832.

Robinson, E., Amburgey, R., Swank, E., & Faulkner, C. (2004). Test cheating in a rural college: Studying the importance of individual and situational factors. *College Student Journal, 38,* 380–396.

Roeser, R. W., Eccles, J. S., & Sameroff, A. J. (2000). School as a context of early adolescents' academic and social-emotional development: A summary of research findings. *Elementary School Journal, 100,* 443–471.

Roeser, R. W., Midgley, C., & Urdan, T. C. (1996). Perceptions of the school psychological environment and early adolescents' psychological and behavioral functioning in school: The mediating role of goals and belonging. *Journal of Educational Psychology, 88,* 408–422.

Ross, L., Green, D., & House, P. (1977). The false consensus effect: An egocentric bias in social perception and attribution processes. *Journal of Experimental Social Psychology, 13,* 279–301.

Ryan, A. M., Hicks, L., & Midgley, C. (1997). Social goals, academic goals, and avoiding seeking help in the classroom. *Journal of Early Adolescence, 17*(2), 152–171.

Ryan, A. M., Keifer, S. M., & Hopkins, N. B. (2004). Young adolescents' social motivation: An achievement goal perspective. In P. R. Pintrich & M. L. Maehr (Eds.), *Advances in motivation and achievement, Vol. 13. Motivating students, improving schools: The legacy of Carol Midgley* (pp. 301–330). Oxford: Elsevier.

Schab, F. (1991). Schooling without learning: Thirty years of cheating in high school. *Adolescence, 26,* 839–848.

Solomon, D., Watson, M., Battistich, V., Schaps, E., & Delucchi, K. (1996). Creating classrooms that students experience as communities. *American Journal of Community Psychology, 24,* 719–748.

Stearns, S. A. (2001). The student-instructor relationship's effects on academic integrity. *Ethics & Behavior, 11,* 275–285.

Tibbetts, S. G. (1997). Gender differences in students' rational decisions to cheat. *Deviant Behavior, 18,* 393–414.

Turner, J. C., & Meyer, D. K. (2004). Are challenge and caring compatible in middle school mathematics classrooms? In P. R. Pintrich & M. L. Maehr (Eds.), *Advances in motivation and achievement, Volume 13. Motivating students, improving schools: The legacy of Carol Midgley* (pp. 331–360). Oxford: Elsevier.

Urdan, T. C., & Maehr, M. L. (1996). Beyond a two-goal theory of motivation and achievement: A case for social goals. *Review of Educational Research, 65,* 213–243.

Voelkl, K. E. (1996). Measuring students' identification with school. *Educational and Psychological Measurement, 56,* 760–770.

Voelkl, K. E. (1997). Identification with school. *American Journal of Education, 105,* 294–316.

Voelkl, K. E., & Frone, M. R. (2000). Predictors of substance use at school among high school students. *Journal of Educational Psychology, 92,* 583–592.

Vowell, P. R., & Chen, J. (2004). Predicting academic misconduct: A comparative test of four sociological explanations. *Sociological Inquiry, 74,* 226–249.

Wentzel, K. R. (1994). Relations of social goal pursuit to social acceptance, classroom behavior, and perceived social support. *Journal of Educational Psychology, 86,* 173–182.

Wentzel, K. R. (1997). Student motivation in middle school: The role of perceived pedagogical caring. *Journal of Educational Psychology, 89,* 411–419.

Whitley, B. E., Jr. (1998). Factors associated with cheating among college students: A review. *Research in Higher Education, 39,* 235–274.

Whitley, B. E., Jr., & Keith-Spiegel, P. (2001). Academic integrity as an institutional issue. *Ethics & Behavior, 11,* 325–342.

Wigfield, A., & Eccles, J. S. (1992). The development of achievement task values: A theoretical analysis. *Developmental Review, 12,* 265–310.

10

IS CHEATING WRONG? STUDENTS' REASONING ABOUT ACADEMIC DISHONESTY

TAMERA B. MURDOCK AND
JASON M. STEPHENS

When a world history teacher brought up the topic of cheating to her public high school students in Fairfax, Virginia, they expressed strong opinions that it was not "okay," and that even minor instances of dishonesty should not be tolerated. Days later, however, when the teacher of these same students submitted their papers to *turnitin.com*, she discovered that over 67 percent of them had lifted some text directly from Internet sources, with a few of them downloading over 80 percent of the material. This seeming contradiction between beliefs and action, as presented on ABC's *Primetime Live* (Velmans & Koppel, 2004), was one of the topics pursued during the one-hour investigation of academic cheating. Like many of the published articles on cheating, this television show revealed that while many students purport to believe that cheating, in the abstract, is wrong, they are also dishonest at some point.

Many students also go to great lengths to make sure their teachers do not suspect them of dishonesty, further underscoring students' awareness that cheating is not an acceptable behavior. In a qualitative study of undergraduates, students described the elaborate impression management strategies they adopted to ensure that their instructors saw them as honest people, such as looking at the ceiling when pondering the answer to a test question, dressing in clothes that have no pockets, and maintaining a serious look on their face during an examination (Albas & Albas, 1993). These students not only appear to recognize that harsh judgments may accompany being suspected of academic dishonesty, but like most of us,

Copyright © 2007 by Academic Press, Inc.
All rights of reproduction in any form reserved.

they also work to avoid negative evaluation from others. How is it then that students explain the high prevalence of their own and others' dishonest behavior?

This chapter focuses on students' reasoning about the acceptability or unacceptability of cheating in different situations and the effectiveness of various contextual variables for buffering negative judgments of people who have been dishonest. After examining the relationship of cheating to moral reasoning, we move into a discussion of the ways in which students neutralize cheating behavior, even if they think it is objectively "wrong." Neutralization is a social cognitive process that obscures or negates personal responsibility for one's conduct (Sykes & Matza, 1957). We begin with a general discussion of neutralizing attitudes and strategies and their relations to academic dishonesty. Subsequent sections detail the psychological mechanisms underlying students' adoption of various neutralizing strategies using an attributional theory of excuse giving (Weiner, 1995) to organize the findings. A specific focus within this section is on the use of neutralizing mechanisms to shift the blame for cheating away from the self and onto the teacher or class where the cheating occurs. Finally, we discuss one of the factors that appears to moderate the effectiveness of students' attributions for dishonesty: the goal they try to achieve by cheating. The chapter concludes with suggestions for educators and for future research and discusses justification within a larger societal context.

MORAL REASONING AND CHEATING

The oft-observed incongruity between belief and behavior is also reflected in the literature on moral reasoning and academic dishonesty. To a large extent, these studies assess morality with instruments that are grounded in Kohlberg's theoretical framework (see Kohlberg, 1984). From Kohlberg's perspective, moral development progresses through six identifiable stages; each higher stage characterizes more mature reasoning. Stages 1 and 2 comprise the "preconventional" level of moral reasoning, in which judgments are based primarily on obedience, avoidance of punishment, and instrumental need and exchange ("I'll help you if you help me"). At Stages 3 and 4, the "conventional" level of moral reasoning, judgments are based primarily on role obligations, social approval, and respect for religious and/or legal rules and authority ("Cheating is wrong because it's against school rules"). Stages 5 and 6 comprise the "postconventional" level of moral reasoning, in which judgments are based primarily on agreed upon contractual or procedural arrangements and universalizable ethical principles ("Do unto others as you would have them do unto you"). Longitudinal studies of moral judgment reveal that very few people reach the highest two stages of development, with most only achieving Stage 4 (Colby, Kohlberg, Gibbs, & Lieberman, 1983; Rest, 1986).

For Kohlberg and other cognitive developmentalists, most notably Jean Piaget (1932) upon whom his model was based, it is the *reasoning* about moral decisions (and not the decision itself) that is of most interest. Any decision, including the decision to rob a store or cheat on an exam, could be seen as justified (the morally "right" thing to do) at any stage of reasoning. Nonetheless, in his early theorizing about the relations between moral development and behavior, Kolhberg (1984) claimed that moral reasoning "can be a quite powerful and meaningful predictor of action" (p. 397). Empirical investigations exploring these relations have produced mixed results.

In his seminal review of the literature on moral cognition and moral action, Blasi (1980) found a clear relationship between moral reasoning and honest behavior in only 7 of 17 experimental studies where cheating was the dependent variable. Similarly, Whitley's (1998) more recent review of literature on academic dishonesty found only small negative correlations between moral reasoning and cheating behavior. Most of the studies in these reviews took place in laboratory settings, where cheating is usually assessed by comparing students' performance on a task (where it is possible to cheat) with the preestablished norms of what is realistic to honestly achieve on that task. Leming (1980), for example, found that while undergraduates high in moral development cheated significantly less in a high-threat/high-supervision situation, they cheated just as often as students low in moral development in the low-threat/low-supervision condition. Similarly, Malinowski and Smith (1985) found that while college students high in moral judgment cheated less overall than their low-reasoning counterparts, they cheated just as often when the temptation to do so became strong (see also Corcoran & Rotter, 1987).

More recently, Bruggeman and Hart (1996) found no differences in moral reasoning or cheating behavior between students enrolled in private religious and public secular high schools, and no relation between moral reasoning and cheating behavior. Similarly, in the only published naturalistic study of the relation between moral judgment and academic dishonesty, Lanza-Kaduce and Klug (1986) found only a weak, nonsignificant correlation ($r = .07$) between moral reasoning and self-reported test cheating in a sample of college freshmen and sophomores.

WHY IS MORAL REASONING NOT CLEARLY RELATED TO CHEATING?

Hypotheses about the lack of congruence between people's level of moral development and their moral behavior can be found in the later work of Kohlberg (see Kohlberg & Candee, 1984) as well as the work of Blasi (1984) and Turiel (1983). As noted earlier, Kohlberg (1969) initially overestimated the extent to which moral reasoning ability influenced actual behavior. Following Blasi's (1980) review, Kohlberg (see Kohlberg & Candee, 1984) clarified that the relation between moral reasoning and moral behavior was mediated by two distinct but related types of moral judgment: a deontic judgment, a "first-order" judgment concerning the rightness or wrongness of given action that was directly deduced from a moral stage or principle, and a responsibility judgment, a "second-order" judgment concerning one's personal obligation to act in accord with that judgment (Kohlberg & Candee, 1984, p. 57). One might, for example, believe that cheating on a test is morally wrong but not feel personally responsible for behaving in a manner consistent with that deontic judgment. This process of externalizing responsibility for one's own behavior is a complex and critical one and will be discussed at length later in this chapter.

CHEATING AS AN ISSUE OF MORALITY VERSUS SOCIAL CONVENTION

Turiel's (1983) domain theory of social reasoning offers another plausible explanation of ambiguous relations between moral reasoning and academic cheating. Domain theory distinguishes between issues that people view as within the realm of morality from those that are seen as matters of social convention. The moral domain encompasses behaviors that concern justice, rights, and human welfare; these behaviors often have intrinsic interpersonal consequences, such as stealing from others, discriminating against them in employment, or otherwise causing physical or psychological harm to them. In contrast, the conventional domain includes socially negotiated uniformities in speech and behavior; these customs or norms, such as addressing doctors by their title, are seen as arbitrary because the behaviors they regulate do not produce inherently positive or negative effects on others. Accordingly, actions in these two domains are presumed to be judged by different sets of criteria, with contextual variables having more of an influence on people's judgments when they are in the social versus the moral domain. Only moral rules or principles are regarded as universal and carry the kind of prescriptive force that would make deviations difficult to rationalize.

The framing of cheating as either a moral or social convention issue has been a focus of recent study. Middle school students were classified as either "morals" (i.e., saw cheating as a moral issue) or "a-morals" (i.e., saw cheating as an issue of social convention) based on their reasoning and emotions about cheating (Eisenberg, 1989). Morals rated cheating for personal gain, called active cheating, and allowing others to copy one's answers, termed passive cheating, as less acceptable than did the students who saw cheating as an issue of social convention. Across both groups of students, passive and active cheating were viewed as more acceptable in the vignettes depicting someone cheating in low-versus high-supervision conditions and when peers' norms were supportive versus unsupportive of cheating. Exam importance also had a small effect on the perceived acceptability of active cheating; dishonesty was slightly more acceptable for high-importance exams than for low-importance exams. In short, framing cheating as a moral issue resulted in harsher judgments than viewing it as a social convention. However, the context in which the cheating occurred affected students' judgments about the severity of the offense, regardless of which lens they used to evaluate the situation.

Context appears to be an important component of the decision to cheat. Since the classic studies of Hartshorne and May (1928), scholars have consistently found that people who are honest in one domain do not exhibit similar behavior in all situations; cheating and reasoning about cheating are affected by the contexts in which they occur. Indeed, Blasi (1980) argues, these contextual variations in situations afford people the opportunity to act in ways that contradict their moral reasoning. Specifically, he argues that people's moral reasoning will be unrelated to their actual actions to the extent that people can find ways of neutralizing or rationalizing the behavior, making it something that no longer reflects on them personally. The extant literature is quite consistent with Blasi's assertion. Academic cheating is strongly related to students' contextualized attitudes about dishonesty. Two reviews of the literature on academic dishonesty report cheating attitudes as one of the strongest predictors of dishonest behavior (Bushway & Nash, 1977; Whitley, 1998). Accordingly, understanding the reasoning processes that students engage in when deciding to cheat is an important area of study.

NEUTRALIZING, JUSTIFYING, AND CHEATING

> *"It perfectly ok to cheat in some circumstances."* High school student on Primetime Live *(Velmans & Koppel, 2004)*

Neutralization was first introduced by Sykes and Matza (1957) as part of a theory of deviant behavior and refers to the use of one of numerous strategies to deflect responsibility for deviant behavior away from oneself, presumably lessening the negative judgments made by oneself and others for the behavior. Common categories of neutralizing strategies include denial of the crime (i.e., the behavior was not wrong), denial of the victim (i.e., it doesn't harm anyone), denial of responsibility (i.e., there were things outside my control that led to this), condemnation of the condemners (i.e., the teacher's fault), and appeal to higher loyalties (i.e., other goals or values were more important).

Data consistently find that the prevalence of neutralizing attitudes toward cheating is stronger among more dishonest students. For example, one study examined cheating in relation to four types of neutralizing techniques: denial of responsibility; appeal to a higher loyalty; condemnation of the condemners, in this case shifting the blame to the instructor based on some aspect of their behavior; and denial of injury in a sample of college athletes and nonathletes (Storch, Storch, & Clark, 2002). After controlling for gender and athletic status, denial of responsibility and appealing to higher loyalties were significant predictors of self-reported cheating. Similarly, in another study of college students, those who reported having cheated in the past year endorsed 11 neutralizing techniques more so than non-cheaters (Haines, Diekhoff, LaBeff, & Clark, 1986). Across both cheaters and noncheaters, the most acceptable techniques were to externalize blame on the teacher, by making statements such as "The instructor doesn't care if I learn" or "The instructor assigns too much work." Most interesting, however, was the relationship between students' endorsement of neutralizing strategies and their perceived effectiveness of various deterrents to cheaters. The more that students neutralized cheating, the less effective they perceived guilt, embarrassment, and negative reactions from friends to be a deterrent to dishonesty and the more they perceived punishments from the university, such as an F for cheating or being dropped from a class, as being effective. In other words, not only do cheaters see themselves as less responsible for cheating than do noncheaters (i.e., neutralize more), but they also report that effective cheating interventions would rely on external rather than internal controls. Reducing dishonesty is not their responsibility!

Actual rather than self-reported cheating was examined in a study of the relationship between gender, excuse-making tendencies, and cheating among undergraduates (Ward & Beck, 1990). In the early part of the semester, students rated the extent to which they approved or disapproved of five reasons for cheating, each one reflecting one of the five categories of neutralizing defined by Sykes and Matza (1957). At a later point in the semester, these same students had an opportunity to grade their own tests, after the instructor had already secretly photocopied and graded them. Students who cheated while grading their own exams could therefore be identified, and 36 of the 128 students did so. Among the men in the class, approximately 35 percent of both the cheaters and noncheaters had high neutralization scores; among women, however, only 30 percent of noncheaters were high in neutralizing in contrast to 73 percent of the cheaters.

Students also perceive that cheaters are more likely to neutralize dishonest behavior when compared to their more honest classmates. Teacher education students rated the perceived frequency with which their peers used various neutralizing techniques (Daniel, Blount, & Ferrell, 1991). The strategies that were endorsed as being used most frequently were all external to the student: the material being too hard, the student having job or family responsibilities, and the instructor assigning too much work. Moreover, there was a strong correlation (.62) between the participants' perceptions of the total number of justifications their peers utilized and the amount of cheating their peers engaged in.

HOW NEUTRALIZING STRATEGIES WORK: AN ATTRIBUTIONAL PERCEPTIVE

From a social psychological perspective, neutralizing strategies can be seen as excuses or accounts of bad behavior, and their effectiveness can be understood through the application of attributional principles (Weiner, 1995, 2001; Weiner, Figueroa-Munoz, & Kakihara, 1991). Attributional theory is a cognitive-emotional theory of motivation that seeks to explain how people's cognitive appraisals of a particular outcome or event influence their emotions and their subsequent behavior. Although its specific application to cheating has been somewhat limited, it has been successfully applied to explaining achievement motivation, helping behavior, and aggression, as well as to explaining people's reactions to behaviors that might be considered moral transgressions.

Attribution theory rests on the assumption that people are innately motivated to make sense of their world; as part of that process, they seek explanations for events around them, particularly for events that are negative, unexpected, or not normative. To the extent that the initial appraisal of the event or behavior is that it is good or positive, positive emotions are elicited, whereas events that are judged as negative elicit similarly valenced emotions. This process occurs regardless of whether one is the actor or the observer. For example, just as I might feel badly after I cheat (negative event), so too might others view me in a negative light.

People's immediate judgment of an event as a success or failure, good or bad, is just the beginning of the attribution process. At this point, the observer and/or the actor asks the question of "why?" The quality of a person's emotional reaction to the event will be modified based on the answer to this question. More specifically, if the respondent judges the cause of the behavior as more internal versus external to the actor, the observer will feel more anger and less sympathy toward the transgressor (Weiner, 1995). Not all internal causes are viewed with equal levels of condemnation: if the actor is seen as having control over the cause, then she will be judged more harshly (eliciting more anger) than if the cause was seen as uncontrollable (eliciting more sympathy).

At the same time, from an interpersonal perspective, if a person has a moral failure (i.e., cheats), the same attributional principles can be applied to his or her self-related feelings. To the extent that the transgressor can externalize the behavior, she or he will suffer less damage to self-esteem. If, however, the person attributes the behavior to something that is internal, then he or she will feel more guilt if the attribution is to something he or she had control of versus shame to the extent that it was uncontrollable.

By applying these attributional principles to academic dishonesty, we see more specifically how the various categories of neutralizing behaviors can be useful in reducing the negative emotions felt toward a cheater or the negative feelings the cheater feels toward himself. The first two strategies mentioned earlier, denial of the crime and denial of the victim, aim to interrupt the attribution process at the point at which a judgment is being made about how negative the behavior is. Rather than dealing with the "why" question, people who adopt this technique reframe the cheating as something that is "not a big deal." As such, there is no need to try to explain why the event occurred in the first place. The remaining strategies seek to deflect responsibility for the behavior away from the cheater by encouraging attributions that are either external or uncontrollable.

In the following pages, we organize the literature on neutralization or justification around attributional principles. We begin with studies that demonstrate how cheating is justified by denying or downplaying its "badness." Subsequently, we discuss the literature on strategies of externalization or uncontrollability. Within that section, we give special attention to methods of externalizing blame to teachers themselves.

STRATEGY #1: REFRAMING THE EVENT: DENIAL OF THE CRIME AND DENIAL OF THE VICTIM

- College student speaking about his term paper writing business.

  "You win out because now you have more money, and he wins out because he has a higher grade on this paper or test, and his parents are happy because their own child is doing well in school, and the school is happy because the student will not drop out or transfer." Primetime Live (Velmans & Koppel, 2004)

- High school student who was caught plagiarizing over 80 percent of his history paper.

  "Cheating is only an embarrassment if you get caught." Primetime Live (Velmans & Koppel, 2004)

The quotes from these two students reflect the attitudes of some people that cheating is not something that is terribly wrong. In fact, Andy, who earns up to $1,000 a week writing papers for others while in college himself, sees it as a "win-win" situation. Across the board, the data suggest that cheaters are less convinced than their more honest peers that cheating is a bad thing. Two types of evidence support this claim. The first type of evidence or group of studies documents the relations between cheating attitudes generally and cheating behavior, and the second demonstrates differences between cheaters and noncheaters in their definitions of dishonesty and in their perceived severity of the behaviors that constitute cheating. We review each of these in turn.

Cheating Attitudes

As used here, general cheating attitudes refer to the extent to which cheating is generally seen as acceptable or unacceptable, right or wrong. In contrast to other forms of neutralization strategies, students using this technique are not arguing for the acceptability of a specific cheating incident in a specific situation, but are rather claiming that cheating is less of a big deal than others might think. These studies differ from studies of moral reasoning, however, because the questions posed of participants are specific to beliefs about cheating, rather than broader questions aimed at classifying people at particular stages of moral development.

Cheating attitudes are consistently strong predictors of cheating behavior (Whitley, 1998). For example, middle school students were asked to respond to a beliefs-about-cheating scale that included questions such as "Is it ok to cheat on your science work?" and "How serious do you think it is if someone cheats on their science work?" (Anderman, Griesinger, & Westerfield, 1998). Students with higher scores on this scale also had higher levels of self-reported cheating in their own science class. More lenient attitudes toward dishonesty were also found among admitted cheaters attending an elite private college than among their peers who did not cheat, when attitudes toward cheating were assessed using three items: the extent to which cheating was justified to pass a course, to help a friend, and sometimes justified (Jordan, 2001). Moreover, tolerance of dishonesty continued to be a significant independent predictor of cheating even after numerous other variables were included in the regression model: extrinsic motivation, norms of cheating, mastery motivation, and knowledge about cheating policies. Finally, other data from undergraduates (Bolin, 2004) using a similar measure also confirmed that students' self-reported cheating behavior increased as a function of their level of permissiveness toward the behavior.

Cheating Definitions

Several studies suggest that while students and teachers agree on the behaviors that constitute cheating, students see some behaviors as less serious than faculty do. Moreover, students exhibit differences in the severity rating assigned by cheaters and noncheaters. When college students and professors rated a list of 17 dishonest behaviors, the two groups generally agreed on what behaviors did and did not constitute cheating, so that the percentage of each group who viewed a behavior as cheating did not differ by more than 5 percent (Graham, Monday, O'Brien, & Steffen, 1994). Interestingly, although students rated all of the 17 behaviors as less severe than the faculty members did, they were accurate in their estimates of the faculty members' ratings. The implication of this finding is that, even though students may be able to rationalize some dishonest behaviors as "not so bad," they are also aware that their professors will not share their views. In contrast, faculty members underestimated the students' severity ratings, suggesting that they are not attuned to students' views on dishonesty. More importantly, however, students' scores of the severity measures were inversely related to their total cheating score, consistent with the notion that cheaters may deflect some self-blame for cheating by diminishing its severity.

Findings from a second study using similar methods suggest that by denying the severity of cheating, dishonest students may not only be neutralizing self-recriminations but also, in their minds, protecting themselves from the harsh evaluations of others: students with more accepting versus intolerant attitudes toward cheating also viewed their professors as having more tolerant attitudes toward cheating as well (Roig & Ballew, 1994). The authors suggest that one of the ways that students justify cheating is by intimating that it is not that much of a big deal, even to professors.

Students clearly have different ideas about the absolute wrongness of cheating. As noted earlier in this chapter, these distinctions may well reflect students' views of cheating as an issue of morality versus social convention. From a motivational perspective, the consequence of this reframing should be to reduce the negative affect that one personally feels for dishonest behavior and inhibit the anger and contempt of others.

STRATEGY #2: DENIAL OF RESPONSIBILITY

Given the generally negative beliefs of teachers and parents about academic dishonesty, efforts to reframe cheating as something that is not a big deal may help the cheater to feel better about herself (and therefore serve as a bridge between her moral beliefs and actual behavior). However, these reframing efforts are probably less effective in changing the evaluations of significant adult others. Research does suggest, however, that a cheater's explanation for his dishonest behavior may change the severity of the judgment that is rendered against him. From an attribution perspective, responsibility can be effectively displaced by describing the behavior as having an external or an uncontrollable cause. As described by Forsyth, Pope, and

McMillan (1985), external attributions serve both a personal and interpersonal goal:

> [c]heating represents less of a threat to students' self-esteem if they can attribute this "immorality" to something external to themselves. . . . such attributions would reduce students' feelings of guilt and immorality after cheating in a classroom situation and allow them to continue to think of themselves as moral persons who simply bent to environmental pressures. In addition, a self-serving attribution pattern will help cheaters maintain an acceptable social image in their educational setting. (p. 73)

Empirical evidence for the utility of external and uncontrollable attributions for cheating derives from several sources: studies that examine the amount of blame people place on cheaters in various situations and studies that show how cheaters themselves most frequently explain their behavior. Both types of studies are reviewed, followed by a detailed examination of studies that focus on a specific target of blame—teachers.

It Is Not My Fault: Externalization and Lack of Control

When people are asked why they cheat, they rarely indicate that it was because of something they could personally control. Five categories of reasons for cheating were identified based on interviews with 12 pharmacy students in London: instructional environment, study skills, assessment employed, personal qualities, and course-specific factors (Ng, Davies, Bates, & Avellone, 2003). Within these categories, 22 specific themes were coded. Note that 3 of these 5 categories cite reasons that are external to the cheaters. Within the two internal categories, study skills and personal qualities, 11 themes were named, of which at least 7 would probably be viewed as uncontrollable by the cheater: time pressure and insufficient support within the study skills category, and peer pressure, ethnic background, pressure to do well, fear of failure, and lack of confidence within the personal qualities category. Of the 22 themes, only 2 were clearly internal and controllable by the individual—laziness, or learning inertia and the desire to save time to allocate to other activites.

Adolescents in grades 7 through college responded to 37 questions about why students cheat (Evans & Craig, 1990). Attribution scores were computed for 3 categories: teacher, classroom, and student characteristics. Across all ages, respondents assigned somewhat more blame to student characteristics than to teacher or classroom characteristics. However, whereas some of the "student" items referred to causes that were internal, and presumably controllable by the student, such as their low effort, others referred to circumstances external to the student, such as the cheating of their friends, and the pressure they received for good grades from their parents. Moreover, of the 13 attributions that were endorsed by at least 75 percent of the students, 10 were external to the student. The three internal attributions that were highly rated were low opinion of their ability, budget study time poorly, and fear of failure.

Does Externalization Work?

Thus far, we have seen that many students who cheat use various strategies to deflect responsibility from themselves and onto others, presumably so that they can avoid the harsh judgments that might otherwise accompany such behavior. It also appears that the mechanisms of deflection that students adopt are effective. That is, we also have evidence that the attributions one makes for cheating *are* effective ways to alter the judgments that others reach about one's dishonest behavior. An attribution perspective on reasoning about cheating was specifically tested in 2 studies by having participants make judgments of hypothetical people who either cheated or helped someone cheat when the motive for cheating involved varying levels of controllability: didn't want to try, didn't have the ability, task was too hard to do well, and the upset from a "breakup" with a girlfriend, leaving the cheater unable to perform well (Whitley, Nelson, & Jones, 1999). In Study 1, the cheater was portrayed as being in need of a C grade to avoid being dropped from the program; in Study 2 the C grade was needed to maintain his current GPA. After reading a scenario, participants rated how much control they perceived the cheater had over cheating, the amount of sympathy and anger they felt toward the cheater and the accomplice, their overall affective liking of the cheater and the accomplice, and the severity of the punishment that should be assigned to them. In Study 1, consistent with attribution principles, higher perceived control over cheating was related to less sympathy for the cheater, but it was not associated with the anger directed at the cheater. Evaluation of the person was inversely related to anger, positively related to sympathy, and negatively related to perceived control. Punishment was related to anger and sympathy ratings but not to control ratings. In Study 2, however, almost all expected correlations were found: perceived control was related to sympathy, evaluation, and punishment, and anger and sympathy were both related to evaluation and punishment severity. Contrary to expectations, in both Study 1 and Study 2 perceived control was not related to anger, anger was not inversely correlated with sympathy ratings, and emotions were not mediators between control ratings and punishment or evaluation rating. The authors suggest that the students' high need for a grade in Study 1 may have moderated the effects that would be predicted by attribution principles. The role of goals in judging others is discussed in more detail on pp. 203–205.

Forsyth et al. (1985) examined the extent to which cheating led to externalization of responsibility using Kelly's (1967) attributional model as the framework. From this perspective, three kinds of attributional information affect people's judgments of responsibility for cheating: the extent to which the behavior is unique to that setting (distinctiveness), the consistency of the behavior over time, and the extent to which the behavior is unique to that individual versus that performed by others (consensus). After being tempted to cheat at an experimental task, participants were asked to rate a series of reasons as to why they had or had not been dishonest. Responses focused on the distinctiveness for the behavior, consensus, and consistency as well as the locus of control for the cheating. As predicted by the theory, students who cheated were more likely to attribute their behavior to aspects of the situation with reference to consensus, distinctiveness, and consistency as compared to those who were observing the task.

The Teacher and/or School is to Blame . . .

"My professor told me that we will only actually ever use 7 percent of what we learn in college." (Primetime Live, Velmans & Koppel, 2004)

One of the frequent ways of externalizing blame for dishonesty is to implicate some aspect of the teacher, classroom, or educational system. Various aspects of the instructional setting are repeatedly cited as circumstances that do or do not make cheating more acceptable. For example, Michaels and Meithe (1989) found a moderate relationship between cheating among undergraduates and pro-cheating attitudes, and an even stronger relationship between these attitudes and intentions to cheat in the future. However, unlike the studies that were reviewed earlier, in this case, six of the eight scale items assessed students' endorsement of circumstances under which cheating is acceptable, such as "Cheating in dumb courses with poor instructors is understandable." Recall also that in the Haines et al. (1986) study cited above, the most frequently used neutralization techniques involved blaming the teacher for the students' deviant behavior.

Pulvers and Diekhoff (1999) investigated whether self-reported cheating and neutralization were higher among students who viewed their classroom less favorably. College students rated one of their classes on seven dimensions: personalization, involvement, student cohesiveness, satisfaction, task orientation, innovation, and individualization. Participants also rated 11 neutralizing statements and indicated whether they had cheated in that class. Cheaters had higher neutralization scores than noncheaters and rated the class as lower on task orientation (operationalized as organized, clear instruction) and on satistifaction (operationalized as interesting presentation). Moreover, correlational data revealed that students who were more critical of the classroom environment were also more likely to say that it was okay to cheat in the class if students needed the grade or if everyone else was doing it.

Whereas these studies all suggest correlations between perceived classroom environment and neutralizing attitudes, they cannot establish if changes in the classroom environment actually affect the acceptability of cheating or if those who cheat are also more likely to neutralize by blaming the teacher. As such, in research conducted with several colleagues, the first author examined the extent to which judgments about blame for cheating could be altered by systematically manipulating aspects of the classroom context. We reasoned that in classroom situations, students have sets of expectations about what constitutes a fair classroom, particularly in relationship to fair grading practices. Moreover, we hypothesized that when teachers use classroom practices that violate this assumed contract, students will shift blame for cheating away from students themselves (external) and toward teachers, thereby increasing the perceived justifiability and likelihood of cheating.

In the first set of studies, we constructed hypothetical vignettes of mathematics classes in which the teacher was described as having either poor or good pedagogy and as emphasizing either performance or mastery goals, yielding a two by two design (Murdock, Miller, & Kohlhardt, 2004). Performance goal structures refer to environments where ability, social comparison, and absolute correctness are emphasized, whereas mastery classrooms focus more on effort, improvement, and mastery. We reasoned that high school students would feel more justified cheating in classroom with poor pedagogy and/or performance goal structures because both of these practices might be seen as unfair in that they reduce the effort $\rightarrow$ achievement linkage. Consistent with our expectations, in these situations students shifted blame toward teachers and away from students; saw cheating as more justifiable; and rated cheating as more likely to occur than in classrooms portrayed as mastery focused and/or with good pedagogy.

In subsequent studies with undergraduate and graduate students, we not only replicated these findings, but we also demonstrated that perceived fairness of the classroom was a mediator between the manipulated classroom context variables and the assignment of blame (Murdock, Miller, & Goetzinger, 2005). Furthermore, these relations were not affected by one's own status as a cheater, suggesting that the effects of context on assigned blame are not limited to students who are themselves dishonest. Although the extent to which these perceived transgressions by the teacher actually motivate versus excuse cheating is unknown, from an attribution perspective one might see how they could serve a motivating function. Just as youths are more apt to engage in retaliatory aggression to the extent that they perceive another student's act as causing them deliberate harm (Graham & Hudley, 1992; Graham, Hudley, & Williams, 1992), so too students who perceive teachers are being unfair may react with anger and retaliate by violating the norms of the school setting.

One of the most interesting aspects of this last set of studies is that graduate students who were themselves middle or high school educators responded no differently than the graduate students in business and psychology who had never taught. In short, they reduced the blame on the student and increased the blame on the instructor as a function of the classroom context. Thus, while teachers are more inclined than their students to make personal versus contextual attributions when their students cheat (Evans & Craig, 1990), when teachers are placed in the situation of being the student, their reasoning processes are identical to those of nonteachers.

MODERATORS OF ATTRIBUTIONS: GOALS AND INCENTIVES IN THE ASSIGNMENT OF BLAME

Our attributions for the behaviors of others also shift based on the perceived incentive for engaging in the behavior: we are more likely to see the cause as dispositional and therefore blame the person when the behavior results in something good (i.e., a reward) than when it will enable him to avoid something undesirable (Grietemeyer & Weiner, 2003; Rodrigues, 1995; Rodrigues & Lloyd, 1998). In a recent series of studies, college students read vignettes in which a faculty member asked a teaching assistant to replace 20 of the student evaluations with forms the instructor had filled out. Participants were then presented with a series of 18 possible incentives that the instructor offered the student for complying, 9 of which were positively valenced (will write you a good letter for grad school, will give you a good grade on your thesis) and 9 of which were negatively valenced (will write you a bad letter for grad school, will give you a poor grade on your thesis). Across 3 experiments, participants rated some combination of responsibility, blame, and expected level of compliance, dispositional causality, and situational causality. Consistent with their hypothesis that the participants' attributions would be asymmetrical across the approach and avoidance conditions; When the incentives were positive, the hypothetical teaching assistant was blamed more and was held more responsible, and her behavior was seen as more dispositional than when she was facing possible negative outcomes for not complying. In addition, the relationship between the valence of the incentive and the level of responsibility was mediated by the extent to which the behavior was seen as dispositional (something about the person). Responsibility judgments were linked to valence in other compliance scenarios as well (e.g., a nurse complying with a doctor's request to fabricate records) as well as in more neutral situations where the behavior was not based on compliance. For example, when the vignette involved going on a hiking trip, liking it, and walking far versus not liking it and walking a short distance, the decision to walk the shorter versus longer was seen as more dispositional in the appetitive condition (i.e., when the person was going to do something he enjoyed doing) and more situational in the aversive condition.

These same principles appear to apply to judgments about academic cheaters. Although no study has specifically focused on judgments and attributions in approach versus avoidance situations, there are data showing that cheaters who are viewed as having their backs against the wall (e.g., needing a specific outcome in order to avoid severe negative consequences such as losing a scholarship) receive more sympathy than those who cheat to achieve something better, such as a 4.00 GPA.

Several studies underscore the application of asymmetrical attributions to students' reasoning when reaching judgments about their dishonest peers. For example, among undergraduate students of varying levels of religious involvement, more students agreed or strongly agreed that cheating was justified when they were portrayed as needing to pass the course to stay in school, to graduate, or to keep a scholarship than when they are pictured as just trying to get a better grade on an exam (Sutton & Huba, 1995). Although highly religious students more often thought cheating was never acceptable as compared to less religious students, the pattern of more and less acceptable reasons was consistent across all groups.

A more recent study examined high school and college students' judgments of the acceptability of cheating in 19 different circumstances (Jensen, Arnett, Feldman, & Cauffman, 2002). Two of the most acceptable motives referred to the negative personal consequences that would come to the student who did poorly (needed to pass the class to get a job that would help her family and would be put on academic probation if she did not pass). In contrast, personal need for a good grade (is competitive by nature, wants to maintain her class rank) were among the six motives ranked as least acceptable. Other motives receiving high acceptability ratings included not wanting to disappoint one's parents (an appeal to higher loyalties), not having enough time to study because of a job, "freezing" and not being able to recall the answers (uncontrollable), and being treated unfairly by an instructor (external). Higher mean acceptability ratings across items were associated with more frequent self-reported cheating behavior.

Cheating to avoid negative outcomes is not only seen as more acceptable than engaging in dishonest behavior in the search a better outcome, but according to at least one study, avoidance is also a stronger motive to cheat. Undergraduate and graduate information technology (IT) students were asked to rate 14 possible reasons for cheating on a scale ranging from *not at all likely* to *highly likely* (Sheard, Markham, & Dick, 2003). At both educational levels, three of the top reasons for cheating had to do with avoidance of failing (will fail otherwise, can't afford to fail, afraid of failing), whereas a desire to get better grades was rated significantly lower and was among the bottom five reasons for academic dishonesty. Thus, we may judge others less harshly for cheating when they are avoiding negative outcomes precisely because we can see ourselves being more likely to transgress in similar situations. However, in the Gietemeyer and Weiner (2003) paper discussed earlier, participants did not expect compliance rates to differ as a function of the incentive to comply. As such, the valence of the incentive may affect not people's likelihood of cheating but rather their willingness to admit that they might be dishonest. In short, their data may reflect the participant's own implicit knowledge of attribution principles and therefore may serve an impression management function.

CONCLUSIONS FOR EDUCATORS

Students appear to have elaborate sets of rules about when and why cheating is more or less wrong; they not only say that they cheat less often when circumstances make cheating more acceptable, but they judge their peers' dishonesty less harshly in those situations. As seen in this review, these reasons are consistent with what we would predict based on attribution theory, and they appear to effectively reduce the negative judgments made by others.

Clearly, both the reasons that students say they cheat and their response to others' cheating are largely determined by the context in which the students engage in the deviant behavior, such as their goals for cheating, their other obligations, the quality of instruction in the class, and the behavior of others around them. This suggests that cheating is viewed not as something within the moral domain, but within the social convention domain. Findings from this literature suggest that teachers openly talk to their students about what is known about high rates of academic dishonesty, including what students use as excuses for the behavior. Emphasizing that cheating is never someone else's fault and that it is not, in fact, a victimless crime may not only deter students from cheating, but may also encourage honest students to be less willing to passively participate in their peers' behavior.

At the same time, teachers themselves need to make sure they are not treating academic dishonesty as negotiable. Research suggests that most teachers are reluctant to take action against known cheaters, which may well contribute to students' attitudes of acceptance (Keith-Spiegel Tabachnick, Whitley, & Washburn 1998). Some scholarship suggests that cheating is less apt to occur in institutions that have formal honor codes (McCabe & Trevino, 1993). Although the mechanisms by which these codes work have not been carefully studied, from an attributional framework ignoring the behavior may be seen as an indicator that it is normative and therefore more acceptable.

At the university level, training professors on the links between learning objectives, instruction, and assessment may reduce the amount of cheating that occurs by increasing students' perceptions that they are being fairly evaluated and are in control of their learning. Even though more universities are considering teaching experience and teaching ability in their evaluation of faculty members, most faculty members outside of schools of education have had only limited pedagogical training.

Finally, students across all levels appear to believe that increased pressure is a justification for cheating. This belief, coupled with a tendency to justify cheating more when failure may have more negative consequences, suggests that cheating may be less apt to occur when there are frequent assessments with opportunities for feedback and improvement than when high-stakes testing is utilized (see also Anderman, and Nichols & Berliner, in this volume).

FUTURE DIRECTIONS FOR RESEARCH

To a large extent, our understanding of the links among context, reasoning, and cheating behavior is based on one-shot correlation designs; we therefore know little about the extent to which justifying circumstances actually motivate cheating rather than simply serving a self-protective mechanism after the fact. Efforts to disentangle cause from effect would be a valuable contribution to the literature, but require a breadth of studies grounded in complementary methodologies. For example, laboratory studies can be excellent vehicles for teasing out some of the specific processes underlying cheating. We might manipulate some of the frequently used excuses for cheating, such as providing poor versus good instruction, and examine its effects on behavior in a controlled situation. The limitation of these studies, of course, is that we know that students' motivation to achieve in that setting will not generalize to an actual classroom. Classroom research on attributions and cheating could be substantially improved by following students over time and collecting data at both the classroom and student level. Such studies might help us tease out the extent to which students' excuses for cheating are actually congruent with what is occurring in the classroom, and the extent to which the excuses versus the actual behaviors of teachers are predictive of dishonesty.

Across these studies, many students justified cheating by citing specific teaching practices, such as unfair grading, poor teaching, or overly difficult subject matter. As noted earlier, these findings suggest that students not only come to a classroom with a set of expectations about how the class should proceed, but they also cheat more often when those expectations are not established. Qualitative investigations to more fully explicate students' understanding of what constitutes fair and just classroom practices (Thorkildsen & Schmahl, 1997; Thorkildsen, Sodonis, & White-McNulty, 2004; Thorkildsen & White-McNulty, 2002) might help elucidate how students reason about cheating in specific contexts.

The higher prevalence of blame on teachers and classrooms intimates that anger may have a motivational role in dishonest behavior. Attributional research on retaliatory aggression shows that people are more likely to be aggressive toward others when they perceive that the other person has behaved negatively toward them, and the behavior was controllable by the person (Graham & Hudley, 1992; Graham, Hudley, & Williams, 1992; Rudolph, Roesch, Greitemeyer, & Weiner, 2004). If students react to teachers' perceived pedagogical failures with similar levels of anger, perhaps cheating may also be an example of retaliatory aggression. Study of the cognitive and emotional processes of students in various classroom settings might prove beneficial to our understanding of dishonesty.

Justifications for cheating can largely be organized through an attributional approach to excuse making. Few of the studies, however, explicitly utilize an attribution framework, and the untapped aspects of this theoretical approach deserve further examination. More specifically, whereas attribution theory posits emotional mediators between cognitions and behaviors, studies of cheating have focused largely on the cognitive components of justification. In fact, across the cheating literature little attention has been paid to the role of emotions in determining cheating behavior. Does the ability to displace blame for cheating behavior actually reduce students' felt emotions of shame, and is it shame rather than externalizing that inhibits or increases a tendency to cheat? Preliminary evidence in this area shows much stronger links between externalizing strategies and cheating than between shame-pronessness and cheating (Poindexter & Murdock, 2005). In addition, students who cheat without being caught may view the event as a success rather than a failure, which may give rise to positive rather than negative emotions. Indeed, one of the students interviewed on *Primetime Live* did not report feeling shame or guilt from cheating; rather, he expressed pride that he had gotten away with it. Some of these differences may reflect the centrality of achievement versus honesty to people's idealized and actual self-images (Stephens, 2004). Further research might help provide an integrated model of how people's self-images and their cognitive and emotional responses affect their propensity to cheat.

CONCLUSIONS: REASONING ABOUT CHEATING IN A LARGER CONTEXT

David Callahan (2004) argues that academic cheating is part of a larger epidemic of societal dishonesty that is fed by decreased opportunities for advancement, coupled with increased individualism and a decline in concern for the welfare of others. Lying, stealing, and deceiving have become the commonplace means that our icons of "success" use to advance their personal wealth and power. Accordingly, the incongruity between students' stated beliefs and self-reported behaviors related to cheating may have a lot to do with the broader forces at play in this "achieving society" (McClelland, 1961). Indeed, the message to our children and adolescents seems clear: material success comes before moral integrity. Cheating on a test may be a small price to pay if it helps one secure admittance into a prestigious college or university. It is not surprising, therefore, that pressure for grades is often the primary reason students cite for cheating (e.g., Calabrese & Cochran, 1990; Newstead, Franklyn-Stokes, & Armstead, 1996; Schab, 1991).

Students' perceived pressure for success is not fictitious: it's very real. According to the Higher Education Research Institute's annual survey, 47 percent of incoming college freshmen in 2003 reported having earned an A average in high school. But students also know that grades are only part of the equation. They must also be involved in a number of extracurricular activities as well as gainful employment. This means time is short, and they sometimes decide to cheat even when they think it is wrong.

The headlines of any newspaper or television support Callahan's arguments that cheating for personal grain is rampant across society, and many of the students in the *Primetime Live* show were familiar with this public ethos. More than one student spoke of the values that are mirrored by the behaviors of our politicians and businessmen. Some went so far as to argue that cheating is so pervasive that the decision not to cheat is a decision to be a "chump." For these and many other students, the question "Is cheating wrong?" appears to be less central to their behavior than the answers to other seemingly more important issues about access to college, competition for scholarships, and presenting the best possible transcript to a future employer. Over and over they insisted that while being moral is a good thing, "everyone else" just cares about their grades. Although teachers and schools can implement honor codes and work to enforce academic integrity, clearly what is happening in schools across the nation reflects broader shifts in our culture and cannot be addressed outside of the macrocontext of students' lives. Until that time, it will become harder and harder to convince students that cheating is wrong, or to persuade them that ethics should be their values standard for their own behavioral decisions.

REFERENCES

Albas, D., & Albas, C. (1993). Disclaimer mannerisms of students: How to avoid being labelled as cheaters. *Canadian Review of Sociology and Anthropology, 30*(4), 451–567.

Anderman, E. M., Griesinger, T., & Westerfield, G. (1998). Motivation and cheating during early adolescence. *Journal of Educational Psychology, 90*(1), 84–93.

Blasi, A. (1980). Bridging moral cognition and moral action: A critical review of the literature. *Psychological Bulletin, 88*, 1–45.

Blasi, A. (1984). Moral identity: Its role in moral functioning. In W. M. Kurtines & J. L. Gewirtz, (Eds.), *Morality, moral behavior, and moral development* (pp. 128–139). New York: John Wiley & Sons.

Bolin, A. U. (2004). Self-control, perceived opportunity, and attitudes as predictors of academic dishonesty. *Journal of Psychology: Interdisciplinary & Applied, 138*(2), 101–114.

Bruggeman, E. L., & Hart, K. J. (1996). Cheating, lying, and moral reasoning by religious and secular high school students. *Journal of Educational Research, 89*(6), 340–344.

Bushway, A., & Nash, W. R. (1977). School cheating behavior. *Review of Educational Research, 47*(4), 623–632.

Calabrese, R. L., & Cochran, J. T. (1990). The relationship of alienation to cheating among a sample of American adolescents. *Journal of Research & Development in Education, 23*(2), 65–72.

Callahan, D. (2004). *The cheating culture: Why More Americans are doing wrong to get ahead*. New York: Harcourt.

Colby, A., Kohlberg, L., Gibbs, J., & Lieberman, M. (1983). A longitudinal study of moral judgment. In *Monographs of the Society for Research in Child Development* (Vol. 48). Chicago: Unviersity of Chicago Press.

Corcoran, K. J., & Rotter, J. B. (1987). Morality-conscience guilt scale as a predictor of ethical behavior in a cheating situation among college females. *Journal of General Psychology, 114*(2), 117–123.

Daniel, L. G., Blount, K. B., & Ferrell, C. M. (1991). Academic misconduct among teacher education students: A descriptive-correlational study. *Research in Higher Education, 32*(6), 703–724.

Eisenberg, N. (1989). Empathy and sympathy. In W. Damon, (Ed.), *Child development today and tomorrow* (pp. 137–154). San Francisco: Jossey-Bass.

Evans, E. D., & Craig, D. (1990). Teacher and student perceptions of academic cheating in middle and senior high schools. *Journal of Educational Research, 84*(1), 44–52.

Forsyth, D. R., et al. (1985). Students' reactions after cheating: An attributional analysis. *Contemporary Educational Psychology, 10*, 72–82.

Graham, M. A., Monday, J., O'Brien, K., & Steffen, S. (1994). Cheating at small colleges: An examination of student and faculty attitudes and behaviors. *Journal of College Student Development, 35*(4), 255–260.

Graham, S., & Hudley, C. (1992). An attributional approach to aggression in African-American children. In D. H. Schunk & J. L. Meece, (Eds.), *Student perceptions in the classroom* (pp. 75–94). Hillsdale, NJ: Lawrence Erlbaum.

Graham, S., Hudley, C., & Williams, E. (1992). Attributional and emotional determinants of aggression among African-American and Latino young adolescents. *Developmental Psychology, 28*(4), 731–740.

Grietemeyer, T., & Weiner, B. (2003). Asymmetrical attributions for approach versus avoidance behavior. *Personality and Social Psychology Bulletin, 29*(11), 1371–1382.

Haines, V. J., Diehoff, G. M., LaBeff, E. E., & Clark, R. E. (1986). College cheating: Immaturity, lack of commitment, and the neutralizing attitude. *Research in Higher Education, 25*(4), 342–354.

Hartshorne, H., & May, M. A. (1928). *Studies in deceit*. New York: McMillan.

Jensen, L. A., Arnett, J. J., Feldman, S., & Cauffman, E. (2002). It's wrong, but everybody does it: Academic dishonesty among high school and college students. *Contemporary Educational Psychology, 27*(2), 209–228.

Jordan, A. E. (2001). College student cheating: The role of motivation, perceived norms, attitudes, and knowledge of institutional policy. *Ethics & Behavior, 11*(3), 233–247.

Keith-Spiegel, P., Tabachnick, B. G., Whitley, B. E., Jr., & Washburn, J. (1998). Why professors ignore cheating: Opinions of a national sample of psychology instructors. *Ethics & Behavior, 8,* 215–227.

Kelley, H. H. (1967). Attribution theory in social psychology. In D. Levine (Ed.), *Nebraska Symposium on Motivation, Volume 15,* Lincoln, NE: Univeristy of Nebraska Press.

Kohlberg, L. (1984). The psychology of moral development: The nature and validity of moral stages. San Francisco: Harper & Row, 1984.

Kohlberg, L., & Candee, D. (1984). The relationship of moral judgment to moral action. In W. M. Kurtines & J. L. Gewirtz, (Eds.), *Morality, moral behavior, and moral development.* New York: John Wiley & Sons.

Lanza-Kaduce, L., & Klug, M. (1986). Learning to cheat: The interaction of moral-development and social learning theories. *Deviant Behavior, 7*(3), 243–259.

Leming, J. S. (1980). Cheating behavior, subject variables, and components of the internal-external scale under high and low risk conditions. *Journal of Educational Research, 74,* 83–87.

Malinowski, C. I., & Smith, C. P. (1985). Moral reasoning and moral conduct: An investigation prompted by Kohlberg's theory. *Journal of Personality and Social Psychology, 49*(4), 1016–1027.

McCabe, D. L., & Trevino, L. K. (1993). Academic dishonesty: Honor codes and other contextual influences. *Journal of Higher Education, 64*(5), 522–538.

McClelland, D. C. (1961). *The achieving society.* Princeton, NJ: Van Nostrand.

Michaels, J. W., & Miethe, T. D. (1989). Applying theories of deviance to academic cheating. *Social Science Quarterly, 70*(4), 870–885.

Murdock, T. B., Miller, A. D., & Goetzinger, A. A. (2005). The effects of classroom context variables on university students' judgment of the acceptability of cheating: Mediating and moderating processes. Unpublished manuscript.

Murdock, T. B., Miller, A., & Kohlhardt, J. (2004). Effects of classroom context variables on high school students' judgments of the acceptibility and likelihood of cheating. *Journal of Educational Psychology, 96*(4), 765–777.

Newstead, S. E., Franklyn-Stokes, A., & Armstead, P. (1996). Individual differences in student cheating. *Journal of Educational Psychology, 88*(2), 229–241.

Ng, H. W., Davies, G., Bates, I., & Avellone, M. (2003). Academic dishonesty among pharmacy students: Investigating academic dishonest behaviours in undergradautes. *Pharmacy Education, 3*(4), 261–269.

Piaget, J. (1932). *The moral judgment of the child* (M. Gabain, Trans.). Glencoe, IL: Free Press.

Poindexter, A. L., & Murdock, T. B. (2005). Guilt and shame proneness as predictors of academic cheating. Unpublished manuscript.

Pulvers, K., & Diekhoff, G. M. (1999). The relationship between academic dishonesty and college classroom environment. *Research in Higher Education, 40,* 487–498.

Rest, J. R. (1986). *Moral development: Advances in research and theory.* New York: Praeger.

Rodrigues, A. (1995). Attribution and social influence. *Journal of Aplied Social Psychology, 25,* 1567–1577.

Rodrigues, A., & Lloyd, K. L. (1998). Reexamining bases of power from an attributional perspective. *Journal of Applied Social Psychology, 28,* 973–997.

Roig, M., & Ballew, C. (1994). Attitudes toward cheating of self and others by college students and professors. *Psychological Record, 44*(1), 3–12.

Rudolph, U., Roesch, S. C., Greitemeyer, T., & Weiner, B. (2004). A meta-analytic review of help giving and aggression from an attributional perspective: Contributions to a general theory of motivation. *Cognition & Emotion, 18*(6), 815–848.

Schab, F. (1991). Schooling without learning: Thirty years of cheating in high school. *Adolescence, 26,* 839–847.

Sheard, J., Markham, S., & Dick, M. (2003). Investigating differences in cheating behaviours of IT undergraduate and graduate students: The maturity and motivation factors. *Higher Education Research & Development, 22*(1), 91–108.

Stephens, J. M. (2004, April). *Beyond reasoning: The role of moral identities, sociomoral regulation and social context in academic cheating among high school adolescents.* Paper presented at the Annual meeting of the American Educational Research Association, San Diego, CA.

Storch, J. B., Storch, E. A., & Clark, P. (2002). Academic dishonesty and neutralization theory: A comparison of intercollegiate athletes and nonathletes. *Journal of College Student Development, 43*(6), 921–930.

Sutton, E. M., & Huba, M. E. (1995). Undergraduate student perceptions of academic dishonesty as a function of ethnicity and religious participation. *NASPA Journal, 33*(1), 19–34.

Sykes, G., & Matza, D. (1957). Techniques of neutralization: A theory of delinquency. *American Sociological Review, 22*, 664–670.

Thorkildsen, T. A., & Schmahl, C. M. (1997). Conceptions of fair learning practices among low-income African American and Latin American children: Acknowledging diversity. *Journal of Educational Psychology, 89*(4), 719–727.

Thorkildsen, T. A., Sodonis, A., & White-McNulty, L. (2004). Epistemology and adolescents' conceptions of procedural justice in school. *Journal of Educational Psychology, 96*(2), 347–359.

Thorkildsen, T. A., & White-McNulty, L. (2002). Developing conceptions of fair contest procedures and the understanding of skill and luck. *Journal of Educational Psychology, 94*(2), 316–326.

Turiel, E. (1983). *The development of social knowledge: Morality and convention.* New York: Cambridge University Press.

Velmans, J. (Producer), & Koppel, T. (Producer). (2004, April 29). The cheating crisis, On *primetime Live*, New York: American Broadcast Company.

Ward, D. A., & Beck, W. L. (1990). Gender and dishonesty. *Journal of Social Psychology, 130*(3), 333–339.

Weiner, B. (1995). *Judgments of responsibility: A foundation for a theory of social conduct.* New York: Guilford Press.

Weiner, B. (2001). Responsibility for social transgressions: An attributional analysis. In Bertram F. Malle, Louis J. Moses, et al. (Eds.), *Intentions and intentionality: Foundations of social cognition* (pp. 331–344). Cambridge, MA: MIT Press.

Weiner, B., Figueroa-Munoz, A., & Kakihara, C. (1991). The goals of excuses and communication strategies related to causal perceptions. *Personality & Social Psychology Bulletin, 17*(1), 4–13.

Whitley, B. E., Jr. (1998). Factors associated with cheating among college students: A review. *Research in Higher Education, 39*(3), 235–274.

Whitley, B. E., Jr., Nelson, A. B., & Jones, C. J. (1999). Gender differences in cheating attitudes and classroom cheating behavior: A meta-analysis. *Sex Roles, 41*(9–10), 657–680.

PREVENTION AND DETECTION OF CHEATING

11

CHEATING ON TESTS: PREVALENCE, DETECTION, AND IMPLICATIONS FOR ONLINE TESTING

WALTER M. HANEY AND MICHAEL J. CLARKE

INTRODUCTION

In a recent book, Steve Levitt and Stephen Dubner described what happened when the Chicago public schools introduced high stakes testing in 1996. Promotion from grade to grade was to be dependent on students' standardized test scores (Levitt & Dubner, 2005). Schools with low test scores could be placed on probation and "face the threat of being shut down, its staff to be dismissed or reassigned" (Levitt & Dubner, 2005, p. 26). Levitt and his colleagues analyzed seven years' worth of test results (1993–2000) to look for two kinds of evidence of possible cheating—namely, unusual patterns of answer concordance and anomalous patterns of score increases and decreases. In one grade 6 classroom, 15 of 22 answer sheets were found to have six or more identical answers, and 9 were found to have four identical wrong answers. The students in this grade 6 class had scored on average at the 5.8 grade level—one full year behind the norms of 6.8 for students tested in the eighth month of grade 6. But when these same students had been in grade 5, they scored at the 4.1 level, and in grade 7 they scored at the 5.5 level—worse than they had apparently scored a year earlier in grade 6. Levitt and Dubner did not explain how the analyses of Chicago test results were undertaken (Levitt & Dubner, 2005),[1] but the

---

[1] The Levitt & Dubner (2005) book provides no explanation as to how analyses of the Chicago test data were undertaken. Jacob and Levitt (2003) provide at least a general account of the analyses, but with little detail, and give no hint at all that they were aware of previous literature on statistical methods for detecting cheating on tests.

Copyright © 2007 by Walter M. Haney and
Michael J. Clarke
All rights of reproduction in any form reserved.

clear implication was that one grade 6 teacher had cheated on the test. Levitt and Dubner were apparently unaware of the long history of using statistical methods to help identify cheating on standardized tests, not just by teachers, but more commonly by students.

One of the most famous incidents of cheating on an exam in the United States involved the "rascal king" of Boston politics, James Michael Curley (Beatty, 1992). In 1902, Curley and his cousin Thomas F. Curley, both leaders of the Tammany Club of Boston Ward 17, took civil service exams for two Irish immigrants, who wanted to become postal workers but had not been able to pass the civil service exam. The Curleys were exposed by a political rival and tried for federal fraud. Eyewitness testimony and handwriting comparisons at their trial showed that Curley and Curley had not only impersonated the Irish immigrants, but had also copied from one another during the exam, for they were found to have given 12 identical wrong answers (Beatty, 1992). These two methods of investigation—handwriting comparisons and analysis of similarity of wrong answers—are still major tools of investigators in their battles against cheating on standardized tests.

One purpose of this chapter is to review the history of using statistical methods to help detect cheating on standardized tests. According to Wesolowsky, "The existence of these methods of detecting copying appears to be very little known among most instructors using multiple choice tests and examinations" (Wesolowsky, 2000, p. 909). Among the traditional uses of statistical detection as applied to academic testing are (1) as a check on exam security and (2) as a way to identify individual suspects of cheating; however, the best use for teachers may be (3) as an assay of the level of independent work among a group of students so as to determine when countermeasures should be taken. While Cizek cautions that "statistical analyses should be triggered by some other factor (e.g., observation)" and maintains that "none of the statistical approaches should be used as a screening tool to mine data for possible anomalies" (Cizek, 2001, p. 12), it is clear that statistical methods are often useful precisely because there is often not an actual observation of the act of copying.

In this chapter, we also summarize evidence on the prevalence and correlates of cheating on tests (and other forms of academic dishonesty). For the most part we focus on the realm of educational tests (as compared, for example, with employment or personality tests). Most of the literature regarding cheating on tests deals with proctored tests, such as high school or college course examinations (proctored by instructors) or college admissions tests (proctored by test administrators). However, we also discuss some historical evidence regarding cheating on self-administered tests, and we discuss the issues of student cheating on online tests taken without a proctor's supervision. Along the way, we also summarize some of what we and others have learned about methods of preventing cheating on tests, including online tests.

Cheating on examinations has a very long history. Miyazaki, for example, notes the variety of forms of cheating on civil service examinations in imperial China, including exchanging papers, copying answers from other examinees, and bribing examination officials (Miyazaki, 1976). Another account of Chinese civil service examinations notes: "The Gest Oriental Library at Princeton University has an example of a 'cribbing garment', which potential examinees could rent for these civil service examinations. Sewn into the coat's lining are 772 essays which are based on Confucian writings" (Hanson, 1990, p. 10; see also Cizek, 1999).

Before turning to summarize research on academic dishonesty and methods of detecting and preventing cheating, let us mention two fairly recent developments illustrating that cheating on tests is a widespread and apparently increasingly severe problem. First in 1999, Gregory Cizek published *Cheating on Tests: How to Do It, Detect It and Prevent It.* In more than 250 pages, Cizek providies a mostly useful summary of literature concerning cheating on tests published between 1970 and 1996 (Cizek, 1999). His chapter on "How to Cheat" presents a remarkable compendium of the dozens of ways students in the United States and elsewhere have found to cheat on tests, ranging from copying from other test-takers to impersonation and even bribing of officials responsible for testing. His "taxonomy of cheating" (pp. 39–53) lists almost 60 different ways of cheating on tests, grouped into three categories (giving, taking, or receiving illicit information, using forbidden materials, and taking advantage of the testing process). Cizek's overall view of the prevalence of cheating on tests is that "Nearly every research report on cheating . . . has concluded that cheating is rampant" (Cizek, 1999, p. 13).

A second recent development indicative of the growing prominence of cheating on tests was the founding in 2003 of a new "test security" company, called Caveon. Founded by a number of veterans of the testing industry, the new firm has the mission of providing "testing programs with some new weapons in their fight against cheating and exam fraud" (Caveon.com, 2005a, p. 1). In a press release issued July 29, 2005, the company announced that it had entered into an agreement to provide its "test cheating analyses" to the Texas Education Agency in addition to contracts with six other states (Caveon.com, 2005b, p. 1).

So, if cheating on tests is rampant and has become so prominent as to be the focus of work by people ranging from an award-winning economist (Steven Levitt) to "test security" consultants, what is known about the prevalence, detection, and prevention of cheating on tests?

STUDIES OF ACADEMIC DISHONESTY

Cheating on examinations clearly predates the twentieth century, but only with the rise of social science did the topic begin to receive scholarly attention (Hanson, 1990). In exploring the literature on academic dishonesty, we have come to appreciate the volume of literature on the subject. Without making any claim to having conducted a thorough literature review on academic dishonesty, let us briefly mention some of the highlights of what has been discovered, giving particular note to studies concerning cheating on tests, issues of statistical methods, and sources of evidence regarding cheating.

First, it should be noted that academic dishonesty literature deals with several kinds of dishonesty other than cheating on tests, including copying of homework assignments, plagiarism, and fabrication of bibliographies (Cizek, 1999; Hanson, 1990; McCabe, 1992; McCabe, Trevino, & Butterfield, 2001). One of the earliest notable studies of dishonesty and deceit in test-like situations was Hartshorne and May's (1928) *Studies in Deceit*, the first of their series of publications on the study of character. Without attempting to describe all of this work, let us simply note two methodological approaches adopted by Hartshorne and May, which foreshadow much subsequent work on the subject of cheating. First, as one strategy for assessing individuals' propensity to cheat, using a variety of different "tests" (copying, speed, and athletic tests, for example), Hartshorne and May gave children each test under supervised conditions with no opportunity for deception. "On the other occasion, the conditions are such as to permit deception" (p. 65)—for example, in violating the rules given for the test by "peeking" or in scoring their own tests. Indices of deception were considered to be the differences in performance between the unsupervised and the supervised performance.

In many of their tests, Hartshorne and May derived indices, which they took to indicate the amount or extent of deceit or cheating (Hartshorne & May, 1928). In addition, as a benchmark for determining the "fact" that cheating had occurred, they compared scores achieved under conditions making cheating possible with the distribution of scores under conditions that ensured honesty. Any scores achieved under the former conditions that exceeded the mean of the latter conditions by three standard deviations were treated as indicative of cheating. In other words, they assumed that scores achieved under conditions allowing cheating and that fell in the top 0.1 percent of the distribution of scores honestly achieved indicated cheating (Hartshorne & May, 1928).

This sort of probabilistic approach to the study of cheating has now become common in investigations of cheating on tests, as recounted later. In a sense this should probably be a cause for comfort; statistical methods for studying cheating seem to have won out over outright deception. Nevertheless, although probabilistic methods may be entirely appropriate in studying patterns among populations for research purposes, they can be quite inappropriate in guiding determinations of cheating in particular cases. Hartshorne and May themselves noted this distinction:

We do not actually do injustice to anyone by this procedure, for we make no use of the results in a personal way, and deal for the most part with groups rather than with individuals. All our relations are worked out on paper for the purpose of discovering facts about human nature. They are not used as a means of "catching" the individual and then confronting him with the "crime." (Hartshorne & May, 1928, p. 75)

Fifty years later, in their 1980 article "The Detection of Cheating on Standardized Tests," Buss and Novick noted the problem of using probabilistic tests in catching individual cheaters: "A statistical test may guarantee that in the long run it will be correct 9999 times out of 10,000. But this is not enough, if *available evidence* pertaining to the 10,000th case is knowingly ignored" (Buss & Novick, 1980, p. 12, italics in original).

In a series of studies in the 1960s, Bowers surveyed college student body presidents, deans, and students about academic dishonesty on campuses nationwide (Bowers, 1964). Bowers found that students' estimates of the prevalence of cheating were considerably higher than estimates of deans and student body presidents (Bowers, 1964, p. 41). Bowers also provides some interesting data on the kinds of cheating on exams that students reported engaging in, and the kinds of exams on which they tended to cheat most commonly (Bowers, 1964).

One section of Bowers' questionnaire asked students whether they had ever engaged in 13 specific acts of cheating. Percentages of over 5,000 respondents admitting various kinds of cheating were as follows: copying a few sentences without footnoting in a paper, 43 percent; getting questions or answers from someone who has already taken the same exam, 33 percent; copying answers from a text or other source instead of doing work independently, 31 percent; "padding" a few items on a bibliography, 28 percent; giving answers to other students during an exam, 17 percent; copying from someone's test or exam paper without his knowing about it, 16 percent; copying from someone's test or exam paper with that person's knowledge, 11 percent; arranging with other students to give or receive answers by use of signals, 2 percent; and, taking an exam for another student, 1 percent (Bowers, 1964).

As to the kinds of tests on which cheating most commonly occurs, Bowers reports the following percentages of students admitting cheating: final exam, 12 percent; midterm test or exam, 7 percent; test or quiz other than final or midterm, 40 percent; paper, 16 percent; lab work, 16 percent; and other, 4 percent (Bowers, 1964).

Since Bowers' study, literally dozens of articles have been published on academic dishonesty, many of which we have not read and most of which are not cited in this paper.[2] Let us mention only works of several investiga-

---

[2] Indeed, even before Bowers' study, far more attention had been given to academic honesty than is indicated in this brief overview of a few key studies. According to Bowers (1964, p. 5), by 1964 a bibliography of over 400 references dealing with problems of academic dishonesty had already been compiled.

tors, which not only are of substantive interest in and of themselves but also provide entrée into the earlier literature. First is Hanson's 1990 dissertation on the impact of student and institutional characteristics on cheating behavior (Hanson, 1990). As far as we know, Hanson's work has received little attention in the published literature (it is not, for instance, included in Cizek's 1999 15-page list of references). Nevertheless, two features make Hanson's study stand out against the backdrop of previous literature on academic dishonesty—or at least what we have read of it. First, unlike most previous studies on academic dishonesty, Hanson's work could make a claim to being a representative national study (though she does not make this claim herself). Her study was based on a followup survey of a random sample of 280,000 college freshmen who responded to a survey of freshmen in the fall of 1985 as part of the Cooperative Institutional Research Program. In the 1987 followup survey, some 3,700 of the random sample of 14,500 students (26 percent) responded, and the colleges they were attending were identified (over 300 in number). These institutions were then surveyed about their academic honesty systems (Hanson, 1990).

This two-pronged approach to data gathering leads to the second notable aspect of Hanson's study. Instead of looking simply at personal correlates of cheating behavior or at institutional characteristics associated with cheating, Hanson looked at how these two sets of features in combination were related to self-reported cheating (Hanson, 1990).

Later we return to more of Hanson's substantive findings, but here we note some of her general findings on the prevalence of cheating. Among the 3,000 responding students, 17 percent reported that they had cheated on an exam or quiz "occasionally" during the past year; less than 1 percent said they had cheated frequently, and 82 percent said they had cheated "not at all" on an exam or quiz in the past year. In contrast, 29 percent said they had copied homework from another student "occasionally" during the past year (pp. 86–87) (Hanson, 1990).[3]

A second body of noteworthy work on academic dishonesty is by Donald McCabe (McCabe, 1992; McCabe & Bowers, 1994; McCabe & Trevino, 1993; McCabe et al., 2001). McCabe surveyed some 16,000 students at 31 selective institutions of higher education in the 1990–1991 academic year, of which 6,100 (or 38 percent) responded. He sampled primarily juniors

---

[3] There appears to be a slight discrepancy between the data cited here and what Hanson's abstract says. The abstract says: "Eighteen percent of the respondents admitted cheating on an examination . . . *during their first two years* of college" (p. xviii, emphasis added). This appears to be incorrect, or at least misleading, in that the actual survey form, included in an appendix, clearly asked respondents to report how often they engaged in activities listed "during the past year" (p. 205). We note this point because the period over which students are asked to report cheating behaviors may help to explain seeming dicrepancies in reported frequencies of cheating behaviors.

and seniors and asked students to indicate whether they had ever engaged in any of 13 cheating behaviors while an undergraduate. McCabe also surveyed faculty members in 16 schools participating in his student survey as to how incidents of cheating are handled on their campuses. McCabe found that around 20 to 25 percent of responding students admitted to having copied from another student on a test or exam.

A third noteworthy recent source on the prevalence of cheating and academic dishonesty is the work of Bernard Whitely (Whitely, 1998; Whitely, Nelson, & Jones, 1999; Whitley & Keith-Spiegel, 2002). A number of previous reviewers, such as Cizek (1999), have conducted narrative reviews of previous research regarding cheating on tests. Whitely's work stands in contrast because he has undertaken meta-analyses of previous studies concerning cheating on tests. Meta-analysis may be described as the quantitative study of previous studies and is widely recognized and employed not only in the social sciences, but also in biomedical research (Leandro, 2005). Whitely conducted a meta-analysis of 107 studies of the prevalence and correlates of cheating among college studies published between 1970 and 1996 (again, by the way, Hanson's 1990 dissertation is overlooked) (Whitely, 1998). The studies Whitely analyzed dealt with one or more types of cheating among college students, namely, cheating on examinations, cheating on homework or other assignments, and plagiarism. More than 40 studies provided estimates of the prevalence of cheating of these three types. The range and means of the prevalence of cheating among college students of these types were: cheating on examinations, 4 to 83 percent, mean 43.1 percent; cheating on homework 3 to 84 percent, mean 40.9 percent; plagiarism 35 to 98 percent, mean 47 percent (p. 238). Whitely found no simple linear relationship between year of data collection and particular types of cheating, but he did find evidence of a curvilinear relationship suggesting that cheating on tests by college students may have diminished somewhat from 1969–1975 to 1976–1985, but increased (to over 45 percent) in the 1986–1995 period (Whitely, 1998).

Whitely's results suggest that cheating on tests by college students is fairly common, about as common as cheating on homework or other assignments, but perhaps slightly less common than plagiarism.[4] Nonetheless, the wide ranges in estimates of cheating of the three types suggest that apparent differences in reported prevalence of dishonesty may well be artifacts of the way in which questions are framed. From her 1987 survey of students two years into their college careers, Hanson found that 17 to 18 percent admitted cheating on an examination during the previous year (Hanson, 1990). In contrast, McCabe found that some 20 to 25 percent of mainly junior-senior-level college students reported copying during an examina-

---

[4]Note, however, that Hanson's study, which used analogous questions on the same population, found that copying homework was more common than cheating on tests.

tion during their undergraduate careers (McCabe & Bowers, 1994; McCabe et al., 2001). The differences in apparent rates of cheating on exams may well be due not just to the populations surveyed but also to how questions were posed. In the survey on which Hanson's analysis was based, students were asked whether they had cheated on a school quiz or exam in the previous year, and the answer options given were "frequently," "occasionally," and "not at all" (Hanson, 1990, p. 205). However, the McCabe survey asked students whether they had "copied from another student on a test or exam" never, once, a few times, or many times, while in college (McCabe & Bowers, 1994; McCabe et al., 2001). Thus, although the McCabe results suggest that the prevalence of exam cheating is somewhat higher than do Hanson's results, it is possible that differences may be entirely due to the manner in which questions were posed.

Numerous observers have pointed out that academic dishonesty occurs far more often than it is detected or punished. For example, Bowers (1964) wrote: "Our data show that the magnitude of the problem is grossly understated by members of the campus community. Moreover, campus authorities say that only a small proportion of those who cheat, even according to their conservative estimates, are caught and punished" (pp. 193–194). Thirty years later, McCabe found that faculty surveyed are reluctant to use formal disciplinary mechanisms to handle incidents of cheating and instead prefer to handle such incidents informally (McCabe & Trevino, 1993).

STATISTICAL INDICES OF ANSWER CONCORDANCE

Having summarized several decades of research regarding academic dishonesty, we now focus more narrowly on cheating on tests and specifically on the use of statistical methods to detect unusual patterns of answer concordance (such as the incident recounted in the introduction of the Curley cousins who had 12 identical wrong answers on a civil service exam). Unusual answer concordance has usually been viewed as evidence of copying or illicit communication during an examination, but sometimes it has been treated as evidence of illicit prior access to test questions. In the next section we describe the early literature regarding studies of answer concordance. Then we focus on more current literature and discuss and compare indices more recently proposed and studied.

EARLY LITERATURE

Saretzky's historical account of how procedures for dealing with cheating evolved at the College Board and the Educational Testing Service (ETS) shows that the problem of cheating was recognized virtually from the start of the College Entrance Examination Board (Saretzky, 1984). He points out that the Board's Document No. 2 from February 1901 warned candidates against use of contraband materials and copying, that exam candidates would be ejected for such cheating by exam supervisors, and that supervisors' judgment in such matters would be "final and without appeal" (Saretzky, 1984, p. 2).

The problem of cheating on college entrance exams clearly predated the introduction of the Scholastic Aptitude Test (SAT) in 1926, as is evident from a passage Saretzky quotes from a transcript of a conference of College Board supervisors in 1926, called to discuss the inaugural administration of the new multiple-choice SAT (Saretzky, 1984):

> In Boston and Cambridge . . . there have been . . . a good many cases of imperson-ation and cheating. One year . . . two boys were expelled from the Boston Latin School . . . and their parents raised a terrible row and said, "Why are you punish-ing our boys, ruining their careers and all they are guilty of is the thing on the basis of which Mayor Curley was elected Mayor of Boston." (Conference tran-script, quoted in Saretzky, 1984, p. 3)

With the increasingly widespread use of multiple-choice tests in the 1920s, including the introduction of the multiple-choice SAT in 1926, copying on tests apparently came to be seen as a more severe problem. After all, in a multiple-choice test, a cheater needs to copy only a single mark or letter—a, b, c, d, or e—rather than anything more extensive. Thus, it could be said that multiple-choice tests, in addition to making standard-ized testing more efficient, also carried the potential for making cheating via copying far more efficient than with the sort of essay or oral tests that predominated before their introduction.[5]

Thus, it hardly seems surprising that a technique similar to that of Harts-horne and May (1928), of identifying outliers as liars, was quickly brought to bear on the problem of copying on multiple-choice tests. In 1927, Charles Bird published an article in *School and Society* entitled "The Detection of Cheating in Objective Examinations" (Bird, 1927). Bird's publication seems to have been the first to set out the basic rationale that is still used today in testing programs for making determinations in cases of suspected copying: "We can tell whether the identical wrong answers in two papers exceed a number which is possible by chance" (Bird, 1927, p. 261).

Bird used an analysis of the frequency of common wrong answers among students in a psychology course examination to investigate students who appeared to proctors to have been copying during the exam. He found that the average number of common wrong answers among random pairs of students was 4.0 (and a standard deviation of 2.38), but each of the students suspected of copying showed in excess of 17 common errors.[6] These

---

[5] However, there is ample evidence that cheating existed on oral and written tests long before the introduction of multiple-choice tests.

[6] This is identical to the EEIC (exact errors in common) index given in Table 11.1.

quantitative results were presented to a disciplinary committee, and though initially denying any dishonesty, when confronted with the evidence, three of the four students confessed guilt. While the fourth maintained innocence, the committee considering the quantitative evidence to be "irrefutable" "unanimously convicted him" (Bird, 1927, p. 262). Without commenting further on Bird's approach, it is worth noting that his approach was based not on theoretical distributions of wrong answers but on empirically determined patterns.

Other relatively early articles concerning statistical analysis of wrong answers on multiple-choice tests as they related to possible copying were Dickenson (1945) and Saupe (1960). Another relatively early publication on statistical indices of copying or collaboration was Anikeef (1954). He relied on theoretical distributions (based on the binomial expansion and the normal approximation to it) to develop his index of collaboration.

What was unique in Anikeef's approach, as compared with all other such literature over the last 60 years, was developing a criterion measure against which to test the validity of his index—a problem of considerable difficulty in efforts to validate statistical indices of cheating (Anikeef, 1954). A group of students was asked to collaborate with other students during an exam in ways not detectable by the instructor and to keep a record of which answers they obtained from which other students. Anikeef then was able to test his collaboration index against these specific records of collaboration.

A key modern work on statistical indices of copying, with particular regard to college entrance exams, is that by Angoff (1972; see also Angoff, 1974). Angoff had collected data on three samples of pairs of SAT takers in order to test the efficacy of eight different indices of possible copying. The first, and Angoff's main analysis sample, was constructed by randomly selecting examinees from the December 1968 administration of the SAT, with care being taken that they all came from different test centers. Since these students were sitting for the SAT in different geographical locations, it was clear that no copying could have occurred among any of the pairs. Thus, Angoff could use pairs of these examinees' answer sheets to develop norms of answer similarity for "honest" examinees, much as Hartshorne and May developed "honest" norms for their quite different tests.

Angoff's other two samples were used mainly as a check on the first. The second was drawn from a test center with "an unblemished security history" in order to check on the possibility that students who might "have studied and learned the same misinformation" would show more answer similarity than students in different geographic locations. The third sample was constructed using procedures analogous to the first, except for a different SAT administration (Angoff, 1972, 1974).

For each of these three samples, Angoff then developed 12 pairs of variables (one each for the SAT verbal and math) having to do with patterns of right, wrong, and omitted answers among pairs of examinees. He then developed eight bivariate distributions among these variables as possible indices of copying. Then, via analyses of covariance relations among pairs of the variables used in the eight indices, Angoff sought to test whether distributions differed in his three samples. Although he did find some statistically significant differences in regression relations in the three samples, Angoff judged the differences to be small and concluded that the data from sample 1 were sufficiently general to use in developing "honest" norms (Angoff, 1972, 1974).

Next, as a means of "validating" the utility of the eight potential copying indices as tools for identifying actual copying, Angoff tested how well the distributions for each of his honest samples might be used to detect "50 cases of known and admitted copiers from recent administrations" (Angoff, 1972, p. 11).[7] In another touch reminiscent of Hartshorne and May (1928), Angoff adopted, in advance, the rule that any instance in which a known copying case equaled or exceeded three standard deviations above the mean of any of the copying indices, for either the SAT-V or SAT-M, "represented a validation of the general procedure" (Angoff, 1972, pp. 11, 13).

Angoff found that every one of the 50 cases was identified by one of the eight indices, but that the most successful indices were five involving counts of wrong or right and wrong answers. Practical and statistical considerations were used to eliminate three of these, and Angoff ended up recommending two, which he called indices B and H, as the most useful indices of copying. Index B was simply the number of questions answered incorrectly, in the same way, by two examinees, relative to the product of the total number of wrong answers by the two examinees (this index B was found by Angoff to identify some 75 to 90 percent of cases of known cheating). The other index, H, was based on the longest run of identically marked incorrect responses and omits between the two examinees, and was shown to identify 80 to 98 percent of the cases of known cheaters. In both his 1972 and 1974 publications, Angoff closes by noting that these two indices had been in successful use at ETS for several years (Angoff, 1972, 1974).

---

[7] After reading Saretzky's (1984) account of how procedures for dealing with questionable scores evolved at ETS, one cannot help but wonder exactly how Angoff's 50 cases of known and admitted copiers were identified. In neither the 1972 nor the 1974 version of his paper does Angoff provide any detail on this point. However, Saretzky's account makes it clear that among the evidence used to confront suspected cheaters in the 1960s was that of copying and that the majority of examinees confronted agreed to cancel their scores. If Angoff's 50 cases were of this sort—that is, examinees who agreed to have scores canceled when confronted with evidence of copying (presumably based on concordance of wrong answers), then Angoff's "validation" of his indices B and H may not be all that they appear to be. That is, it may not be validation of statistical indicies of copying in terms of cases *independently* identified as being true copying cases.

It is not clear what statistical methods may have been used at ETS in cases of suspected copying prior to Angoff's work leading to use of indices B and H. However, it seems clear that some kinds of statistical comparisons were used at least as early as the mid-1960s, since a 1966 ETS memo, entitled "Statement of Existing Procedures for Disposition of Discrepant Scores," states that "the resume of evidence" used to confront a suspected cheater might include "the score comparison, the indication of impersonation, or of copying, as the case may be."[8]

From around 1980 through the 1990s, ETS relied on an index of suspected copying called index K. It was developed by Frederick Kling in 1979 according to Saretzky (1984, p. 13). The index as of 1993 is given in Table 11.1. As a result of Holland's 1996 report, the K index was slightly modified by the ETS and, presumably, remains in use today (Holland, 1996).

Houston conducted a series of studies on copying during college course exams in order to investigate the kind of copying that occurs, characteristics of apparent copiers, and methods of reducing its prevalence (Houston, 1976, 1977, 1978, 1983, 1986). Later, we summarize some of Houston's substantive findings, but here let us simply describe his statistical method for measuring copying.

> Cheating was assessed by comparing the number of wrong answers held in common by a target individual and his adjacent neighbors (a number which is inflated by answer copying). The greater the difference between adjacent and distant overlap, the greater the answer copying. (Houston, 1976a, p. 730)

The Houston procedure for measuring copying on multiple-choice exams was not used to identify individual cases of cheating. Also, in light of the "delicate ethical issues" involved in some of Houston's experiments (e.g., giving some exams under conditions that made it relatively easy to cheat via copying), Houston dealt with ethical issues in his research both by having his research reviewed by a campus review board and by allowing subjects, after a debriefing as to the nature of the experiment, to withhold their data from analysis (Houston, 1983).

Two other, largely theoretical, articles relevant to indices of suspected copying were published in the 1970s. Frary, Tideman, and Watts reported on the development of a variety of indices dealing with the probability of two examinees' answers corresponding merely by chance (Frary & Tideman, 1997; Frary, Tideman, & Watts, 1977). However, unlike most previous analyses, their indices were based on correct and incorrect

---

[8] Memo "Statement of existing procedures for disposition of discrepant scores," July 25, 1966, quoted in Saretzky (1984, p. 6). Saretzky also refers to a 1965 review of ETS test security procedures, including "the use of comparisons of wrong answers in copying cases" (p. 7). Unfortunately, however, he gives no detail as to the nature of the procedures.

TABLE 11.1 Methods for Detecting Unusual Answer Similarity in Multiple Choice Tests

Identifier	Description
H	Longest run of identical incorrect answers (EEICs) between a student pair (Haney, 1993; Saretzky, 1984)
HH	Harpp-Hogan. The number of EEICs divided by the number of different responses (DR) between a student pair, HH = EEIC/DR (Harpp & Hogan, 1998; Harpp et al., 1996).
Chance	Application of the binomial expansion formula to the number of EEICs (Bellezza & Bellezza, 1989; Rizzuto & Walters, 1997). $$P_{ch} = \frac{n!}{(n-x)!x!} p^x (1-p)^{n-x}$$ where the variables are the same as defined for the ESA index in order for this index to appear on the same scale as others. Used here as $-\log(P_{Ch})$ in order to correlate it more easily with other indices.
K	$$P_K = \sum_{i=x}^{m} \frac{m!}{(m-i)!i!} p^i (1-p)^{m-i}$$ where P_K = the probability that the subject copied from the source x = the number of EEICs between subject and source (Holland, 1996) m = the number of wrong answers by the source p = the probability that a subject with a given grade would answer the question in the same way as the source; for $x/t \le 0.3$ this is $p = 0.085 + 0.45*(0.18 + 0.4 \, x/t)$, where t is the number of items on the test. For $x/t > 0.3$, $p = 0.085 + 0.45*(0.18 + 0.4 \, x/t)$ (Haney, 1993; Holland, 1996; Sotaridona & Meijer, 2002). Used here as $-\log(P_K)$ in order to correlate it more easily with other indices.
ESA	Error Similarity Analysis. Number of identical incorrect answers (exact errors in common) between a student pair. As the quizzes may have different numbers of multiple-choice questions, a z-index is used (Bellezza & Bellezza, 1989), which is $$ESA = \frac{x - \bar{x}}{\sigma} = \frac{x - np}{\sqrt{np(1-p)}}$$ where n = number of incorrect answers in common (EEIC) x = number of EEICs between subject and source $\bar{x}$ = average number of EEICs between all student pairs p = probability of choosing a response randomly, i.e. 1/4 for questions with five possible responses, i.e., 4 incorrect responses.
G2	Sum of all identical responses between a student pair (C_{ab}) minus the probability that a student in the class would provide that answer divided by the covariance of the measure (Frary et al., 1977; Wesolowsky, 2000). $$C_{ab} = \sum_{ij} y_{ij} + \sum_{ij} x_{ij}$$ C_{ab} = EEICs (x_{ij}) + identical correct answers (y_{ij}) for students *a* and *b* summed over *i* questions and *j* answer choices. The pairwise index is then given by the following equation, where P_{jj} is the probability that subject *a* would have the same response as the source *b* on the ith question as determined from the overall responses of all students taking the test. $$g2 = \frac{C_{ab} - \sum_i P_{jj}}{\sqrt{\sum_i P_{jj}(1 - P_{jj})}}$$ The *Pjj* values are weighted according to the student's relative performance on the quiz in a series of equations given in reference (Frary et al., 1977) that depend on whether the response was correct or incorrect.

responses of examinees. Their two approaches differed mainly in terms of whether or not the response vector of one of the two examinees compared was considered to be fixed. Having explored the theoretical distribution of the two answer correspondence indices (which, when standardized in terms of appropriate standard errors, were called g statistics), Frary, Tideman, and Watts applied their correspondence indices to both simulated data (including simulated cheating) and real data (including two data sets—one in which pairs of examinees took the same form of a test in the same room and the other in which examinees in different rooms took different forms). The Frary G2 index given in Table 11.1 was determined to be the better of the two indices for practical applications (Frary & Tideman, 1997; Frary et al., 1977).

Frary, Tideman, and Watts (1977) provide an eminently clear and direct statement of the limits and potential use of copying indices such as their own. In the introduction they note:

> It should be noted at the outset that any index based on response similarity could take on a high value for a given pair of examinees due purely to chance, however unlikely. Therefore, it would never be feasible to *prove* that cheating occurred based *only* on the size of the response similarity index for a specific pair of examinees, just as it is never possible to prove, by citing statistics, that a scientific hypothesis is true. (Frary et al., 1977, p. 236)

Results of their simulations suggested that in order to identify as many as 80 percent of the cases of cheating, limits on response similarity indices would have to be set such that 10 to 20 percent of the pairs of examinees snared would be innocent. Thus, Frary, Tideman, and Watts closed their paper with the following suggestion:

> The fact that statistical evidence alone could never absolutely confirm an individual's cheating highlights the need for prevention rather than detection. It is in monitoring cheating to evaluate methods of preventing it that the methods of this paper have the greatest potential. (Frary et al., 1977, p. 255)

An even more unusual article relevant to indices of suspected copying was that of Levine and Rubin (1979). Perhaps the most notable aspect of Levine and Rubin's article is that they considered indices of suspected copying in the broader context of "inappropriate" patterns of examinee response. Inappropriate response patterns were considered to be indicative of both spuriously high scores (for example, a "low ability examinee copies answers to several difficult questions from a much more able neighbor") and spuriously low scores (e.g., a "very able examinee, fluent in Spanish, but not yet fluent in English, misunderstands the wording of several relatively easy questions") (Levine & Rubin, 1979, pp. 269–270). Using item response theory (Linden & Hambleton, 1996), Levine and Rubin explore the possibility of identifying spuriously high and low scores on the SAT by identifying aberrant patterns of examinee ability item performance. In the case of spuriously high scores (as with copying), examinees of low ability would get some very difficult items correct. In the reverse case, that of spuriously low scores, high-ability examinees would get easy items incorrect. As far as we have been able to determine, no operational college admissions testing program has yet taken seriously what Levine and Rubin's approach suggests—namely, devoting as much attention to suspected cases of spuriously low as to those of spuriously high scores (Levine & Rubin, 1979).

Buss and Novick's article, "The Detection of Cheating on Standardized Tests: Statistical and Legal Analysis" (1980) remains a key work on the subject. The article makes five points that are directly relevant to statistical analyses concerning suspected copying. First, since answer similarity indices are often coupled with evidence of unusually large score gains in investigating cases of suspected cheating, Buss and Novick argue that in the use of unusually large score gains as triggers to cheating inquiries, hypotheses alternative to cheating, and any independent evidence bearing on them, ought to be considered. Second, they suggest that any statistical analysis of answer concordance ought to be based not just on the numbers of identically incorrect answers but more broadly, on "at the very least, the number of items for which one examinee gave the correct answer, and the other an incorrect answer" (Buss & Novick, 1980, p. 11). Third, the assumption that examinee responses to successive items are independent (commonly implicit in indices of copying) may not be valid in particular applications. Fourth, as Buss and Novick themselves write, "perhaps most crucially, no formal hypothesis test can consider the wide range of alternative hypotheses, other than the primary alternative hypothesis of cheating, that might account for any unusually large concordance in responses of two examinees" (Buss & Novick, 1989, p. 13). Fifth, when testing agencies undertake multiple comparisons of alternative pairs of answer sheets, a great many comparisons will be made. As Buss and Novick comment, "Failure to take this problem into account in an analysis may be to employ something less than accepted professional practice" (Buss & Novick, 1989, p. 13).

In the 1980s and 1990s, several more publications dealt with statistical methods for detecting copying on multiple-choice tests. In something of a throwback to earlier approaches, Bellezza and Bellezza suggested using the binomial expansion for analyzing error similarity on multiple-choice tests (Bellezza & Bellezza, 1989). Their index suffers from some of the same defects of cheating indices noted by Buss and Novick (Buss & Novick, 1980—for example, by focusing only on similarity of wrong answers), though Bellezza and Bellezza do explicitly acknowledge some such limitations and take some account of them by suggesting a liberal probability for two examinees picking the same wrong answer by chance (0.4) (Bellezza & Bellezza, 1989).

In 1986, two dissertations were completed which dealt with indices of copying. In one of the dissertiations, Hartman studied two methods for detecting likely cheating in 170 fifth grade classrooms in a large public school system (Hartman, 1986). Her two indices were what she called the "wrong-in-common" index (which was the total number of wrong answers that a pair of students have in common, as a function of the total number of wrong answers for both) and the "run" index (which was defined in terms of the number of items answered incorrectly in the same way and in succession by a pair of examinees). Hartman concluded that cutoff scores for the two indices must vary depending on the total of wrong answers for pairs of students examined. Nevertheless, she concluded that "test irregu- larities" occurred in 5 to 8 percent of the fifth grade classrooms studied (Hartman, 1986).

In the other 1986 dissertation, Gutauskas sought to test two approaches for reducing copying on standardized tests used in eighth grade science classes (Gutauskas, 1986). She called the two approaches the Houston (after the work of J. Houston mentioned earlier) and Tauber methods. In the Houston approach, students received alternate test forms in which both the test questions and alternate multiple-choice answers were rearranged. In the Tauber approach, students received a test form in which each ques- tion was preceded with multiple numbers and, depending on directions they received, had to use different sets of those numbers in responding on a separate answer sheet. Both approaches were found to be effective in reducing copying from the side but not as effective in reducing copying from the front.

In 1987, Hanson, Harris, and Brennan published a research report enti- tled "A Comparison of Several Statistical Methods for Examining Allega- tions of Copying" (Hanson, Harris, & Brennan, 1987). Among the indices they tested were Angoff's B and H, and like Angoff's, their work was based on actual distributions of answers among pairs of examinees on an actual standardized test (in their case, a licensure exam). To test the efficacy of different indices, they employed five types and five levels of simulated copying. "For the type of copying thought to be most realistic, the methods do not differ greatly in performance, and approximately 5%, 20%, 50%, 85% and 95% of the simulated copiers who copied 10%, 20%, 30%, 40% and 50% of the items, respectively, could be detected with a false positive rate of .001" (Hanson et al., 1987, p. 3).

More recently, Harpp and colleagues have reported on their use with college classes of an index based on joint distributions of pairs of students' errors in common (EIC) and their exact errors in common (EEIC) (Harpp & Hogan, 1998; Harpp, Hogan, & Jennings, 1996). As a conservative cutoff for detecting copying, they suggest using outliers varying from the means of their distributions by more than 5.0 standard deviations. They report using their approach successfully in 75 different college exams from six different institutions over a period of four years.

In two followup studies, Harpp and colleagues investigated evidence bearing on a common defense when pairs of students are accused of cheating based on unusual numbers of identical wrong answers on a test. The defense often offered by students is that they had similar patterns of answers because they had studied together in preparation for the test. In one study, Harpp and Hogan (1993) examined answer patterns among identical twins who had studied together. They found that, while the pairs of twins scored similarly on the exam, they exhibited no more than the normal number of exact errors in common. In another study ("Crime in the Classroom III"), Harpp and Hogan (1998) recounted a case in which two students, who had been identified as having an very unusually high number of exact errors in common, admitted cheating (via stored memory on a calculator). Upon retest, three months after the first exam, the students were given the same test previously taken but with questions and answers scrambled. In the case of what they termed the "ultimate identical twin," they examined how the source student performed on the same exam, finding that while the student achieved essentially the same grade, he did so by a different path, that is, without having more than the normal number of exact errors in common with himself on the two occasions.

A number of other studies have been done on indices of copying on multiple-choice exams, but rather than review more of this literature here, let us mention two points made in many of these studies. First, in developing statistical indices of copying, it is preferable to rely not on theoretical but on empirically derived distributions [this point has been made repeatedly, for example, by Bird (Hanson et al., 1987), Angoff (Angoff, 1972, 1974), and Hanson, Harris, and Brennan (Hanson et al., 1987)]. The last-named authors comment, for example: "False positive rates based on theoretical assumptions were not found to agree well with false positive rates produced using benchmark data, and it is suggested that benchmark data be used whenever possible" (Hanson et al., 1987, p. iii).

Second, because probabilistic statistics can never conclusively prove that copying took place in a particular instance, such statistical methods may be more useful in preventing copying than in addressing particular cases of suspected copying. Although he was slightly inaccurate in what he suggested about certainty of detection, Bird made this point in 1927: "If we are correct in assuming that *certainty* of detection is one of the most important factors in the *prevention* of dishonesty, this method may be used not merely to detect cheating, but to prevent it" (Bird, 1927, p. 262, italics in original). Although they do advocate use of their procedure to pursue individual cases of suspected cheating, Harp and Hogan also suggest that by following fairly simple procedures, "cheating will be very substantially decreased, (if not eliminated), in multiple-choice exams and likely greatly reduced in other methods of examination" (Harpp & Hogan, 1993, p. 311; see also Harpp & Hogan, 1998; Harpp, Hogan, & Jennings, 1996). Specifically, Harpp and Hogan suggest that use of randomly assigned seating and scrambled or multiple test forms can dramatically reduce the extent of cheating via copying on college course examinations. Houston's work also suggests the use of multiple test forms as a means of reducing cheating via copying (Houston 1976a, 1976b). Hanson's (1990) research also suggests that student honor codes through which students pledge not to engage in academic dishonesty, including cheating on tests, can have some impact in reducing cheating. Finally, our own experience indicates that simply informing students that statistical methods to detect copying are being employed can dissuade students from engaging in such behavior.

COMPARISON OF CURRENT INDICES

As previously noted, the traditional methods of statistically detecting unusual instances of similarity in answers all depend largely on corresponding incorrect responses, which Harp and Hogan term *exact errors in common* (EEIC—that is, identical wrong answers). Table 11.1 shows that there are three general types of indices: (1) indices that are based on the number of EEICs truncated or divided in some way—these include the H and HH indices; (2) binomial expansions that calculate the theoretical probability that two students would choose the same set of incorrect responses—these include Chance and K; and (3) z-indices, where the difference between the number of observed student pair matches and the expected number of student pair matches is divided by the standard deviation—these differ in how they estimate the probability of each possible response to a question and include the ESA, G2, ω (Wollack, 1997, 2003, Wollack, 2004; Wollack & Cohen, 1998; Wollack, Cohen, & Serlin, 2001), and Scrutiny!, which is similar to ESA (*Scrutiny!*, 2005; Wollack, 2003) and Zb (Wesolowsky, 2000, 2005) (see Table 11.2 below).

As pointed out by Frary and others, the g2, ω, and Zb indices have an inherent advantage (see Table 11.2) because they consider both correct and incorrect responses in common and scale for the test performance of the examinees. However, they do not explicitly include differing responses. The EEIC, Chance, and Scrutiny! indices include only EEIC responses, as do the H, HH, and K indices, but the latter set of indices are also controlled

TABLE 11.2   Listing of Statistical Indices for Detecting Unusual Answer Similarity According to General Type and Factors Included

Index	EEIC	CRIC	Scaled	DR	Ref.
H	x			x^a	(Saretzky, 1984)
HH	x			x^b	(Harpp et al., 1996)
Ch	x				(Bellezza & Bellezza, 1989)
K	x		x	x^c	(Holland, 1996)
S₂	x	x	x	i	(Sotaridona & Meijer, 2003)
ESA	x				(Bellezza & Bellezza, 1989)
Scrutiny!	x				(*Scrutiny!*, 2005; Wollack, 2003)
g2	x	x	x	i	(Frary et al., 1977)
ω	x	x	x	i	(Wollack, 1997)
Zb	x	x	x	i	(Wesolowsky, 2000)

EEIC = exact errors in common.

CRIC = correct responses in common.

Scaled = the probability (p) of responding to a given question in a particular way is scaled for student performance.

DR = different responses.

x indicates the factor is explicitly included.

i indicates that DR is implicitly included in some way because the total number of responses N = EEIC + CRIC + DR, so that DR is a function of EEIC + CRIC (see (Holland, 1996)).

[a] See the definition of H in Table 11.1.

[b] HH = EEIC/DR.

[c] The K index includes only those different responses that were simultaneously incorrect, but different.

in some way by different responses. For example, if a student pair has two different responses, it may not be flagged by the H and HH indices, even though the pair has a relatively high number of EEICs because the two different responses will halve the HH index and may decrease the H index by as much as two-thirds. The K index, whose first term is very similar in form to the Chance index, has an advantage over the other indices that only include EEIC responses because K is not only scaled on the basis of student performance, but is also substantially decreased by different responses, albeit in a complex manner that depends on subsequent terms.

As indicated by Wollack (Wollack et al., 2001), g2 has a greater tendency to flag student pairs, even though they may have a large number of different responses. Values above which a significant amount of unusual answer similarity has been suggested to occur for the indices listed in Table 11.1 are approximately 3.3, 3, 4, and 4.5 for the ESA, Ch, K, and g2 indices, respectively (Frary et al., 1977; Harpp et al., 1996; Holland, 1996; Rizzuto & Walters, 1997). Values of the HH (Harpp et al., 1996) statistic greater than one and more than five standard deviations are said to be significant. HH values of six or more are touted as "nearly 100% accurate in finding highly suspicious pairs."

Although Holland points out that there is no information in different responses beyond that in the sum of correct, incorrect and omitted responses, the K index, in effect, includes different responses in a manner that decreases the index. This occurs because K is based on the binomial equation (Chance index in Table 11.1), but is modified to focus on the upper tail by adding terms dependent on the difference between the number of EEICs and the number of wrong answers marked by the source (Holland, 1996). Thus the likelihood that a student pair will be flagged decreases with the number of differing incorrect (but not identical) responses to the same questions.

Because most students will have a number of differing responses, the HH index has a very narrow distribution relative to that of the other indices, with a consequently much smaller standard deviation, thereby isolating the outliers. On the other hand, dividing by the number of different responses rapidly decreases this index as the number of different responses increases, which would be expected to decrease the power of this index to detect copying relative to other indices.

The inclusion of different responses should cause the HH and K indices to correlate fairly well, at least at the high end. Similarly, the inclusion of different responses by the HH index and their exclusion by the g2 index would suggest a lower level of correlations, despite the anecdotal claim to the contrary (Harpp et al., 1996).

The g2 and K indices are directional in that they depend on whether student a is considered the subject and student b the source or the reverse. When used as a general monitoring index, the effect of direction is relatively unimportant. On the other hand, when used as indicators of individual student behavior, the direction may be an important consideration (Wollack et al., 2001).

The ESA, H, HH, and Chance indices assume that the probability is the same for all EEIC responses; in the case of the Chance index, this is $1/n$, where n is the number of possible *incorrect* responses to a question (Angoff, 1974; Rizzuto & Walters, 1997). In contrast, the g2 index incorporates the actual probabilities for the class applied to both correct and EEIC responses weighted by student performance, and the K index calculates the probability by using an empirical formula that varies with student performance alone (Frary, 1993; Frary et al., 1977). (As noted previously, Bellezza utilized a probability of 0.4 rather than 0.25 for questions with five possible answers.) Consequently, if a student is wildly guessing or using an algorithm, such as "when in doubt, choose c," a number of highly improbable responses may be accumulated, which makes that student more likely to be flagged by the g2 index. On the other hand, the Chance index can either over- or underestimate the probabilities of EEIC pairwise responses. The ESA, H, and HH indices do not correct for the probabilities of answer responses at all (Harpp & Hogan, 1998; Harpp et al., 1996). Indices may show false positives if two students are using the same method of guessing.

Because the HH and K indices have been used not only to monitor the reliability of high-stakes tests, but also to detect individual students, these indices are somewhat conservative. On the other hand, the g2 index tends to yield more false positives than the K, ω, and S_2 indices (Sotaridona & Meijer, 2002; Wollack, 1997). Statisticians compare the various indices on the basis of their minimizing Type I errors (false positives) while at the same time exhibiting a high power to detect copying. However, these comparisons are often based solely on simulated data. Owing to the strengths and deficiencies of each index, it may be useful to look at several indices in assessing general trends in real testing situations. Indeed, several of the software packages available include several of the indices listed in Table 11.1 (Harpp & Hogan, 1998; *Integrity*, 2005; *ScanExam-II*, 2003; *Scrutiny!*, 2005; Wesolowsky, 2005).

Other z-type indices include Wesolowsky's Zb index, which is more self-consistent in the weighting of the student's ability (Nathanson, Paulhus, & Williams, 2006; Wesolowsky, 2000), and Wollack's ω index (Wollack, 1997, 2003, 2004; Wollack & Cohen, 1998; Wollack et al., 2001), which weighs the probability of an answer match using a measurement model known as the nominal response model. This model provides probabilities that examinees of different ability levels will select each of the choices. On the basis of simulated data, Wollack showed that ω generated fewer Type I (false positive) errors than g2 (Wollack, 1997), while both Wollack and Sotaridona state that the power of ω in detecting copying is greater than that of the K index for small and medium sample sizes of 100 and 500 simulated students, respectively (Sotaridona & Meijer, 2002).

Overall, it seems that g2, K, and ω can be used on sample sizes as small as 100 students (Sotaridona & Meijer, 2002; Wollack, 2003). The inherent advantages of g2 and ω are that they utilize more data (i.e., both identical correct and incorrect responses), whereas the K index is based only on incorrect responses. On the other hand, the g2 and K are easier to use, for they do not require item response parameters. While the empirical adjustment for student ability appears to be better for the K index than the g2 (Wollack, 1997), this is based on responses to SAT math and verbal sections derived by an undisclosed method and are different for both (Holland, 1996). Thus, they can only be used as approximations for responses on other tests.

A new index based on the Poisson distribution, rather than the binomial, is the S_2, which, like g2, Zb, and ω, utilizes both correct and incorrect responses. However, a weighting factor of one is used for EEIC responses, a factor of zero for different responses, and a factor of δ if both the source and the copier respond correctly to an item (Sotaridona, 2003a; Sotaridona & Meijer, 2003). The weighting factor (δ) is high when the estimate of the probability of a correct response is low, and low when the estimate of the probability of a correct response is high (Sotaridona, 2003b).

The processing power needed to calculate these indices varies not only with the sophistication of the probability factors, but also roughly with the square of the number of students (n) taking the test. More precisely, the number of student pairs that must be calculated for nondirectional indices is $n(n-1)/2$ and for directional ones $n(n-1)$. For large classes, this can lead to significant processing times. Nevertheless, all of the indices listed in Table 11.1 could be readily calculated using Visual Basic™ subroutines in Excel™ on a typical laptop[9] for class sizes of up to about 150 students. For in-class tests, CPU time can be conservatively used by comparing only those student pairs who are sitting within a reasonable visual distance of one another. Of course, this assumes that there is no electronic communication between examinees. While computationally more efficient than ω, the S_2 is less accurate on small sample sizes (Sotaridona, 2003a; Sotaridona & Meijer, 2003).

ONLINE TESTING

Most of the research on answer similarity on tests discussed so far deals with proctored tests administered as part of college or precollege courses or as part of national testing programs, such as the SAT or ACT college admissions testing programs. Nonetheless, answer similarity analyses can also be useful in monitoring online testing in which individuals take tests via computers and telecommunications. Online testing is increasingly used in a variety of situations ranging from weekly quizzes to examinations, for both regular classroom courses and distance learning. Commercially available and widely used courseware, such as Blackboard™ and WebCT™, allow comprehensive online tests and quizzes to be administered outside of class at the student's convenience. These can be automatically graded, and the correct-answer feedback can be provided to the student immediately following the quizzing period.

In an effort to promote learning by encouraging students to keep pace with the course through frequent quizzes, one of us (MJC) administered

---

[9] PC using XP Professional 2002 operating system, 512 M RAM, 1.6 GHz CPU and 30 G hard drive.

a series of online quizzes over a first-semester general chemistry course. Quizzes included a mixture of multiple-choice and algebraic "calculated" questions that provided different input values, which we thought might discourage answer copying. We had intended to check for how independently students were taking the quizzes by using the g2 index to detect for unusual answer similarity, but the program was no longer available at our institution and other computer programs were not easily interfaced with our data from WebCT. Consequently, we were unable to check for unusual answer similarity until we had written our own programs in Excel™. Although there are commercially available programs, as well as some freely provided to academic researchers, we found them either too expensive (*Integrity*, 2005; *Scrutiny!*, 2005) or cumbersome to use (Harpp & Hogan, 1998; *ScanExam-II*, 2003; Wesolowsky, 2005). Our experience illustrates how educators might promote learning through monitoring conveniently administered online quizzes.

Quizzes generally contained at least 20 questions, and the number of calculated questions depended on the relevance of algebraic problems to the material covered during the week the quiz was administered. The ground rules were that students could take the quiz anytime within a 24-hour period and use any source material, except each other.

Six recognized methods (Angoff, 1972; Bellezza & Bellezza, 1989; Frary et al., 1977; Harpp et al., 1996; Holland, 1996), which are described in Table 11.1, were chosen as indicators of answer copying and their algorithms were written as a Visual Basic™ subroutine interfaced with Excel™ spreadsheets. While the H and often the Chance index are considered obsolete, we included them for comparison purposes. Excel subroutines were written so that necessary lists, tables, and arrays could be easily entered through cursor selection of downloaded data. As shown in Figure 11.1, the results could be readily displayed as a 3-D bar graph indexed by student identifiers along horizontal axes, and the copying index could be plotted along the vertical axis. This individual pair data could be revealed by placing the cursor over the corresponding bar on the plot (Billo, 2001).

Several months after the course was over, students were surveyed to ascertain how independently they took the quizzes. The purpose of utilizing both the indices and survey was to determine whether it was possible to assay the general level of independent quiz taking. Data were not used to detect or implicate individual students.[10]

By most indices, the incidence of copying appeared to increase slowly, but it remained at perhaps acceptable levels until the last half of the

---

[10]The test and survey protocols were approved by the Boston College Institutional Review Board.

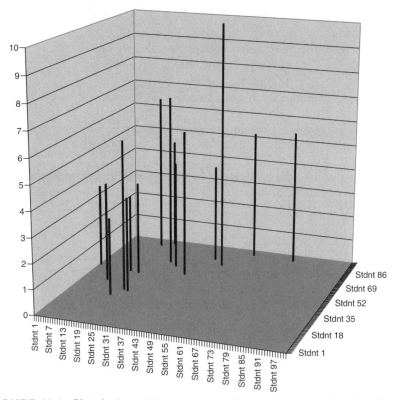

FIGURE 11.1   Plot of values of K2 student pair indices versus student identifiers. Values less than 3 are not plotted in order to more easily reveal suspicious student pairs. For example, when their responses are compared by using the K2 index, quiz takers with the identifiers Stdnt 54 and Stdnt 71 have a K2 index value of about 10, which is highly significant.

semester. As might be expected at the high end, where copying was strongly suspected, the indices correlated fairly well with each other. Indices such as the ESA and Chance, which depend essentially on the same variable, were highly correlated. On the other hand, the HH and g2 indices were the least well correlated both at the high end (0.67 when only the 20 highest student pairs by any index were considered), but particularly over all student pairs (0.25).

Notably, an overlap of student pairs (i.e., triples, quartets, etc.) appeared to develop over the semester, suggesting that collaborative groups as well as pairs had formed. This corresponds well with results of the student survey data, which indicated that at least a plurality of the students felt that the quizzes were a learning exercise. When coupled with the results that 79 percent of those responding were aware that collaboration or answer exchange was occurring by the end of the course, it may be that students felt increasing justification and/or pressure to respond in kind. Indeed, the results illustrated in Figure 11.2 show that the number of students using these methods increased over the semester and may have included over half the class.

Curiously, the responses to questions about whether the online quizzes assisted in learning or keeping abreast of the material were somewhat negative, while the responses of those answering a question concerning whether taking the test in collaboration with others assisted in learning were overwhelmingly positive.

While the average Pearson correlation coefficients between the number of students flagged by the indicators and the number of calculated questions on each quiz was -0.31, this varied with the cutoff values used. A permutation p-test indicated no statistically significant correlation regardless of the cutoff values used. Nevertheless, when queried as to whether calculated questions helped keep students "honest," students strongly responded "yes." When queried as to the number of calculated questions necessary to keep students honest, students answered that three to five were adequate.

Students found ways around taking the tests independently by a variety of methods. Notably, using electronic means of exchanging answers was far less prevalent than the more personal means of taking the quiz together or in groups. Only 4 percent reported having someone else take a quiz for them. Taking the quiz with others may account for the lack of statistical

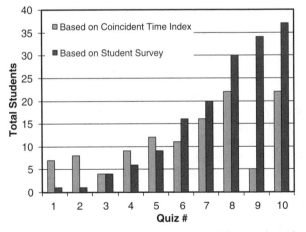

FIGURE 11.2 Weekly quiz number versus the cumulative number of students self-reported as using some form of collaboration (uncorrected for the approximately 60 percent sample size), and the number of student pairs with coincident response times over the number calculated as having a probability of greater approximately 0.001 (three to five) as calculated on the basis of answering the questions in random order. The anomaly in Quiz 9 is likely due to the "Thanksgiving effect" (see text).

evidence for the effect of calculated questions on answer similarity, in that collaboration was readily available on both multiple-choice and calculated questions. Responses also indicated that answer exchange networks increased slowly in the first few quizzes and more markedly in the second half of the semester. Overall 37, or 54 percent, of the respondents self-reported engaging in some form of answer exchange.

The answers to the calculated questions for each student were also extracted from the WebCT online data. Traditionally, teachers often feel that it is diagnostic of copying when students give an incorrect numeric answer for their own input data, but it happens to be the correct number for another student's input data. Unfortunately, it is very tedious to extract calculated questions from the online test responses, and, in our limited study, the data yielded relatively few clearcut indications of copying. Nevertheless, such data, if readily available, might provide strong corroboration in some cases. The location of the IP addresses of the student computers used at the time the quizzes were taken might also provide corroborating data in cases of an unusual answer similarity or temporal correspondence in responding to questions. However, obtaining this information raises strong ethical issues concerning privacy.

UNUSUAL SIMILARITIES IN TIME

A possible problem in applying traditional methods of detecting answer similarity to quizzes is the limited number of test items that can be reasonably posed. Studies involving statistical methods seldom use fewer than 40 test items (Sotaridona, 2003b; Sotaridona & Meijer, 2003; Wollack, 2003). Consequently, large changes in any index from one quiz of 20 to 30 items to another may not reflect the true overall level of answer exchange, while trends over several quizzes are more likely to be valid.

Peterson et al. suggested monitoring students taking online tests in real time as the test was administered (Peterson, Gordon, Elliott, & Kreiter, 2004). Unusual coincidences in the times of student responses were considered to be an index of suspicion of answer copying. In some ways, the analyses of temporal proximity may be viewed as analogous to evidence of spatial proximity (i.e., seating proximity), frequently undertaken in past studies of answer concordance. Specifically, in past studies seating proximity has often been viewed as possibly corroborative of answer similarity. For instance, if two students were found to have a highly unusual number of identical wrong answers *and* they were seated in adjacent seats during an examination, the inference that they cheated was strengthened.

With online testing, physical proximity may be less relevant, because students may take online tests from their homes or dormitories via telecommunications. But temporal proximity in test-taking may provide evidence of collaboration. Fortunately, courseware programs such as WebCT record the times that students answer each question. The coincidence of these times provides an independent measure of answer exchange, so that combining this with one or more of the traditional indices should increase the confidence level of determining trends in answer exchange, at least for those collaborating by taking the quiz together. Moreover, since it is independent of whether the responses are correct, the unusual similarity in answer times can potentially reveal collaborating students who do well on a test.

A difficulty in this method is that students who happen to be taking the quiz at the same time may reasonably be expected to have a certain number of simultaneous responses to the same questions, even though they are not collaborating. This number will be higher if students are answering the questions more or less in order than if the timing of their responses is random. Also, this number will increase with the number of questions administered within a given time period or with decreasing test time for a set number of questions. For the case of students answering questions in random order, which may underestimate the expected number of temporal overlaps in answer responses, this number can be estimated by a standard statistical equation (Feller, 1968).[11]

---

[11] The expected number for the random case can be given by the equation, which was adapted from reference (Feller, 1968) by Dr. Jenny Baglivo, Department of Mathematics, Boston College:

With **m** questions and **n** time slots, the probability on a random basis for **k** matches between two students of answering the same questions in identical time slot (Feller, 1968) is:

$$p_{[k]} = S_k - \binom{k+1}{k}S_{k+1} + \binom{k+2}{k}S_{k+2} - \binom{k+3}{k}S_{k+3} \ldots \pm \binom{m}{k}S_m$$

where the S values are the sum of the probabilities of overlapping events

$S_0 = 1$

$S_1 = \sum_i p_i = m\left(\frac{1}{n}\right)$

$S_2 = \sum_{ij} p_{ij} = \binom{m}{2}\left(\frac{1}{n(n-1)}\right)$

$S_3 = \sum_{ijk} p_{ijk} = \binom{m}{3}\left(\frac{1}{n(n-1)(n-2)}\right)$

etc., and

$p_i = \frac{1}{n}$

$p_{ij} = \frac{1}{n(n-1)}$

$p_{ijk} = \frac{1}{n(n-1)(n-2)}$

etc.

For the quizzes shown in Figure 11.2, which were 30 to 70 minutes in duration and involved 17 to 22 item responses, plus the times the quiz was started and submitted, the expected number of coincident times on the basis of answering the questions in random order was three to five. While this may underestimate the number of coincident responses, as students were inclined to answer the questions more or less in the order given, the number of student pairs with coincident times above this number is plotted in Figure 11.2 for a series of quizzes. Notable aspects of this plot are that the cumulative number of students admitting to having used some means of collaboration (which was overwhelmingly by taking a quiz in conjunction with one or more other students) correlates reasonably well with the number of student pairs having more than three to five coincident times— except for Quiz 9.

What happened on Quiz 9? This quiz was administered during Thanksgiving week, when students were traveling (sometimes before the university holiday started) and were allowed four days instead of the usual one to take the quiz. Consequently, students were spread out in both time and space so that the number of those collaborating by taking the quiz at the same time appears to have dropped precipitously, only to pick up again on the last quiz of the semester. Though based on limited data, this suggests that a more useful coincident time index could be developed with an improved statistical approach.

The outlying results from the last quiz of the semester, which was taken by 111 students, are shown in Figure 11.3 and are uncorrected for the

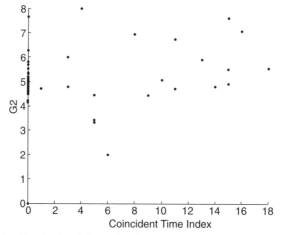

FIGURE 11.3 Graph of outlying points for Quiz 10 for the g2 index versus the number of coincident times for item responses, including the start time of the quiz and the time it was submitted. Shown are 86 outlying student pairs out of approximately 6,005. Student pairs for which the difference in grade was greater than 5 question responses (out of 32) were excluded. Item response times were taken as coincident if they occurred within 48 seconds of the time of the other member of the student pair.

expected number of simultaneous responses for students taking the quiz at the same time. Of the 6,005 unique student pairs, only 86 were outlying by any of the indices given in Table 11.1 or the coincident time index. Those lying high along the vertical axis exhibit some reasonable probability of collaborating on answers on the basis of the g2 index, and those lying to the right on the horizontal axis are more likely to have collaborated on the basis of the coincident time index. Those residing in the upper right quadrant might reasonably be expected to have a much higher probability of collaborating overall.

Although the degree of unusual answer similarity on quizzes may be relatively low for high statistical significance, combining these with a separate index for unusual similarities in response times may provide sufficient statistical reliability to monitor the level of answer exchange for unproctored students taking quizzes online and may reveal collaborative efforts (i.e., in the case of few incorrect responses) where the traditional indices fail.

The development of more sophisticated statistical methods for determining the probability of coincidences in the times during which individual questions were answered, possibly taking into account the degree to which students answered the questions in the order they were posed, would likely be very useful in detecting collaboration in online tests.

Since unproctored online quizzes seem to promote collaboration among students, some may view this as an opportunity to foster interactive learning groups. For example, it may be possible to turn this to an advantage by allowing students to form defined study/quiz groups, but then emphasizing that additional learning is expected by providing an extra point incentive for these students on in-class tests for questions that were either identical to or of the same basic type as those that had previously appeared on the online quizzes.

CONCLUSIONS

In this chapter, we have reviewed literature on the prevalence of academic dishonesty among students in general and the detection and prevention of cheating on tests in particular. Studies dating back to the 1960s indicate that cheating on tests is fairly common, being admitted to by some 15 to 45 percent of college students surveyed. While numerous forms of cheating on tests have been documented, for example, by Cizek (1999), copying or illicit communication during testing appears to be one of the most common forms of cheating. At the same time, numerous observers have noted that the problem of cheating on tests is addressed far less commonly than it occurs. This is unfortunate because, as Wesolowsky has noted, the existence of statistical methods for detecting copying "appears to be little known among instructors using mutiple choice tests and examinations (Wesolowsky, 2000, p. 909).

As a result, in this chapter we reviewed literature on statistical indices of answer concordance, which have been used for more than 80 years to detect possible cheating on tests. Although determination of cheating in individual cases, by almost all accounts, requires evidence beyond statistical evidence of answer concordance (such as anomalous patterns of score gains and loses, evidence of proximity during testing, and eyewitness testimony from exam proctors), there is considerable evidence that answer similarity analyses can be used to monitor the extent of cheating and methods, such as use of multiple test forms and randomized seating, to prevent cheating.

We compared a number of different answer similarity indices, variously taking into account wrong answers in common, correct answers in common, and different answers. We showed that at Boston College, six of the most commonly used indices correlate reasonably well (in the range of 0.70 to 0.95), at least when used to identify the most aberrant cases.

Finally, we summarized the experience of one of the authors (MJC) in using answer similarity analyses in monitoring how students behave while taking unproctored online quizzes. We found that students adapt to unproctored online testing over time by collaborating. While various forms of collaboration in online tests are used, the easiest and perhaps most common way is for students to take the test together (e.g., congregating with laptops in a dorm room). We also found that the use of calculated questions had no statistical effect on the level of answer exchange. In addition, we found it helpful to employ a variety of indices available for monitoring answer concordance in tests. Finally, we suggest that time-based correlations, such as plots of unusual answer similarity versus an index of unusual temporal similarity, may be useful in detecting collaboration in online testing and can potentially detect collaboration when students have few incorrect responses (i.e., where statistical indices of answer similarity typically fail). Finally, courseware vendors might facilitate data capture of test response items, including calculated questions, time of response, and possibly IP address, which would facilitate using the various types of indices in monitoring for collaboration.

ACKNOWLEDGMENTS

We thank our colleagues Professors E. Joseph Billo (Department of Chemistry) and Jenny Baglivo (Department of Mathematics) for their help in writing the computer subroutines and statistical analysis, respectively. Part of MJC's contribution to this work was supported while he was serving at the National Science Foundation under the IR/D program.

REFERENCES

Angoff, W. H. (1972). *The development of statistical indices for detecting cheaters* (CEEB-RB-72-26, ED069687). Princeton, NJ: Educational Testing Service.

Angoff, W. H. (1974). The development of statistical indices for detecting cheaters. *Journal of the American Statistical Association, 69*, 44–49.

Anikeef, A. M. (1954). Index of collaboration for test administrators. *Journal of Applied Psychology, 38*, 174–177.

Beatty, J. (1992). *The rascal king: the life and times of James Michael Curley (1874–1958)*. Reading, MA: Addison-Wesley.

Bellezza, F. S., & Bellezza, S. F. (1989). Detection of cheating on multiple-choice tests by using error-similarity analysis. *Teaching of Psychology, 16*(3), 151–155.

Billo, E. J. (2001). *Excel for chemists: A comprehensive guide* (2nd ed.). New York: Wiley-VCH.

Bird, C. (1927). The detection of cheating in objective examinations. *School & Society, 25*, 261–262.

Bowers, W. J. (1964). *Student dishonesty and its control in college*. New York: Columbia University, Bureau of Applied Research.

Buss, W. G., & Novick, M. G. (1980). The detection of cheating on standardized tests: statistical and legal analysis. *Journal of Law and Education, 9*(1), 1–64.

Caveon.com. (2005a). *Caveon.com Newsletter*. Retrieved July 22, 2005, from the World Wide Web: http://www.caveon.com/articles/newsltr_10_03_1.htm.

Caveon.com. (2005b). *Caveon™ test security to apply its test cheating detection services to the Texas Assessment of Knowledge and Skills (TAKS™)*. Retrieved August 7, 2005, from the World Wide Web: http://www.caveon.com/pr/press7-19-05.htm.

Cizek, G. (1999). *Cheating on tests: How to do it, detect it and prevent it*. Mahwah, NJ: Lawrence Erlbaum.

Cizek, G. (2000). An overview of issues concerning cheating on large–scale tests. Paper presented at the National Association of Test Directors 2001 Symposia, New Orleans La.

Cizek, G. J. (2001, April). An overview of issues concerning cheating on large-scale tests. Paper presented at the National Council on Measurement in Education, Seattle, WA.

Dickenson, H. F. (1945). Identical errors and deception. *Journal of Educational Research, 38*, 534–542.

Feller, W. (1968). *An introduction to probability theory and its applications* (3rd ed., Vol. I). New York: John Wiley & Sons.

Frary, R. B. (1993). Statistical detection of multiple-choice answer copying: Review and commentary. *Applied Measurement in Education, 6*(2), 153–165.

Frary, R. B., & Tideman, T. N. (1997). Comparison of two indices of answer copying and development of a spliced index. *Educational and Psychological Measurement, 57*(1), 20–32.

Frary, R. B., Tideman, T. N., & Watts, T. M. (1977). Indices of cheating on multiple-choice tests. *Journal of Educational Statistics, 2*, 235–256.

Gutauskas, S. (1986). A comparison of the Houston and Tauber methods to reduce cheating on multiple-choice tests. Unpublished Ed.D. Dissertation, Temple University, Philadelphia.

Haney, W. (1993, April 15). Cheating and escheating on standardized tests. Paper presented at the Annual Meeting of the American Educational Association, Atlanta, GA.

Hanson, A. C. (1990). *Academic dishonesty: The impact of student and institutional characteristics on cheating behavior*. Unpublished Ph.D. Dissertation, University of California, Los Angeles.

Hanson, B. A., Harris, D. J., & Brennan, R. L. (1987). *A comparison of several statistical methods for examining allegations of copying (ACT Research Report Series 87–12)*. Iowa City, IA: American College Testing Program.

Harpp, D. N., & Hogan, J. J. (1993). Crime in the classroom—detection and prevention of cheating on multiple-choice exams. *Journal of Chemical Education, 70*(4), 306–311.

Harpp, D. N., & Hogan, J. J. (1998). Crime in the classroom. Part III: The case of the ultimate identical twin. *Journal of Chemical Education, 75*(4), 482–483.

Harpp, D. N., Hogan, J. J., & Jennings, J. S. (1996). Crime in the classroom. Part II: An update. *Journal of Chemical Education, 73*(4), 349–351.

Hartman, A. G. (1986). Detecting cheating on multiple-choice examinations. Unpublished Ph.D. Dissertation, Tulane University, New Orleans.

Hartshorne, H., & May, M. A. (1928). *Studies in the nature of character* (Vol. 1). New York: Macmillan.

Holland, P. W. (1996). *Assessing unusual agreement between the incorrect answers of two examinees using the K-Index: Statistical theory and empirical support.* (ETS RR-96-7). Princeton, NJ: Educational Testing Service.

Houston, J. P. (1976a). Amount and loci of classroom answer copying, spaced seating, and alternate test forms. *Journal of Educational Psychology, 68,* 729–735.

Houston, J. P. (1976b). The assessment and prevention of answer copying on undergraduate multiple-choice examinations. *Research in Higher Education, 5*(4), 301–311.

Houston, J. P. (1977). Cheating behavior, anticipated success-failure, confidence and test importance. *Journal of Educational Psychology, 69,* 55–60.

Houston, J. P. (1978). Curvilinear relationships among anticipated success, cheating behavior, temptations to cheat and perceived instrumentality of cheating. *Journal of Educational Psychology, 70,* 758–762.

Houston, J. P. (1983). Alternate test forms as a means of reducing multiple-choice answer copying in the classroom. *Journal of Educational Psychology, 75*(4), 572–575.

Houston, J. P. (1986). Classroom answer copying: Roles of acquaintanceship and free versus assigned seating. *Journal of Educational Psychology, 78*(3), 230–232.

Integrity. (2005). Castle Rock Research. Retrieved September 1, 2006, from the World Wide Web: http://integrity.castlerockresearch.com/.

Leandro, G. (2005). *Meta-analysis in medical research: The handbook for the understanding and practice of meta-analysis.* Malden, MA: BMJ Books, Blackwell Publishers.

Levine, M. V., & Rubin, D. B. (1979). Measuring the appropriateness of multiple-choice test scores. *Journal of Educational Statistics, 4*(4), 269–290.

Levitt, S., & Dubner, S. (2005). *Freakanomics: A rogue economist explores the hidden side of everything.* New York: William Morrow.

Linden, W. J. V. D., & Hambleton, R. K. (1996). Item response theory: Brief history, common models and extensions. In W. J. V. D. Linden & R. K. Hambleton (Eds.), *Handbook of modern item response theory.* New York: Springer.

McCabe, D. L. (1992). The influence of situational ethics on cheating among college students. *Sociological Inquiry, 62*(3), 365–374.

McCabe, D. L., & Bowers, W. J. (1994). Academic dishonesty among males in college: A thirty year perspective. *Journal of College Student Development, 35*(1), 5–10.

McCabe, D. L., & Trevino, L. K. (1993). Academic dishonesty: Honor codes and other contextual influences. *Journal of Higher Education, 64*(5), 522–538.

McCabe, D. L., Trevino, L. K., & Butterfield, K. D. (2001). Cheating in academic institutions: A decade of research. *Ethics and Behavior, 11,* 219–232.

Miyazaki, I. (1976). *China's examination hell.* New York: Weatherhill.

Nathanson, C., Paulhus, D. L., & Williams, K. M. (2006). Predictors of a behavioral measure of scholastic cheating: Personality and competence but not demographics. *Contemporary Educational Psychology, 31*(1), 97–122.

Peterson, M. W., Gordon, J., Elliott, S., & Kreiter, C. (2004). Computer-based testing: Initial report of extensive use in a medical school curriculum. *Teaching and Learning in Medicine, 16*(1), 51–59.

Rizzuto, G. T., & Walters, F. H. (1997). Cheating probabilities on multiple choice tests. *Journal of Chemical Education, 74*(10), 1185.

Saretzky, G. (1984). *Treatment of scores of questionable validity: The origins and development of the ETS Board of Review* (ETS Archives Occasional Paper ED254538). Princeton, NJ: Educational Testing Service.

Saupe, J. L. (1960). An empirical model for the corroboration of suspected cheating on multiple-choice tests. *Educational and Psychological Measurement, 20,* 475–489.

ScanExam-II (2003). University of Western Ontario. Retrieved September 1, 2006, from the World Wide Web: www.ssc.uwo.ca/sscl/softwaredistribution.html.

Scrutiny! (2005). Assessment Systems Corporation. Retrieved September 1, 2006, from the World Wide Web: http://www.assess.com/software/scruting.htm

Sotaridona, L. S. (2003a). Cheating detection using the s2 copying index. *The Philippine Statistician, 52*(1–4), 59–67.

Sotaridona, L. S. (2003b). Statistical methods for the detecting of answer copying on achievement tests. Unpublished, University of Twente, Twente, The Netherlands.

Sotaridona, L. S., & Meijer, R. R. (2002). Statistical properties of the K-index for detecting answer copying. *Journal of Educational Measurement, 39,* 115–132.

Sotaridona, L. S., & Meijer, R. R. (2003). Two new statistics to detect answer copying. *Journal of Educational Measurement, 40,* 53–69.

Wesolowsky, G. O. (2000). Detecting excessive similarity in answers on multiple choice exams. *Journal of Applied Statistics, 27*(7), 909–921.

Wesolowsky, G. O. (2005). *S-Check.* McMaster University. Retrieved, September 1, 2006, from the World Wide Web: http://www.business.mcmaster.ca/msis/profs/wesolo/ReadMeScheck.pdf; http://www.business.mcmaster.ca/msis/profs/wesolo/wesolo.htm.

Whitely, B. E. (1998). Factors associated with cheating among college students. *Research in Higher Education, 39*(3), 235–274.

Whitley, B. E., & Keith-Spiegel, P. (2002). *Academic dishonesty: An educator's guide.* Mahwah, NJ: Laurence Erlbaum.

Whitely, B. E., Nelson, A. B., & Jones, N. (1999). Gender differences in cheating attitudes and classroom cheating behavior: A meta-analysis. *Sex Roles, 41*(9–10), 657–680.

Wollack, J. A. (1997). A nominal response model approach to detect answer copying. *Applied Psychological Measurement, 21,* 307–320.

Wollack, J. A. (2003). Comparison of answer copying indices with real data. *Journal of Educational Measurement, 40,* 189–205.

Wollack, J. A. (2004). Detecting answer copying on high-stakes tests. *The Bar Examiner, 73*(2), 35–45.

Wollack, J. A., & Cohen, A. S. (1998). Detection of answer copying with unknown item and trait parameters. *Applied Psychological Measurement, 22,* 144–152.

Wollack, J. A., Cohen, A. S., & Serlin, R. C. (2001). Defining error and power for detecting answer copying. *Applied Psychological Measurement, 25,* 385–404.

12

The Pressure to Cheat in a High-Stakes Testing Environment

Sharon L. Nichols and
David C. Berliner

INTRODUCTION[1]

A common error of attribution is to believe that an individual's character is stable across settings. Qualities such as virtue, honesty, and integrity are often used to describe stable and generalizable dispositions of individuals, both adults and children. But from the classic, groundbreaking studies of Hartshorne and May (1928, 1929, 1930), we learned that students who cheat do so as a function of the situations in which they find themselves rather than because of some general flaw in character. Hartshorne and May found that students who recognize virtue, honesty, integrity, and truthfulness as desirable goals were also often found to be "cheating," that is, peeking at answers, changing answers, inflating scores, and stealing coins. Eventually, psychology recognized that a good deal of human *B*ehavior was the result of an interaction between *P*ersonological variables and *E*nvironmental variables, or as is commonly stated, $B = (P \times E)$. Stability of behavior across situations is far less frequent than the general public would believe.

In Hartshorne and May's lengthy series of assessments of school children to determine propensity for cheating, the stakes or consequences associated with the outcomes of the assessments were generally

---

[1] The authors would like to acknowledge Michelle Shuler for her valuable editorial help with this chapter.

Copyright © 2007 by Academic Press, Inc.
All rights of reproduction in any form reserved.

inconsequential. Yet even under these conditions it was hard to find students who never cheated, and it was equally hard to find students who cheated all the time. Individual acts of cheating, the researchers concluded, were a function of the person in a situation, an interaction. In this series of studies each apparently similar situation was perceived differently by each individual, and thus each situation was responded to differently.

Unlike the assessment exercises of Hartshorne and May, the No Child Left Behind (NCLB) law of 2002 creates a highly competitive environment, one in which student test scores matter at least as much to teachers and administrators as they do to the students themselves. Under these conditions, teachers and administrators who are being pressured to improve test scores for all subgroups of children every year may well be in pressure-inducing situations that encourage behavior we ordinarily call cheating. The integrity, honesty, and truthfulness of America's teachers and administrators may well be compromised if these individuals find themselves in situations that they see as unfair, and the result could be humiliation, school closure, and loss of employment.

The passage of NCLB was a response to the widespread belief that teachers and students are often lazy or unmotivated and that each holds expectations for performance that are much too low. The theory of action behind NCLB is that teachers and students will change when enticing incentives and/or punishing consequences are attached to rigorous programs of testing (cf. Raymond & Hanushek, 2003). The assumption is that under threat of aversive punishments or enticing rewards everyone will work harder. But working "harder" does not equal more effective, and ultimately, this simplistic agenda for school reform ignores a lengthy and rich literature on what constitutes effective teaching and learning. Student motivation, for example, is often severely compromised when students are enticed to perform by extrinsic forces rather than pushed to perform by their own intrinsic reasons for learning (Deci, Koestner, & Ryan, 1999). In addition, many educators argue that extrinsic rewards alone cannot overcome the range of background experiences and individual differences in learning style and motivation brought to school by students (e.g., Deci & Ryan, 1985; Deci, Koestner, & Ryan, 2001; Good, 1999, 2000; McCaslin & Good, 1996). Furthermore, family and youth poverty, along with racial and ethnic segregation, may be the real culprits behind low school performance; these are characteristics of American life that NCLB never addresses (Berliner, 2005; Kozol, 2005). For these reasons NCLB may not improve the schools, and many will attribute its failure to the very teachers and administrators who are working so hard to meet the laws' many, often impossible (and underfunded), demands. Thus, with pressure both high and continuous, the honesty and integrity of America's teachers and administrators may be compromised in ways unprecedented in our nation's history.

ORIGINS OF THE PRESSURES
EXERTED IN NCLB

Originating partly from concerns that America might lose its economic leadership of the world (National Commission for Excellence in Education, 1983) and partly from concerns about the adequacy of education for America's disadvantaged students, NCLB mandates that administrators, teachers, and students all should be held accountable for what students learn. Learning more than in the past, and more quickly than in the past, is what every district and school in the nation must now do, though the amount and speed of achievement growth are determined by local conditions. Students' failure to learn as much and as quickly as required by NCLB results in serious consequences for everyone—districts, schools, teachers, and, of course, students. This is the origin of the pressure educators feel.

The practice of high-stakes testing—attaching important consequences to standardized test performance—is not new. Students in New York have had to pass the prestigious Regents exams to obtain a high-status high school diploma for over 100 years, and other states such as Texas and Kentucky had been increasing the stakes associated with testing for the past 20 years. In fact, testing with stakes is familiar to anyone who has ever taken the SATs, ACTs or GREs, in which test scores determine whether one is admitted to or rejected by the college of one's choice. What is new with NCLB is the national and systematic application of this kind of testing pressure on all students in virtually every grade and during every school year. By law, students in grades 3–8, and 10, must take high-stakes tests, but in practice, significant numbers of first and second grade students are also being tested, and many high school students are required by their districts to take high-stakes achievement tests in grades preceding and following grade 10. Thus, almost no one in the public schools is immune to the pressures associated with these tests.

Failure on these tests results in humiliating consequences for educators and students, including very young students, special education students, and language minority students. Although the range and impact of these consequences can vary widely, what is common to everyone under NCLB is that the focus of education is now on the schools as well as the students. This public attention to "passing" and "failing" schools creates a climate in which success and failure on tests are synonymous with "good" and "bad" teachers, administrators, and districts. Although this is not all bad (focus on dysfunctional districts, inadequate principals, and weak teachers is overdue), when achievement testing takes place under such conditions, the honesty, integrity, and trustworthiness of all adults in the educational system are likely to be tested as often as are the honesty, integrity and trustworthiness of the students in the system.

PRESSURE IS LINKED TO CHEATING

The theoretical link between the pressure accompanying high-stakes testing and cheating comes from the eminent social scientist Donald Campbell. Campbell's law (1975, p. 35) states, "The more any quantitative social indicator is used for social decision-making, the more subject it will be to corruption pressures and the more apt it will be to distort and corrupt the social processes it is intended to monitor." The very act of measuring an outcome that is highly valued, say an indicator from whose values consequences ensue (amount of oil reserves, accounts receivable, predicted sales, scores on achievement tests), leads us to question the trustworthiness of the outcome.

Campbell claims that when the stakes are high it isn't only the measure that could be corrupted, but corruption of the social processes may also be associated with the measure. Instances of the ubiquity of Campbell's law have been described elsewhere, including the fields of business (Enron profits and loss), athletics (grades of college players), the military (recruitment practices, body counts during combat), government (reported deficits, percent in poverty) and education (Nichols & Berliner, 2005). In our contemporary system of education, Campbell's law predicts that because of high-stakes testing there will be corruption of the indicators used to measure student achievement (i.e., test scores), as well as greater corruption of the educators working in that educational system.

For example, studies show that as test-related pressure goes up, sound instructional decision making goes down. Teachers report that the tests drive instruction, forcing them to teach to the test, narrow the curriculum, restrict their instructional creativity, and compromise their professional standards (Pedulla et al., 2003). A more pernicious effect is that as a result of high-stakes testing pressure, many administrators and teachers now view learning as a product, a simple test score, rather than as a process. This commodification of learning, where knowledge becomes merely a valued test score, fosters a culture of doing whatever it takes to get an acceptable number, and thus cheating becomes more likely, more acceptable, and more justifiable.

ADULT CHEATING

Sufficient evidence exists that high-stakes testing environments foster the corruption of indicators and educators, as predicted by Campbell (Nichols & Berliner, 2005). But, as we discuss below, corruption comes in many forms—some of which are blatantly deceitful and others of which suggest that there is more of a moral grey area. For example, a teacher's "help" of a struggling student with one or two challenging test items may seem a small and forgivable infraction when compared to the potential motivational and psychological costs of a struggling student failing yet another test or the costs to a teacher's family if her job is lost. The dilemma of teachers and administrators under NCLB is similar to those posed by Kohlberg (1984) in his studies of moral development. One of those dilemmas posed the question of whether it is right or wrong for a person who can't afford the medicine needed to save a dying spouse to steal it. When faced with the dilemma of either (a) cheating to help a struggling, demoralized student or to keep their own job for the financial protection of their own family, or (b) not cheating and watching a student falter or their family's economic security harmed, teachers' moral choice appears clear: stealing the medicine can be justified.

The culture of cheating created by the pressures of high-stakes testing under NCLB is pernicious on many levels. Although we articulate how this plays out among adults throughout this chapter, we are equally worried by how this climate affects students. We know from the early Hartshorne studies that student cheating is a function of person variables in interaction with environmental variables. More recent research has shed light on what aspects of person–environment interactions are more likely to predict student cheating. For example, an emphasis on extrinsic goals is more likely to lead to student cheating (Anderman, Griesinger, & Westerfield, 1998; Evans & Craig, 1990). Similarly, students' perceptions of their teachers as incompetent make it more likely that they will cheat (Murdock, Miller, & Kolhardt, 2004). As we explore the range of ways high-stakes testing pressure leads to the corruption of adults, we must be equally mindful of how this "cheating" culture (Callahan, 2004) is socializing our youth in American classrooms.

It may be difficult for some people to ever defend cheating, especially forms of cheating that improve one's income through bonuses awarded for higher student performance. Examples of those forms of cheating abound, and they present no moral quandary. But later in this chapter we offer examples of cheating that may not be as easy to condemn—and we shouldn't. When lives and livelihoods are at stake, we might want to forgive minor rule infractions in support of a greater good. And we might want to inquire about the value of instilling a system that leads to dilemmas infused with moral confusion in the first place, especially when cheating (blatant or otherwise) leads to an invalid test score.

ADULT CHEATING AND OTHER DUBIOUS BEHAVIOR

In this section we provide a range of examples of adult cheating, dubious behavior, misdirection and fraud, all as a function of contemporary high-

	"Blatant" examples of cheating	Grey areas of cheating
Nontest indicators	CELL A	CELL B
Tests as indicators	CELL C	CELL D

FIGURE 12.1   Matrix for organizing examples of cheating.

stakes testing. The dictionary definition of cheating is to deceive or mislead, so cheating includes more than a concern with student test scores, which is often the first thing most people think of when the term *cheating* is used in proximity to the word *testing*. If a state fudges its high school dropout numbers, as Massachusetts has done (Rothstein, 2004), or if a district lies about its college entrance figures, as Houston has done (Schemo & Fessenden, 2003), or if a test is made patently easier, as Arizona has done (Gassen & Sterba, 2003), the public is deceived. Deliberate attempts to deceive are acts of cheating, and so we use that term to describe a large variety of incidents we find dubious.

We have organized these incidents into a 2 × 2 matrix differentiating those incidents by type of behavior (apparently blatant examples of cheating—or direct attempts to deceive—versus grey area activities) and whether or not the incident was directly related to testing and/or the test score or not. Difficult judgments necessarily had to be made about what might be a dubious incident, what might be blatant, and what might be a grey area of cheating. Nevertheless, to organize what follows, we will use the 2 × 2 matrix provided as Figure 12.1.

CELL A: EXAMPLES OF BLATANT CHEATING—
WITH NONTEST INDICATORS

Example A-1

Birmingham, Alabama, provides an egregious example of how high-stakes testing pressures and the lure of cash can corrupt educators. In this case, educators not only deceived the public but blatantly violated the rights of their own city's youth (Orfield, Losen, & Wald, 2004).

Birmingham's administrators knew that if students predicted to do poorly on upcoming tests could be prevented from taking those tests, then test scores would be higher. Dropouts need only to occur selectively, and in a timely fashion, for test scores to be manipulated and the public misled. Birmingham officials implemented such a policy when they had 522 young people "administratively withdrawn" from high school, just before administration of their annual high-stakes tests. In general, parents were not included in the withdrawal meetings; some parents were never even notified their children had been withdrawn; and requests by concerned parents and guardians to have their children allowed back in school were denied. One mother fighting for her child's reinstatement was informed that her daughter's standardized test scores were low and that she probably wouldn't graduate, so the school officials refused to readmit her.

Pushing students out of school is not an uncommon way for educators to mislead the public about test scores. In New York City, for example, students had been pushed out for years (Lewin & Medina, 2003). Apparently, low-performing students were often counseled into taking a high school equivalency exam, the GED, and were therefore pushed out of school. Since these students are not recorded as dropouts because they were supposedly studying for the GED, this activity could mislead the public about the actual dropout rate. In fact, after these students left school, there were no records of what they did. Knowledgeable educators estimate that if pushout rates were added to dropout rates, the dropout figure would be 5 to 10 percent higher than it is. Pushing out the poor-performing students also raises a school's test scores, so the New York City schools cheated the public twice (misleading dropout rates and misleading test scores) with its extensive pushout program.

After months of not commenting about charges that pushout programs were widespread in the New York schools, Chancellor Joel I. Klein conceded that the practice occurred, that it was not merely a few instances but a systematic program of deceit, and that the problem was a tragedy. Promises were made to halt these practices. But two years later, in a recent lawsuit, it is contended that students at Boys and Girls High School in Brooklyn were still being pushed out (Herszenhorn, 2005). Allegedly, a common practice was to warehouse many low-performing students in a school auditorium, have them just fill out worksheets for a few hours each day and then assign the same worksheets for them to do as homework. If that wasn't enough to have them drop out on their own, after a while they were told they had accumulated no credits, which became the grounds for dropping them from the school rolls. Pushing out and dropping students from school, if they are the "right" students, increases test scores in a deceitful way. It has become an all too common practice under the NCLB law (Nichols & Berliner, 2005).

Example A-2

Another incident that illustrates cell A involves manipulation of dropout rates to show evidence of success by a district, cheating the public of accurate data about the holding power of the district's schools. This incident comes from the Houston Independent School District, apparently a national leader in systematic programs of cheating and misdirection by its educators (see Example A-3).

In this case a Sharpstown High School dropout and his mother noticed that the high school he should have been attending had recorded no dropouts for that year (Schemo, 2003a). That was obviously not true. The mother, with the help of a local newspaper and a whistleblower on campus, provoked an investigation. She found that 462 students had left Sharpstown and all were reported to be in charter schools or other schools, though Sharpstown administrators had not asked these students where they were going and had no knowledge of what they were doing. Sharpstown had started with about 1,000 freshmen, but its senior class was about 300 when the story was written. Yet not a single student was recorded as having dropped out. In 2000–2001, the year that Houston said it had a 1.5 percent dropout rate, about 5,500 left the schools and over half of those should probably have been counted as dropouts but were not (Schemo, 2003a).

Previous to the announcement that it was deceiving the public, Sharpstown high school staff received bonuses based on the good attendance, low dropout rates, and increased test scores of its students. For his leadership of the Houston School District, Superintendent Rod Paige was honored by McGraw-Hill Publishers, and on the basis of his record as a school leader, he was elevated to the highest educational position in the land, secretary of education, under President G. W. Bush (Schemo, 2003a). In addition, for their outstanding urban education programs Houston received $1 million from the Broad Foundation. Sharpstown and Houston, home of Enron, are examples of the ubiquity of Campbell's law: If an indicator takes on excessive value, the indicator and the individuals who work with it are likely to be corrupted.

Example A-3

The Houston Independent School District had another indicator of success that it was proud of—the rate of college attendance by its high school graduates (Schemo, 2003b). Jack Yates High School, for example, from 1998 to 2002, reported that 99 to 100 percent of its graduates planned to attend college. Yet at Jack Yates High School students had to make do without a school library for more than a year, and the principal replaced dozens of experienced teachers with substitutes and uncertified teachers, who cost less. These were not likely to be the kinds of events that would give rise to high rates of college attendance in the future. In 1998, Davis High School reported that every one of its graduates planned to go to college. But that year at Davis High students averaged a combined verbal and quantitative score on the SAT of only 791 (out of a possible 1600!). This state of affairs was also not likely to have led many students to college. Sharpstown High School, which had falsely claimed zero dropouts in 2002, reported in 2001 that 98.4 percent of its graduates expected to attend college (Schemo, 2003b).

After investigations began, one of the principals in Houston said that it was customary for administrators to inflate their college attendance figures in the hope of attracting the children of active, involved parents. And more students also generate more money for their schools from the state. At this principal's own school the claim was made that almost all of its graduates were headed for college. But in fact, the principal admitted, most of her students "couldn't spell college, let alone attend" (Schemo, 2003b, p. 14).

College attendance rates in Houston were patently false, designed to mislead. As with the dropout scandal, when asked about this deception of the public, the former superintendent of Houston schools Rod Paige, then serving as secretary of education, refused to comment.

CELL B: EXAMPLES OF GREY AREAS OF CHEATING—WITH NONTEST INDICATORS

Example B-1

Narrowing the curriculum is a subtle, but insidious effort to cheat students out of a quality educational experience. Although not an example of test cheating per se, denying students opportunities to explore a range of topics represents an unconscionable form of cheating students. Furthermore, if a narrowing of curricula choices for students occurs in some districts and schools, but not in others, that raises problems of validity for the construct being measured. So the highly valued indicator is corrupted when extensive narrowing of the curriculum occurs, as predicted by Campbell's law.

The pressure to do well on tests that focus primarily on literacy, mathematics, and science forces educators whose jobs are on the line to spend more time on these subjects and less time on others. For example, the Council on Basic Education sought examples of the ways in which the spirit of liberal arts education in K–12 settings atrophies in a test-driven culture, and they found them (Zastrow & Janc, 2004). According to their multistate survey of principals from Illinois, Maryland, New York, and New Mexico, 25 percent reported decreases in instructional time for the arts, with 33 percent reporting anticipation of future decreases. Principals serving schools with greater proportions of ethnic minority students report the greatest amount of decreases in arts education. Overall, 29 percent of elementary principals reported decreases in time for social studies, with 46 percent of all high-minority elementary schools reporting decreases.

This seems to be a national trend. In Massachusetts, there has been a disturbing and not so subtle trend to eliminate arts education from schools. Over the past two years Massachusetts' public school systems have lost 178 teachers of the arts (visual arts, dance, music, and theater), going from 3,996 in 2002–2003 to 3,818 in 2004–2005. In pockets around the state there were noticeable changes, including the Plymouth public schools, where three of its five middle school art teachers were laid off so that the schools could increase time spent in English and math; the Stoneham public schools where students receive elementary art and music lessons for half-year periods of time, rather than for the whole year (starting in Fall 2005); and layoffs in the Southbridge area that forced schools to give elementary students art and music once every three weeks starting in the fall of 2005 rather than once a week (Vaishnav, 2005). These decisions not only cheat students and their families of what is generally considered to be an appropriate education, but compromise the validity of the tests as well.

The state of California is also noticeably squeezing the arts programs out of the public school system. Currently, California ranks fiftieth in the nation in the ratio of music teachers to students. And still, music offerings continue their significant decline. The greatest decline came in general music, which suffered an 85 percent decrease in student enrollment, with the number of music teachers declining by 26.7 percent during the same time period (1999–2000 through 2003–2004). And in one San Jose district the music program was being severely cut back, going from nine classes and two teachers to five classes and one teacher (Posnick-Goodwin, 2005).

Curriculum narrowing is a way to improve test scores through means that are not entirely proper. It is a grey area of deceit that deserves more attention.

Example B-2

It isn't just whole curriculum areas that are being cut or modified; the delivery of instruction within curriculum areas is being compromised as well. Later, we discuss explicit examples of the ways in which teachers are forced to "teach to the test" under our discussion of grey areas of cheating related directly to testing (see discussion for cell D). Here we focus on the more subtle ways in which test validity is compromised and students are cheated out of opportunities to learn when instruction becomes narrowed and more rote learning is emphasized. For example, teachers from higher-stakes states report that the pressure they feel is directly tied to decisions about time allocation (Pedulla et al., 2003), and their instructional decisions are also affected, with negative impacts on the range and types of activities they can engage in such as field trips, organized play, or class enrichment. In Tennessee, a newspaper-sponsored survey of teachers (asking how high-stakes testing is affecting them) found that many teachers have narrowed their curriculum. One middle school teacher complained of being unable to teach children about America's first moon landing because of the demands a heavy testing schedule places on classroom time (Editorial, 2003).

Tragically, even our youngest students are being cheated. In the elementary grades, recess, play time, and physical education are being severely cut or minimized, and even students in kindergarten are being denied the opportunity to rest and rejuvenate (Ohanian, 2002). This is especially alarming in light of research showing the benefits of being well rested for learning among students of all ages (Johnson, Chilcoat, & Breslan, 2000; Randazzo, Muehlbach, Schweitzer, & Walsh, 1998; Siren-Tiusanen & Robinson, 2001; Tietzel & Lack, 2002; Wolfson & Carskadon, 2003). For example, nap time is typically a daily ritual for the prekindergarten students at countless schools across the country. But many, feeling the pressures to get students ahead as early as possible, agree with Prince George's County schools chief Andre J. Hornsby who said, "Nap time needs to go away . . . we need to get rid of all the baby school stuff they used to do" (Trejos, 2004, p. A1). Naps, recess, and any opportunities to break from the academic side of schooling are increasingly being denied our youngest students. The Waltham, Massachusetts, newspaper reports that recess for elementary students has been cut one day a week to make time for the high-stakes testing and to be sure that they look good on NCLB (*Waltham Daily News Tribune*, 2004). The report also says that some nearby districts have done away with recess at the elementary grades completely—a trend that is occurring all over the country (Ohanian, 2002). The educators around St. Louis have also cut back on recess and physical education (Aquilar, 2004).

Even lunch is being threatened. A teacher at a district near Waltham expressed concern that lunch had been reduced at her elementary school to less than 15 minutes on many days so that more time could be put in on the rigorous curriculum areas, meaning the areas that are tested. The school had started serving finger food—wraps and chicken nuggets—to get the students in and out of the cafeteria faster.

Example B-3

In the quotes presented next, it is clear that teachers are sacrificing curriculum for test preparation activities and greater concentration on those subjects being tested (Taylor, Shepard, Kinner, & Rosenthal, 2003).

> Teacher A: "We only teach to the test even at second grade, and have stopped teaching science and social studies. We don't have assemblies, take few field trips, or have musical productions at grade levels. We even hesitate to ever show a video. Our second graders have no recess except for 20 minutes at lunch." (2003, p. 31)
>
> Teacher B: "I eliminated a lot of my social studies and science. I eliminated Colorado History. What else? Electricity. Most of that because it's more stressed that the kids know the reading and the math, so, it was pretty much said, you know, you do what you gotta do." (2003, p. 30)
>
> Teacher C: "Those things (science and social studies) just fall to the backburner and quite frankly, I just marked report cards for the third grading period and I didn't do science at all for their third grading periods. Same for the social studies." (2003, p. 30)
>
> Teacher D: "We don't take as many field trips. We don't do community outreach like we used to like visiting the nursing home or cleaning

up the park because we had adopted a park and that was our job, to keep it clean. Well, we don't have time for that any more." (2003, p. 30)

Teacher E: "I had to cut out some things in order to do the CSAP stuff. It's not the number of days. I think it would be more accurate to say the number of labs. I think what is more significant is that I have had to cut the number of hands-on investigations. I would say I have had to cut one quarter of the labs." (2003, p. 31)

Teacher F: "[I eliminated] projects, [I] eliminated curriculum such as novels I would teach, we didn't have time to go to the library, we didn't have time to use the computer labs because they had to cut something. [I] Cut things I thought we could live with out. [I] Cut presentations, anything that takes very much time, I cut film. We have been cutting like crazy." (2003, p. 31)

The general public does not understand that achievement tests associated with NCLB were not designed to measure literacy and numeracy under conditions of near full-time study. The norms for these tests and the various cut scores determining a students' competency were almost surely based on assumptions of an ordinary, varied, and complete curriculum. When that is not the case, as is well documented here, the scores obtained on the tests are more difficult to interpret. Validity has been compromised because a measure that was designed to assess learning in typical classroom environments is used in classrooms that are no longer typical. Students have overlearned the curriculum and have been drilled often on similar measures to those used for assessing their learning under ordinary, typical classroom conditions. In these situations the issue of cheating arises because the scores obtained on the measures used are artificially inflated. The public, then, is cheated out of a valid assessment of how its school children are doing. Furthermore, we should not forget that at the same time as such cheating occurs, teachers' professional judgments have been ignored and the arts and humanities have been downgraded as subjects of importance in the public's thinking. The deceit associated with the inflation of scores, therefore, has side effects of some significance.

CELL C: EXAMPLES OF BLATANT AREAS OF CHEATING—WITH TESTS AS INDICATORS

Example C-1

Conditions created by high-stakes testing have apparently led increasing numbers of educators to engage in blatant acts of test-related cheating. These acts are "blatant" because of the obvious attempt to mislead others by manipulating the test indicator directly. One example includes acts where administrators "order" their teachers to cheat, thereby involving all adults as active co-conspirators in making the school look better on the high-stakes accountability measure. One of the most notorious examples of these incidents took place in New York. A year after Jon Nichols, a mathematics teacher, arrived at Community Elementary School 90, he was approached by the principal, Richard Wallin, with what seemed at first an unusual pep talk. "I was taken into the office at the beginning of the school year and told that the students were expected to do well—no matter what it takes . . . at the time, I didn't know what that meant exactly" (Archibold, 1999, p. 8). He learned soon enough. It meant cheating, and Nichols and other teachers were provided with detailed instructions, down to palm-sized crib notes to check against students' answer sheets as they took city and state examinations, as reported by Edward F. Stancik, the special investigator for New York City schools.

The report identified 32 schools but cited Community Elementary Schools 88 and 90 in District 9 in the Bronx, and Public School 234 in District 12 in the Bronx, as particularly egregious cases. Nichols, 33, who investigators confirmed had participated in the case, described how an administrator approached him before city reading and math exams in April 1994. "Keep this handy, he said, and he gave me a piece of paper," Nichols recalled. "It was a 2-by-3 sheet with a list of numbers and letters on it. It was the answers to the test. Some of the numbers had asterisks next to them," Nichols continued. "These were the hard questions, and I was told not to help with those. The kids were expected to get them wrong" (Archibold, 1999, p. 8).

Unfortunately, this is not an isolated incident. In Kentucky, one principal's certification was suspended owing to allegations of cheating on a statewide assessment test. According to the story, this Milken Family Foundation Award winner (an award worth $25,000 that is tied to academic improvement) was investigated after statewide test scores spiked sharply one year. "Investigators found that among some 80 violations, the principal had encouraged inappropriate practices, such as teachers developing tip sheets, and students had been encouraged to seek help from teachers, according to investigation and court documents." As a result, the principal was stripped of all his teaching and administrative certificates for 18 months (Blackford & Mueller, 2004).

Example C-2

Other ways administrators and teachers cheat include blatantly changing test scores or identifications associated with test scores as a way to render test results completely misleading. For example, Helen Lehrer, the former principal at Pacific High School in Boerum Hill, Brooklyn, was accused of erasing and changing as many as 119 answers on 14 Regents competency tests in global studies and U.S. history given one spring (Gendar, 2002).

In Texas, where the pressure to do well on statewide tests has been increasing significantly since at least 1990 (Nichols, Glass, & Berliner,

2006), officials employed one of the more creative strategies for altering test scores. According to one report, the Austin School District manipulated test results to make it appear as if several schools performed better than they did, the Texas Education Agency says. "There must be 50 ways to cheat on TAAS tests, and administrators at several Austin schools employed one of the more clever chicaneries in manipulating history" (Stinson, 1998, p. 3). Commissioner of Education Mike Moses explained the trickery in a letter to the school district: "Student identification number changes were submitted for students tested at (the schools), which resulted in the exclusion of those students from the accountability subset of TAAS results used to determine the 1998 accountability ratings" (Stinson, 1998, p. 3). In plainer English, administrators changed poor-performing students' test ID numbers so that they did not agree with previous numbers, knowing that the inconsistencies would cause the TEA to eliminate the students' scores from ratings calculations (Stinson, 1998).

In Arizona, the state's eighth-largest school district is auditing all of its schools after officials alleged that a principal changed test scores so that her teachers could get incentive money (Ryman, 2003). In Louisiana, a suspiciously high number of erasures on standardized-test answer sheets prompted the State Department of Education to throw out scores from 10 Orleans Parish public schools. Regulators were keeping an eye on erasures because they can help identify classrooms where teachers are improperly helping students. Infractions can range from a teacher coaching a student during the test to a principal erasing students' wrong answers and filling in the right ones. According to New Orleans Schools Superintendent Morris Holmes, "We are investigating as if there were improprieties. We have a whole lot of erasures that statistically could not just happen by chance. So we are trying to determine how that happened" (Meitrodt, 1997, p. A1).

Example C-3

Other kinds of blatant acts of cheating involve the efforts of teachers to directly prompt, guide, and help students prior to or during test-taking. For example, in Wisconsin, one elementary school principal and two teachers were accused of giving students answers to questions on the statewide assessment (the Wisconsin Knowledge and Concepts Examination and TerraNova), both taken in November 2003. It was believed that the stakes attached to the Wisconsin assessment provided the motive for the alleged aid. The allegations are that the teachers "had students memorize the sequence of answers" (Carr, 2004, p. 1A) on a multiple-choice section of the test. Allegations involved students in third, fourth, and fifth grades. The school in which the allegations occurred has low test scores. In 2000, only 23 percent of third graders were in the top two brackets, and 32 percent drew the lowest ranking ("minimal"). By comparison, 4 percent of students statewide were ranked as "minimal" (Carr, 2004).

In Chicago, one elementary school's principal and curriculum coordinator were accused of giving teachers copies of the Iowa Test of Basic Skills (ITBS) and telling them to use it to help students prepare for the exam. Although the chief executive officer said he did not believe the alleged cheating was linked to stakes attached to that year's ITBS, the timing couldn't have been more coincidental as that year was the first year stakes were attached to the ITBS (Rossi, 1996). Unfortunately, this was not an isolated incident as researchers, using a newly developed algorithm to detect suspicious test score patterns across time, found over 1,000 separate instances of classroom cheating representing 4 to 5 percent of all Chicago public schools (Jacob & Levitt, 2001, 2002). Chicago schools were just beginning to implement the high-stakes testing system, and thus, Jacob and Levitt's analysis, conducted prior to NCLB enactment, seems a gloomy foreshadowing of what was to come when the federally mandated high-stakes testing practice began nationwide.

In Massachusetts, a Worcester elementary school principal submitted her letter of resignation to the superintendent of schools a year after allegations that she helped students to cheat on the Massachusetts Comprehensive Assessment System (MCAS) exam. The principal, Irene Adamaitis, was alleged to have distributed the test to teachers days before they were administered to students. The dramatic improvement in MCAS scores at this school had prompted the investigation by the Worcester school system and the state's Department of Education (LeBlanc, 2004).

Since the statewide testing program began five years ago, more than 200 California teachers have been investigated for allegedly helping students on state exams. Cheating behavior included whispering answers into a student's ear during an exam; photocopying test booklets so students would know vocabulary words in advance; and erasing answers marked with wrong answers and changing them to correct ones. Teachers receive enormous pressure from principals to work on raising scores—not just for bragging rights, but because federal funding can be withheld. Some incidents in the last five years include the following (Hayasaki, 2004):

- In the San Joaquin Valley's Merced County, a third grade Planada School District teacher gave hints to answers and left a poster on a wall that also provided clues.
- In the Inland Empire, a Rialto Unified School District third grade teacher admitted telling students, "You missed a few answers; you need to go back and find the ones you missed" (2004, p. 4). A student reported that the teacher looked over pupils' shoulders and told them how many questions were wrong.
- Near the Mexican border, in the El Centro Elementary School District, a principal asked a student why he had erased so many

answers. The student responded that the teacher had told him to "fix them" (2004, p. 4).

- In El Monte, a Mountain View School District eighth–grade teacher admitted using the board to demonstrate a math problem and saying, "This is a silly answer. If you marked this one, erase it and pick another" (2004, p. 5). Records stated that the teacher "said she was very sorry and wept during the interview" (2004, p. 5).
- In the Ontario-Montclair School District, a student told investigators that a teacher read 10 math answers. One student said he handed his test booklet to that teacher and then went back to change five answers after the teacher said, "Why don't you try again?" (2004, p. 5).
- Near Salinas, a Hollister School District teacher admitted changing about 15 answers.

Beverly Tucker, California Teachers Association chief counsel for 16 years, said the number of teachers her office defended against allegations of cheating had risen. She could recall one or two cases stemming from the decade before the current testing began. Since 1999, she estimated the union has defended more than 100 (Hayasaki, 2004).

CELL D: GREY AREAS OF CHEATING— TESTS AS INDICATORS

Example D-1

It is plausible that teachers and administrators sometimes try to resist a system they see as corrupt and unfair, as do tax, religious and civil rights protesters across this nation. One example is the North Carolina principal who will not test what she calls "borderline kids," her special education children, despite the requirement to do so. She says, "I couldn't. The borderline children experience enough failure and do not need to be humiliated by a test far beyond their abilities" (Winerip, 2003, p. 9).

By not testing all the children in the school the principal is cheating. The test scores at her school are inflated, and thus she is deceiving the public. But this is also an act of human kindness. At the same time it is an act of resistance to laws made by policymakers in some other community. It is not easy to judge this principal's behavior harshly.

Example D-2

In this incident, the subjectivity of essay scoring provides the grey area in thinking about a controversy in New York. There is a practice used in scoring the statewide New York Regents examinations called teacher "scrubbing"—the process of "tweaking" student scores on tests (Campanile, 2004, p. 15). Scrubbing is most often done on the English and history exams because they have essay questions that are subjectively scored. In this process, at least two different teachers review a student's exam. The scores these two teachers assign are averaged (required under state rules). Then, these papers are sent back to respective departments, where some teachers, often under the guidance of principals, set aside tests that are just a few points shy of passing. Then, teachers review exam responses and "find" extra points to assign to students. This process is heavily debated. Some say it is not cheating. According to one teacher, "I'm sorry if it's shocking for laymen to hear. Scrubbing is something we do to help the kids get their asses out of high school" (Campanile, 2004, p. 15).

But other educators argue that it is a blatant form of cheating. And there are plenty of incentives to do so with merit pay bonuses and diplomas on the line. According to one high school staffer, "The students of the school benefit because they pass. The school benefits because the pass rate is up ... but it is grade inflation. It's cheating. You're falsifying exams. It's totally corrupt" (Campanile, 2004, p. 15). We leave readers to decide whether scrubbing is a humane and sensible practice, or one that deserves to be called cheating of some sort.

Example D-3

Test preparation has become a big business. Test preparation companies provide tutoring for individual students and also offer large, formal curricula packages for test preparation to schools and districts. These private enterprises are driven by profit, and that means their test preparation strategies need to "work." To work well, the test preparation company needs items and an understanding of the test format that are very close to the actual items and formats used on the tests themselves. Thus there are incentives to get close to preparing students too directly for the test.

For example, in Tacoma in 1995, Comprehensive Test of Basic Skills (CTBS) scores were at the 42nd percentile and rose to the 63rd in a few months. Superintendent Rudy Crew was then hailed as a miracle worker, received a large bonus, and on the basis of this work was promoted to the chancellorship of the New York City public schools. But to help get scores up Dr. Crew had hired a "test prep" consulting firm, and the students were drilled on tests very much like those on the CTBS. Many practice tests with suspiciously similar items to the CTBS were used. The Tacoma miracle stopped (i.e., scores dropped) after Dr. Crew left Tacoma and the test prep company stopped working there. When the test preparation program developer was questioned, he stated that wealthy children throughout the country get coaching from tutors to improve their SAT scores. He wondered what the difference was between what he did in Tacoma and what is done for the wealthy children all the time (Sacks, 2001). For us, this defense of his actions moved what looked at first like a suspiciously blatant incident of cheating into a much greyer area.

Teaching to the test certainly gets a state's own test scores up, but it may not involve learning that is transferable to other tests. Teachers recognize this as deceit, as is reflected in the following comments by some Florida teachers (Jones & Egley, 2004):

> Teacher A: "I can say one thing, if my kids learn one thing in third grade, it is this: how to pass a standardized test even if you are not familiar with the material. Now is that what our goal is? Perhaps we should revisit it." (2004, p. 17)
>
> Teacher B: "I have seen that schools are teaching to the test (how can you not?) and that is not a true reflection of student abilities. This is only a reflection of the abilities of each school to teach effective test-taking strategies, not academics." (2004, p. 17)
>
> Teacher C: "Schools aren't improving their academics as students score better on the FCAT. They are just taking more time to teach to the test and unfortunately, away from real learning. We aren't getting smarter students, we are getting smarter test takers. That is NOT what we are here for! . . . The schools who score well are focusing on teaching to the test at a very high cost to their students." (2004, p. 17)

Another dubious incident we classified in the grey zone was the report of a school that, in an effort to increase student test performance, focused on using test-taking strategies (Dewan, 1999). Teachers there emphasized such things as underlining key words and drawing arrows when rounding numbers. Children who used the strategies on practice tests were rewarded with, among other things, movies and popcorn. But that wasn't all they got, according to another teacher who was instructed to "give out candy during the test to children who were using the strategies. If the child wasn't using them, the teacher whispered to the child 'use the strategies'" (Dewan, 1999, online).

Thus, teaching to the test and other forms of helping students become problems in interpreting test scores, probably by inflating them, and thus compromising the validity of the examinations. A sadder note about the ubiquity of test preparation programs, however, comes from the research of Harold Wenglinski (2004). Wenglinski looked at 13,000 students who had taken the fourth grade National Assessment of Educational Progress (NAEP) tests in 2000. He found that frequent testing actually reduced scores on NAEP and that emphasizing facts (over reasoning and communication) also reduced students' scores. Since these two characteristics are prevalent in most of the test preparation programs we have examined, it is quite likely that many of the activities engaged in during test preparation are counterproductive.

CONCLUSION

We believe that the high-stakes testing environment creates pressures on educators that were never before present in our nation's schools. These summative high-stakes tests have too much riding on them: bonuses to modestly paid people, school closures and loss of jobs, shame and humiliation for lack of progress, and so forth. These conditions induce some educators to engage in blatant cheating, or impel them toward acts that are morally ambiguous. The circumstances now affecting education are the same as those that have affected many fields of endeavor, and there is even a social science law to account for this phenomenon—Campbell's law. Whether in finance or in education, both the blatant and the greyer acts of deceit mislead the public. When indicators take on too much value, as when stakes are high, both indicators and educators are corrupted. When the public is misled through outright chicanery or through compromised test validity, the reputation of the whole system and its entire workforce is damaged.

Although there is no way to track the corruption of educators historically, there does seem to be a dramatic contemporary rise in the incidents of cheating and morally suspect behavior by teachers and administrators. We believe that NCLB and its associated accountability model have sown these seeds of corruption in a once morally trustworthy, highly respected profession. For self-protection from humiliation, for job protection, for help to their students, and sometimes out of simple greed, educators have been put into a situation where crossing the moral line seems tempting and excusable. We know from the Hartshorne and May studies with which we began this chapter, that teachers like others in our society are neither honest nor dishonest all the time. Teachers, like others, weigh the situations they are in and decide how they should act under the particular conditions in which they find themselves.

If there is a link between the particular conditions in which teachers now find employment and increases in cheating and other unsavory behavior, as we suspect, then we should consider changing the conditions of employment. Tests could be formative, not summative, for teachers and their schools. Assessments could be the source of information and discussion by the professional staff, rather than the source of job loss or monetary gain. The results of poor school performance could be high-quality professional development rather than humiliation or job loss. Investment in neighborhood efficacy and health care, as well as job training and income support, could be targeted for schools that do not do well on tests, since almost all of the schools that show low achievement serve impoverished communities (Berliner, 2005). A program of school improvement that concentrates on helping families in schools that do not do well has as much evidence to support it as any other model of school improvement, but such evidence is often ignored (Berliner, 2005). In short, a system of school improvement that does not rely on humiliation, threats, or cash bonuses could be developed as easily as the one we have now were we to become more concerned with annual family income than annual yearly progress, more concerned about teacher professional growth than test score growth, and more concerned about school improvement than school bashing.

The NCLB accountability program appears to foster conditions that will change the public's perception of educators as being trustworthy and virtuous to one in which teachers are seen as more corruptible; such perceptions make our educators more common. This cannot be good for the nation. Teachers may never have had the respect they deserve, but respect has been there nonetheless. The calling to be a teacher was strong, and because of that, many teachers endured low status and modest pay so that they could be special persons in the lives of children. Professional educators had some dignity, and their work provided them feelings of self-worth. If high-stakes testing corrupts the profession as it seems to be doing, the nation may lose from the profession those who endured its difficulties for the small honors it bestows. This is too great a price to pay for an accountability system that appears not to work and for which there are alternatives.

REFERENCES

Anderman, E. M., Griesinger, T., & Westerfield, G. (1998). Motivation and cheating during early adolescence. *Journal of Educational Psychology, 90,* 84–93.

Aquilar, A. (2004, September 28). Many Illinois districts opt out of PE requirement. *St. Louis Post-Dispatch,* A01.

Archibold, R. C. (1999, December 8). Teachers tell how cheating worked. *New York Times,* B8.

Berliner, D. C. (2005, August 2). Our impoverished view of educational reform. *Teachers College Record.* Retrieved November 8, 2005 from www.tcrecord.org.

Blackford, L. B., & Mueller, L. (2004, March 22). Bell may hire controversial educator. *The Herald-Leader,* A1.

Callahan, D. (2004). *The cheating culture: Why more Americans are doing wrong to get ahead.* New York: Harcourt.

Campanile, C. (2004, January 26). Teachers cheat; Inflating regents scores to pass kids. *New York Post,* 15.

Campbell, D. T. (1975). Assessing the impact of planned social change. In G. Lyons (Ed.), *Social research and public policies: The Dartmouth/OECD Conference* (Chapter 1, pp. 3–45). Hanover, NH: Public Affairs Center, Dartmouth College.

Carr, S. (2004, January 30). MPS looks into claims staff helped on tests. *Milwaukee Journal Sentinel,* A1.

Deci, E. L., Koestner, R., & Ryan, R. M. (1999). A meta-analytic review of experiments examining the effects of extrinsic rewards on intrinsic motivation. *Psychological Bulletin, 125,* 627–668.

Deci, E. L., Koestner, R., & Ryan, R. M. (2001, Spring). Extrinsic rewards and intrinsic motivation in education: Reconsidered once again. *Review of Educational Research, 71*(1), 1–28.

Deci, E. L., & Ryan, R. M. (1985). *Intrinsic motivation and self-determination in human behavior.* New York: Plenum Press.

Dewan, S. (1999, February 25). The fix is in: Are educators cheating on TAAS? Is anyone going to stop them? *Houston Press.* Retrieved August 26, 2006, from http://www.houstonpress.com/issues/1999-02-25/news/feature.html.

Editorial. (2003, September 24). High-stakes testing turns the screw. *The Commercial Appeal,* p. B4.

Evans, E. D., & Craig, D. (1990). Teacher and student perceptions of academic cheating in middle and senior high schools. *Journal of Educational Research, 84,* 44–52.

Gassen, S. G., & Sterba, J. (2003, September 3). State talks tweak for AIMS test. *Arizona Daily Star,* A1.

Gendar, A. (2002, November 13). Three state tests on hold till HS erases cheat label. *Daily News,* p. 3.

Good, T. (Ed.) (1999). *Elementary School Journal (Special Issue), 99*(5).

Good, T. (Ed.) (2000). *Elementary School Journal (Special Issue), 100*(5).

Hartshorne, H., & May, M. A. (1928). *Studies in deceit,* vol. 1 of *Studies in the nature of character.* New York: Macmillan/Maxwell Macmillan.

Hartshorne, H., & May, M. A. (1929). *Studies in self-control,* vol. 2 of *Studies in the nature of character.* New York: Macmillan/Maxwell Macmillan.

Hartshorne, H., & May, M. A. (1930). *Studies in the organisation of character,* vol. 3 of *Studies in the nature of character.* New York: Macmillan/Maxwell Macmillan.

Hayasaki, E. (2004, May 21). One poor test result: Cheating teachers. *Los Angeles Times,* A1.

Herszenhorn, D. (2005, October 12). Brooklyn high school is accused anew of forcing students out. *New York Times,* B1.

Jacob, B., & Levitt, S. D. (2001). *Rotten apples: An estimation of the prevalence and predictors of teacher cheating* (Working paper 9413). Cambridge, MA: National Bureau of Economic Research.

Jacob, B., & Levitt, S. D. (2002). *Catching cheating teachers: The results of an unusual experiment in implementing theory* (Working paper 9414). Cambridge, MA: National Bureau of Economic Research.

Johnson, E., Chilcoat, H. D., & Breslan, N. (2000). Trouble sleeping and anxiety/depression in childhood. *Psychiatry Research, 94,* 93–102.

Jones, B. D., & Egley, R. J. (2004, August 9). Voices from the frontlines: Teachers' perceptions of high-stakes testing. *Education Policy Analysis Archives, 12*(39). Retrieved December 2, 2004 from http://epaa.asu.edu/epaa/v12n39/.

Jones, M. G., Jones, B. D., Hardin, B. H., Chapman, L., Yarbrough, T., & Davis, M. (1999). The impact of high-stakes testing on teachers and students in North Carolina. *Phi Delta Kappan, 81,* 199–203.

Kohlberg, L. (1984). *The psychology of moral development: Essays on moral development, 2.* San Francisco: Harper and Row.

Kozol, J. (2005). *The shame of the nation: The restoration of apartheid schooling in America.* New York: Crown Publishers.

LeBlanc, S. (2004, April 8). Worcester school principal resigns in wake of MCAS cheating probe. Associated Press Wire.

Lewin, T., & Medina, J. (2003, July 31). To cut failure rate, schools shed students. *New York Times.* Retrieved December 20, 2005 from http://www.pipeline.com/~rgibson/shedstudents.html.

McCaslin, M., & Good, T. (1996). The informal curriculum. In D. C. Berliner & R. Calfee (Eds.), *Handbook of educational psychology* (pp. 622–673). New York: Macmillan.

Meitrodt, J. (1997, September 17). LA voids scores in 19 schools: Erasures blamed. *Times-Picayune,* A1.

Murdock, T. B., Miller, A., & Kohlhardt, J. (2004). Effects of classroom context variables on high school students' judgments of the acceptability and likelihood of cheating. *Journal of Educational Psychology, 96*(4), 765–777.

National Commission for Excellence in Education. (1983, April). *A nation at risk: The imperatives for educational reform.* Washington, DC: U.S. Department of Education, National Commission for Excellence in Education.

Nichols, S., & Berliner, D. C. (2005, March). *The inevitable corruption of indicators and educators through high-stakes testing.* EPSL-0503-101-EPRU. Retrieved March 23, 2005 from http://www.greatlakescenter.org/pdf/EPSL-0503-101-EPRU.pdf.

Nichols, S., Glass, G., & Berliner, D. C. (2006). High-stakes testing and student achievement: Does accountability pressure increase student learning? *Educational Policy Analysis Archives*. 14(1). Retrieved August 24, 2006 from http://epaa.asu.edu/epaa/v14n1.

Ohanian, S. (2002). *What happened to recess and why are our children struggling in Kindergarten?* New York: McGraw-Hill.

Orfield, G., Losen, D., & Wald, J. (2004). *Losing our future: How minority youth are being left behind by the graduation rate crisis.* Cambridge, MA: Civil Rights Project at Harvard University.

Pedulla, J. J., Abrams, L. M., Madaus, G. F., Russell, M. K., Ramos, M. A., & Miao, J. (2003, March). *Perceived effects of state-mandated testing programs on teaching and learning: Findings from a national survey of teachers.* Boston: Boston College, National Board on Educational Testing and Public Policy. Retrieved January 7, 2004 from http://www.bc.edu/research/nbetpp/statements/nbr2.pdf.

Posnick-Goodwin, S. (2005, June). Is it curtains for the arts in California's public schools? *California Educator, 9*(9). Retrieved online: http://www.cta.org/CaliforniaEducator/v9i9/Feature_1.htm accessed September 27, 2005.

Randazzo, A. C., Muehlbach, M. J., Schweitzer, P. K., & Walsh, J. K. (1998). Cognitive function following acute sleep restriction in children ages 10–14. *Sleep, 21*, 861–868.

Raymond, M. E., & Hanushek, E. A. (2003, Summer). High-stakes research. *Education Next*, pp. 48–55. Retrieved September 1, 2006, from http://www.educationnext.org/.

Rossi, R. (1996, May 5). Grade school under investigation as teachers allege test cheating. *Chicago Sun-Times*, p. 13.

Rothstein, K. (2004, February 26). Minority dropout skyhigh in Mass. *The Boston Herald*, 1.

Ryman, A. (2003, October 8). District audits schools after test deceit. *The Arizona Republic*, B4.

Sacks, P. (2001). *Standardized minds: The high price of America's testing culture and what we can do to change it.* New York: Perseus Publishing.

Schemo, D. J. (2003a, July 11). Questions on data cloud luster of Houston schools. *New York Times*, A14.

Schemo, D. J. (2003b, August, 28). For Houston schools, college claims exceed reality. *New York Times*.

Schemo, D. J., & Fessenden, F. (2003, December 3). Gains in Houston schools: How real are they? *New York Times*, A1.

Seattle Post-Intelligencer. (2004, September 21). Schooldays: Leave time for recess. B6

Siren-Tiusanen, H., & Robinson, H. A. (2001). Nap schedules and sleep practices in infant-toddler groups. *Early Childhood Research Quarterly, 16*, 453–474.

Stinson, R. (1998, September 17). TAAS cheaters meet national standard. *San Antonio Express-News*, p. 3A.

Taylor, G., Shepard, L., Kinner, F., & Rosenthal, J. (2003, February). A survey of teachers' perspectives on high-stakes testing in Colorado: What gets taught, what gets lost. (CSE Technical Report 588: CRESST/CREDE/University of Colorado at Boulder). Los Angeles: University of California.

Tietzel, A. J., & Lack, L. C. (2002). The recuperative value of brief and ultra-brief naps on alertness and cognitive performance. *Journal of Sleep Research, 11*(3), 213–218.

Trejos, N. (2004, March 14). Time may be up for naps in pre-k classes. *Washington Post*, p. A01. Retrieved November 27 2004 from: http://www.washingtonpost.com/wp-dyn/articles/A58706-2004Mar14.html.

Vaishnav, A. (2005, July 24). Art teachers fade in Bay State: Educators cite shifting focus. *Boston Globe*. Retrieved August 16, 2005. from http://www.boston.com/.

Waltham Daily News Tribune. (2004, February 27). Class requirements cut into recess time.

Wenglinsky, H. (2004, November 23). Closing the racial achievement gap: The role of reforming instructional practices. Education Policy Analysis Archives, 12(64). Retrieved November 27, 2004, from http://epaa.asu.edu/epaa/v12n64/.

Winerip, M. (2003, October 8). How a good school can fail on paper. *New York Times*, B9.

Wolfson, A. R., & Carskadon, M. A. (2003). Understanding adolescents' sleep patterns and school performance: A critical appraisal. *Sleep Medical Review, 7*, 491–506.

Zastrow, C., & Janc, H. (2004, March). *Academic atrophy: The condition of the liberal arts in America's public schools*. A report to the Carnegie Corporation of New York. Washington, DC: Council for Basic Education.

Epilogue
Moving an Integrity Agenda

David Callahan

Senior Fellow, Demos and author of
The Cheating Culture: Why More Americans Are Doing Wrong to Get Ahead.

Over recent decades, academic dishonesty has emerged as one of the most exhaustively researched aspects of education. Now, with the research collected in this fine volume, our understanding of the cheating phenomenon has grown even deeper and richer. Scholars know more than ever about who cheats and why, what kind of classroom settings may encourage cheating, what sort of solutions work best to deter cheating, and much more. Knowledge is growing about cheating in all educational settings, even in middle school, but most of the research on academic cheating focuses on higher education. Today, no leader in higher education could possibly suggest that more research is needed before their institution can effectively address the problem of cheating. Nor could any educator who cares about their profession deny a simple truth: the high incidence of cheating among college students strikes at the very soul of higher education. Widespread academic dishonesty, it is increasingly clear, ranks as one of the most profound challenges facing colleges and universities. And yet on most campuses this problem rarely makes it to the top of the agenda. Numerous schools do not even have basic honor codes. Many schools that do have such codes do not effectively promote or enforce them. An ongoing, energetic commitment to academic integrity remains the exception on college campuses, not the rule. The situation is even worse in secondary schools, where cheating habits are often first formed.

This must change. The research reported in this volume and elsewhere provides the ammunition for sweeping new efforts to promote academic integrity. A psychological perspective allows us to better understand the motives and rationalizations of cheaters, making human the alarming

---

statistics that emerge regularly from large surveys of students. Such insights can lead to better anti-cheating policies. But ultimately these efforts depend on mustering more political will among educators. In concluding this volume, it seems appropriate to reflect on how such will could emerge and lead to a much broader movement to promote academic integrity. While the research in this book looks at students in many settings, the thoughts that follow are confined to higher education and draw on numerous conversations with faculty and administrators over the past two years.

Clearly, nearly all the key players at any university have a major interest in reducing cheating. Faculty members often take cheating personally and loath the role of cop more than just about any other part of their job. Administrators fear cheating scandals and dislike the burden of adjudicating integrity violations. Many student leaders want to see more integrity on campus and a level playing field. But while all these players have a strong interest in less cheating, each also tend to have more urgent concerns on their minds and, in some cases, have direct incentives to ignore the issue of academic integrity.

Start with faculty members. To prevent, detect, or punish cheating is a time consuming activity for professors—as is pushing one's institution to change policies in this area. Preventing cheating requires, among other things, changing how writing assignments are administered to make plagiarism more difficult, creating new exam questions every year, and more closely proctoring exams. Detecting cheating requires reading written assignments with vigilance and submitting suspicious material to Google or Turnitin.com. Punishing cheating is also time consuming and unpleasant, entailing confrontations with students, documentation of cheating, and involvement in administrative processes for adjudicating charges of academic dishonesty. Finally, championing the cause of academic integrity on campus can mean involvement in special committees and event planning.

The incentives for faculty to spend their time on any of these activities are virtually non-existent. A typical academic, whether tenured or not, has little reason to make academic integrity a top priority. At research universities, faculty are evaluated mainly on their scholarship and publishing. To the extent that they spend time going after cheaters they are distracting themselves from career building. At teaching colleges, faculty are often judged by their student evaluations—evaluations which rarely assess a professor's effort to enforce an honest and level playing field in their courses. On all campuses, faculty are also judged by how well they attend to various administrative duties and few have an incentive to dream up new, additional responsibilities. These basic facts of academic life may explain why so many faculty do not energetically take up the cause of integrity, whether in regard to their own students or their university at large.

Nor do college administrators have compelling reasons to put this issue on the front burner. These leaders are judged by any number of criteria: how much money they raise; what ranking their school achieves from *U.S. News & World Report*; whether their sports teams do well; how many entering students complete their degrees; how much diversity they are able to achieve among the student body and faculty; whether new buildings get built or new programs flourish or star professors are recruited; and on and on. Rarely, however, do administrators face tough questions about the integrity climate on campus. The severity of the cheating problem at different schools may be known by researchers like Donald McCabe, but this information is often not widely or publicly available. Nor is the integrity climate an indicator of educational quality that outsiders—e.g., prospective students or college counselors or donors—tend to factor into their assessments of a school.

Meanwhile, taking on the cheating problem on a campus requires a significant expenditure of administrative energy and resources. Developing and implementing an honor code from scratch can take several years. Effectively promoting and enforcing such a code requires money and staff. In short, greater integrity on a campus doesn't come cheap or without major hassles. And yet, for most administrators, there is no urgent imperative to focus attention in this area.

Student leaders also have other priorities. While many student leaders are concerned about integrity, this doesn't tend to be a first-tier issue for such leaders—compared to say, housing or financial aid or even parking availability.

Academic dishonesty is difficult to address under any circumstances. But it is especially difficult when most of the players on a university campus lack a strong incentive to tackle the problem. Given this, what can be done to move the issue of academic integrity further up the ladder of priorities at colleges and universities?

Part of the answer to this question lies in the successes so far of the growing movement of scholars and educators who are focused on academic integrity. For all the disincentives to tackle the cheating problem, individuals exist throughout higher education who are determined to reduce cheating. The best hope for progress may lie in better supporting these change-makers. The Center for Academic Integrity, founded in 1992, has played a key role in this regard. With nearly 400 member institutions, the Center helps schools develop or strengthen integrity policies by conducting assessments, disseminating best practices, and providing a national—indeed, international—network through which scholars, educators, and students can exchange ideas and find support. In effect, CAI is in the business of empowering anti-cheating activists by giving them the tools to design and push effective integrity policies. The organization operates like advocacy networks in other fields, with a national office, an annual conference, and a variety of materials designed for members.

Looking ahead, CAI illustrates one effective strategy for tackling the cheating problem. The more that faculty, administrators, and students can look outside their institution for guidance and support, the less they have to reinvent the wheel in developing effective policies or struggle against feelings of isolation. However, for all its successes, CAI remains small and underfunded. Even as the premier national anti-cheating organization, it has struggled to raise funds on a large scale and to develop adequate capacity. Fewer than 10 percent of U.S. colleges are members. A top priority of the integrity movement must be ensure that CAI grows much larger and reaches more widely onto campuses. One way to achieve this goal is for motivated faculty and administrators across the U.S. to get their institutions to become dues-paying members of CAI.

Other national organizational resources must also be brought to bear on the cheating problem. There are several dozen higher education associations in the United States. Many have significant financial and institutional capacity. Were these associations to focus much more attention on the cheating problem—spotlighting the issue at conferences, disseminating best practices, training members, etc.—those faculty and administrators committed to reducing cheating would have yet more resources and support. Just as importantly, educators who may give little thought to the problem would be exposed to its dimensions and solutions. Moving forward, those committed to promoting academic integrity should seek to gain influence in the major higher education associations and shape the agenda of these groups.

Two other strategies can help move the cheating issue higher up the agenda at colleges and universities. The first is the savvy dissemination of research findings to national and local media. Over recent years, the media has shown a strong interest in the subject of cheating, with front page stories appearing in the *New York Times* and television news shows like *Primtime Live* devoting segments to the issue. Scholars like Donald McCabe have been prominently quoted in these stories, and survey data have provided an empirical foundation to this reporting that might not otherwise exist. On a local or campus level, data on cheating that is specific to a school can attract significant media attention. For instance, a 2006 survey by two doctoral students at Ohio University Athens that documented widespread cheating led to extensive press attention and sparked a campus-wide conversation about how to foster more integrity. This same tactic has been used elsewhere and can be an effective way to force the cheating issue onto a university's agenda. In particular, negative press attention regarding cheating is a sure-fire way to stir up alums and trustees. The caveat here is that this public tactic is not suitable for scholars who need to maintain the trust of university officials over time to conduct large-scale student surveys.

A final strategy for elevating the cheating issue is to find ways to get federal and state government officials involved in this problem. Billions and billions of dollars flows annually into higher education. Various legislative committees and executive branch agencies oversee the appropriation of these funds. Such oversight and leverage could be used to demand more proactive efforts by universities—especially public universities—to reduce cheating. At the very least, government officials could use the power of public hearings and the bully pulpit to demand action. The best recent analogy is the successful effort by John McCain and other U.S. Senators to push Major League Baseball to implement a more effective drug-testing policy. Needless to say, federal and state officials have far more power over the higher education institutions that receive government funding than the U.S. Senate has over Major League Baseball, a private entity.

To elicit government action on the cheating problem, advocates of academic integrity could take the familiar path of other advocacy groups: lobbying specific legislators or officials to devote attention to the cheating problem and then supporting the efforts of such officials with research and policy analysis. Getting a U.S. Senator such as McCain to orchestrate national hearings in Congress about cheating—subpoenaing university presidents if necessary—would be an important symbolic step forward. Much more could be achieved at the state level, where the legislative oversight of higher education is greater.

Academic dishonesty is more than a problem on many campuses. It is a crisis, and one that deserves a central place on the agenda of university leaders and the world of higher education more broadly. While few incentives exist for administrators, faculty, or students to make this issue a priority, integrity activists can make substantial headway using different strategies. Many of these strategies have been extensively employed by advocates in other fields who have successfully done battle with large and obstinate institutions. Currently, the academic integrity movement remains small, underfunded, and comparatively unsophisticated when it comes to advocacy. There is enormous potential for this movement to grow more rapidly, mobilize new resources, and professionalize its advocacy efforts.

One thing is certain: thanks to the scholars who have contributed to this volume, along with other researchers, we now have a better understanding of exactly what we are up against when it comes to student cheating.

INDEX

Page numbers with "t" denote tables; those with "f" denote figures